Studies in International Performance

Published in association with the International Federation of Theatre Research

General Editors: Janelle Reinelt and Brian Singleton

Culture and performance cross borders constantly, and not just the borders that define nations. In this new series, scholars of performance produce interactions between and among nations and cultures as well as genres, identities and imaginations.

Inter-national in the largest sense, the books collected in the *Studies in International Performance* series display a range of historical, theoretical and critical approaches to the panoply of performances that make up the global surround. The series embraces 'Culture' which is institutional as well as improvised, underground or alternate, and treats 'Performance' as either intercultural or transnational as well as intracultural within nations.

Titles include:

Patrick Anderson and Jisha Menon (*editors*)
VIOLENCE PERFORMED
Local Roots and Global Routes of Conflict

Elaine Aston and Sue-Ellen Case
STAGING INTERNATIONAL FEMINISMS

Christopher Balme
PACIFIC PERFORMANCES
Theatricality and Cross-Cultural Encounter in the South Seas

Helen Gilbert and Jacqueline Lo
PERFORMANCE AND COSMOPOLITICS
Cross-Cultural Transactions in Australasia

Judith Hamera
DANCING COMMUNITIES
Performance, Difference, and Connection in the Global City

Alan Read
THEATRE, INTIMACY & ENGAGEMENT
The Last Human-Venue

Joanne Tompkins
UNSETTLING SPACE
Contestations in Contemporary Australian Theatre

S. E. Wilmer
NATIONAL THEATRES IN A CHANGING EUROPE

Forthcoming titles:

Elaine Aston and Sue-Ellen Case (*editors*)
PERFORMING GLOBAL FEMINISMS

Adrian Kear
THEATRE AND EVENT

Palgrave Studies in International Performance
Series Standing Order ISBN 978–1–403–944566 (hardback)
978–1–403–944573 (paperback)
(*outside North America only*)

You can receive future titles in this series as they are published by placing a standing order. Please contact your bookseller or, in case of difficulty, write to us at the address below with your name and address, the title of the series and the ISBN quoted above.

Customer Services Department, Macmillan Distribution Ltd, Houndmills, Basingstoke, Hampshire RG21 6XS, England

Violence Performed

Local Roots and Global Routes of Conflict

Edited by

Patrick Anderson

and

Jisha Menon

Introduction, selection and editorial matter © Patrick Anderson &
Jisha Menon 2009
Individual chapters © contributors 2009
Chapter 8 © Johns Hopkins University Press 2007
Chapter 9 © Taylor & Francis 2004
Chapter 12 © Johns Hopkins University Press 2002

First published 2009 by
PALGRAVE MACMILLAN

Palgrave Macmillan in the UK is an imprint of Macmillan Publishers Limited,
registered in England, company number 785998, of Houndmills, Basingstoke,
Hampshire RG21 6XS.

Palgrave Macmillan in the US is a division of St Martin's Press LLC,
175 Fifth Avenue, New York, NY 10010.

Palgrave Macmillan is the global academic imprint of the above companies
and has companies and representatives throughout the world.

Palgrave® and Macmillan® are registered trademarks in the United States,
the United Kingdom, Europe and other countries.

ISBN-13: 978–0–230–53726–2 hardback
ISBN-10: 0–230–53726–X hardback

This book is printed on paper suitable for recycling and made from fully
managed and sustained forest sources. Logging, pulping and manufacturing
processes are expected to conform to the environmental regulations of the
country of origin.

A catalogue record for this book is available from the British Library.

Library of Congress Cataloging-in-Publication Data
Violence performed : local roots and global routes of conflict / [edited by]
    Patrick Anderson and Jisha Menon.
      p.   cm. — (Studies in international performance)
    Includes index.
    ISBN 978–0–230–53726–2 (alk. paper)
    1. Violence. I. Anderson, Patrick, 1974– II. Menon, Jisha, 1972–
    HM1116.V53 2009
    303.6—dc22   2008029976

10   9   8   7   6   5   4   3   2   1
18   17   16   15   14   13   12   11   10   09

Printed and bound in Great Britain by
CPI Antony Rowe, Chippenham and Eastbourne

Contents

List of Figures

Notes on Contributors

Patrick Anderson is an Assistant Professor in the Department of Communication at the University of California, San Diego, where he is also affiliated with the Critical Gender Studies Program and the Ethnic Studies Department. He is currently completing a manuscript called *"So Much Wasted": Hunger, Performance, and the Morbidity of Resistance*, which examines the role of self-starvation in clinical, aesthetic, and carceral domains. A former Fulbright Scholar and a graduate of the PhD program in Performance Studies at Berkeley, Anderson's recent writing has appeared in *Cultural Studies*, *TDR*, the *Radical History Review*, *Women and Performance*. He has served on the board of directors for Performance Studies International (PSi) since 2002.

Catherine M. Cole is Professor in the Department of Theater, Dance, and Performance Studies at the University of California, Berkeley. She is the author of *Ghana's Concert Party Theatre* (2001), which received a 2002 Honorable Mention for The Barnard Hewitt Award for outstanding research in theatre history from the American Society for Theatre Research, and was a finalist for the Herskovitz Prize in African Studies. Cole is the co-editor of *Africa After Gender?* (2007), and she is currently writing *Stages of Transition: Performing South Africa's Truth Commission* (forthcoming, Indiana University Press). Her dance theatre piece *Five Foot Feat*, created in collaboration with Christopher Pilafian, toured North America in 2002–05. Cole's research has received funding from the National Endowment for the Humanities, the Fund for US Artists, the American Association of University Women, the ELA Foundation, the University of California Institute for Research in the Arts, and the National Humanities Center.

Mary Karen Dahl is a Professor of Theatre at Florida State University. She has a longstanding interest in the relationship between performance and politics. She writes on the ethics of violence, theatre and terror, and theatre and political theories of citizenship and the state. Her book *Political Violence in Drama: Classical Models, Contemporary Variations* was selected as a Choice Outstanding Academic Book for 1987. It uses theoretically informed close readings of plays to address a central question, how/if killing to save the world can be justified, using texts by Aeschylus, Bertolt Brecht, Howard Brenton, Albert Camus, Max Frisch, Eugene Ionesco, Slawomir Mrozek, Sophocles, and Ernst Toller. Related essays include "Postcolonial British Theatre: Black Voices at the Center," in *Imperialism and Drama* for Routledge; "Stage Violence as Thaumaturgic Technique," in *Violence in*

Drama, for Cambridge University Press; and "State Terror and Dramatic Countermeasures," in *Politics and Terror in Modern Drama*, for Edinburgh University Press. Her work in progress poses questions about political theory and citizenship: a book manuscript, provisionally titled *Theatre for New Times*, draws on British political thinkers such as Stuart Hall and David Held and contemporary British playwrights including David Edgar, Hanif Kureishi, Maria Oshodi, Winsome Pinnock, and Timberlake Wertenbaker. "Mediating Truth: The Competition to Shape Reality" is the umbrella concept for a series of essays on media and theatrical treatments of contemporary events such as the "Behzti Affair" and the case of Stephen Lawrence in the United Kingdom.

Maya Dodd is Assistant Professor in the School of Liberal Education at FLAME (Foundation for Liberal and Management Education) in Pune, India. She has previously been a Post-Doctoral Research Fellow of the Committee in South Asian Studies and a Lecturer in the Department of Anthropology at Princeton University. Her PhD in Modern Thought and Literature at Stanford University was completed in 2006. As examined in the doctoral thesis, her current research interests continue to focus on questions of justice and political theory in the wake of the Emergency in India. Following an interest in deliberative democracy, she has also worked in talk radio.

Laura Edmondson is an Assistant Professor of theatre studies at Dartmouth College. Her articles on East African theatre and performance have appeared in *Theatre Journal*, *Theatre Research International*, *TDR*, and the anthology *African Performance Arts* (Routledge, 2002). Her book, *Politics and Performance in Tanzania*, is forthcoming from Indiana University Press.

Susan Haedicke is Associate Professor in the School of Theatre, Performance, and Cultural Policy Studies at the University of Warwick in the UK. In addition, Dr Haedicke works as a professional dramaturg, most recently with Friches Théâtre Urbain, a street theatre company in Paris, France. Her adaptation of *Macbeth* for a promenade performance on stilts has been performed at numerous street theatre festivals in Europe and Asia and at the National Theatre in London in August 2005. She worked on a new creation with Dody diSanto to be presented at various East Coast Fringe Festivals in 2006. Dr Haedicke has published several essays on community-based theatre and she co-edited, *Performing Democracy: International Perspectives on Community-Based Performance* (University of Michigan Press, 2001). Her current research looks at European street theatre, particularly in France. Dr Haedicke established Inside French Theatre, an annual summer program for American students in Paris, France in 1999, and has directed the program for the past nine years. Inside French Theatre offers students intensive training in physical theatre, stilts, and aerial acrobatics and the opportunity to perform a play in the parks of Paris

Ketu H. Katrak is a Professor of Asian-American Studies at the University of California, Irvine. Her current research explores the intersections of ethnicities, multiple locations, diasporic politics in Asian American Literature, and the impacts of new immigration on ethnic formation in the Unites States. Professor Katrak's many books include *Politics of the Female Body: Postcolonial Women Writers of the Third World* and *Wole Soyinka and Modern Tragedy: A Study of Dramatic Theory and Practice.*

Eddy Kent is a post doctoral fellow in the Department of English at Rutgers, the University of New Jersey. His research interests include Victorian literature, colonialism, social theory, and ethics. From 2009 he will be an Assistant Professor in the Department of English and Film Studies at the University of Alberta.

Suk-Young Kim is Assistant Professor of Theatre and Dance at the University of California at Santa Barbara. Her research interests include East Asian and Russian theatre, and the questions of gender and nationalism in performance. She is the recipient of the International Federation for Theater Research New Scholar's Prize (2004), ASTR Research Fellowship (2006), and the Library of Congress Kluge Fellowship (2006–07). She is currently working on a book project entitled *Popular Theater and Film in North Korea.*

Sonja Arsham Kuftinec is an Associate Professor of Theatre Arts and Dance at the University of Minnesota. Her research and teaching interests include performance and social change, community-based theatre, and the intersection between theatre and identity. She has published close to 20 articles on theatre and facilitation (*Theater Topics*), Balkan theatre (*Journal of Dramatic Criticism, South East European Performance, Text and Performance Quarterly*) and on Cornerstone and community-based theatre (*Theater Journal, Brecht Yearbook*) featured in her 2003 book *Staging America: Cornerstone and Community-Based Theater.* She received honorable mention for the Bernard Hewitt Award for the 2003 best book in theatre history. Professor Kuftinec also works professionally as a director and dramaturg. Since 1995 she has been creating original theatre and leading workshops with young people across ethno-religious boundaries in Bosnia, Romania, Croatia, Serbia, and Germany. Her production *Where Does the Postman Go When all the Street Names Change?* won an ensemble prize at the 1997 International Youth Theatre Festival in Mostar. In 2002 her production *Between The Lines*, created with Balkan youth from seven countries, opened at Berlin's House of World Cultures. *There is a Field*, an original, collaborative production meditating on the Middle East, premiered on the University of Minnesota main stage in February 2003. Professor Kuftinec also works as a conflict resolution facilitator for Seeds of

Peace, an organization bringing together youth from the Middle East and Balkan regions. Her current research derives from Boal-based workshops with youth in Afghanistan, Pakistan, India, and Jerusalem. She is at work on a new book project, *From Ethnos to Ethics*, which examines the intersections among theatre, facilitation, and nation-formation.

Barbara Lewis, Associate Professor, is the Director of the William Monroe Trotter Institute for the Study of Black Culture at the University of Massachusetts-Boston, where she holds a joint appointment in the Departments of Africana Studies and English. As a theatre historian, she has published on lynching and performance, minstrelsy, and the black arts movement of the sixties. As a playwright, her work has been presented at festivals and on professional stages nationally and internationally. As a Francophone scholar, she co-translated *Faulkner, Mississippi* by Edouard Glissant, which was published by Farrar, Straus & Giroux (1999). Dr Lewis has taught at City College, Lehman, and New York University. Prior to being named Director of the Trotter Institute, she was Chair of the Department of Theatre at the University of Kentucky.

Jon McKenzie is an Associate Professor in the Department of English at the University of Wisconsin. He is the author of *Perform or Else: From Discipline to Performance* and is currently completing *Contesting Performance: Global Genealogies of Research* (co-edited with Heiki Roms). His research explores a wide range of topics, including performance and globalization, new media and teaching, and emerging forms of electronic civil disobedience. Professor McKenzie's most recent essays include "StudioLab UMBRELLA," which traces the development of the pedagogy he calls "StudioLab"; and "Global Feeling: (Almost) All You Need is Love," on the ability of performance and media to help "design experience" on a global scale.

Jisha Menon is an Assistant Professor in the Department of English at the University of British Columbia. She will begin her appointment as Assistant Professor in the Department of Drama at Stanford University in January 2009. Her articles on subcontinental partition, and transnational feminist theatre have appeared in *Modern Drama*, and *Feminist Review*. She is currently working on a book manuscript entitled *Bordering on Drama: Performing Community and Nation in Postcolonial India*; the research for this project is supported by a grant from the Social Sciences and Humanities Research Council of Canada.

Tony Perucci is a theatre director, activist and writer living in Chapel Hill, North Carolina. He is Assistant Professor of Performance Studies in the Department of Communication Studies at the University of North Carolina.

He is currently revising his book manuscript, *Tonal Treason: Paul Robeson and the Politics of Cold War Performance.*

Mark Phelan is a Lecturer in Drama at Queen's University, Belfast. His research focuses on Irish Theatre, specializing in theatre and performance in the North of Ireland. He has published a number of articles on Irish theatre and photography and is currently working on two book projects: a commissioned collection of essays on Stewart Parker and a monograph on the Northern Revival and the Ulster Literary Theatre. He has taught at Trinity College, Dublin, and has been an invited speaker at the Royal Irish Academy for both the 2003 "Representing the Troubles" conference and the inaugural conference of the Irish Theatrical Diaspora (2004).

Peggy Phelan is the Ann O'Day Maples Chair in the Arts and Professor of Drama and English at Stanford University, California. She is the author of *Unmarked: The Politics of Performance*; *Mourning Sex: Performing Public Memories* (honorable mention Callaway Prize for dramatic criticism 1997–99); the survey essay for *Art and Feminism*; the survey essay for *Pipilotti Rist*; and the catalog essay for *Intus: Helena Almeida*. She is co-editor, with the late Lynda Hart, of *Acting Out: Feminist Performances* (cited as "best critical anthology" of 1993 by American Book Review); and co-editor with Jill Lane of *The Ends of Performance*. She has written more than 60 articles and essays in scholarly, artistic, and commercial magazines ranging from *Artforum* to *Signs*. These essays have been cited in the fields of architecture, art history, psychoanalytic criticism, visual culture, performance studies, theatre studies, and film and video studies. She has edited special issues of the journals *Narrative* and *Women and Performance*. She has been a fellow of the Humanities Institute, University of California, Irvine; and a fellow of the Humanities Institute, The Australian National University, Canberra, Australia. She served on the Editorial Board of *Art Journal*, one of three quarterly publications of the College Art Association, and as Chair of the board. She has been President of Performance Studies International, and has been a fellow of the Getty Research Institute and a Guggenheim Fellow.

Freddie Rokem is Professor of Theatre Studies at Tel Aviv University and has served as the Dean of the Faculty of the Arts (2002–06). He is also a permanent visiting Professor (Docent) of Theatre Studies at Helsinki University. His book *Performing History: Theatrical Representations of the Past in Contemporary Theatre*, published by University of Iowa Press, received the ATHE Prize for best book in theatre studies for 2001. His most recent book, *Strindberg's Secret Codes* was published by Norvik Press in 2004. He has also published numerous articles in scholarly journals and chapters in books. Rokem is the editor of *Theatre Research International* (2006–09). He is also a translator and a dramaturg, and serves as Vice-President of Performance Studies International (PSi) and is a member of the executive committee of The International Federation for Theatre Research (IFTR).

Leslie A. Wade is currently Associate Professor of Dramatic Literature, Theory, and Criticism at Louisiana State University. A doctoral graduate of the University of California at Santa Barbara, Wade has presented his work at national and international theatre conferences and has published essays in numerous journals including *Theatre Studies, Text and Performance Quarterly, Theatral Annuaire, Journal of Dramatic Theory and Criticism, Theatre Symposium,* and *Western European Stages*. His book, *Sam Shepard and the American Theatre* was published by Greenwood Press (1997). He serves on the editorial board of *Theatre History Studies* and the *American Theatre* series of Southern Illinois University. Recipient of a Louisiana State Arts Council's fellowship in playwriting, Wade has also received writing awards from the Association for Theatre in Higher Education, the Los Angeles Arts Council, and the American College Theatre Festival. He has twice directed the grant-project "Native Voices and Visions," a script-writing competition for Louisiana playwrights. Wade currently serves as the LSU Theater Department's director of graduate studies and co-director of LSU in London.

Acknowledgements

No book is written in absolute isolation. This is especially true for an edited volume, and we are grateful to our contributors for their diligent work and constant attention to detail. *Violence Performed* began as a seminar at the 2003 conference of the American Society for Theater Research called "Documenting Violence/Violating the Document," chaired by Janelle Reinelt and Jisha Menon, and we would like to thank all the participants at that original seminar. Janelle Reinelt and Brian Singleton, the editors for the *International Performance* series at Palgrave Macmillan, have been exceedingly supportive as we nurtured this project into a book – as have Paula Kennedy, Steven Hall, and Penny Simmons. We are profoundly grateful to Katrina Hoch and Mobeley Luger, our graduate research assistants at UCSD and UBC respectively, for their rigorous copy proofing and commentary.

We are grateful to the Johns Hopkins University Press for permission to reproduce Chapter 8, originally published in *Theatre Journal* 59:2 (2007): 167–87; and for permission to reproduce Chapter 12, originally published in *Theatre Journal* 54:4 (2002): 555–73. We are grateful to Taylor & Francis, Ltd, for permission to reproduce Chapter 9, originally published in *Cultural Studies* 18:6 (2004): 816–46.

Our colleagues and students in the Department of Communication at UCSD and the Department of English at UBC, also deserve special thanks for providing the institutional and intellectual context in which we were able to work. In particular, we would like to thank Robert Horwitz, Dan Hallin, Giovanna Chesler, Nitin Govil, and David Serlin at UCSD; and Patricia Badir, Richard Cavell, Mary Chapman, Sian Echard, Sneja Gunew, Tina Lupton, and Laura Moss at UBC.

Our own mentors are responsible for driving us to ask the questions at the heart of this book: Shannon Jackson, Kaja Silverman, Ruthie Gilmore, and W. B. Worthen at Berkeley; Harry Elam, Peggy Phelan, Akhil Gupta, and Purnima Mankekar at Stanford; Della Pollock at Chapel Hill; Ania Loomba at the University of Pennsylvania; and D. Soyini Madison and the late Dwight Conquergood at Northwestern.

We would also like to thank our families for their enduring support, especially Haridas Menon, Sanjay Rajagopalan, and George Iwaki, Finally, it is to the women who gracefully and selflessly taught us how to truly *be* in the world – Chandrika Menon and Janis Haynie Anderson – that we owe our largest debts.

P. A. and J. M.

Introduction: Violence Performed

Patrick Anderson and Jisha Menon

> *Violence remains the founding language of social representation.*
> Allen Feldman, *Formations of Violence*

In the relatively few years since 11 September 2001 – and beginning precisely with the repetition of images from that day – we have seen an unprecedented increase, if not in the actual performance of state-sanctioned violence, in the representational prominence of violence both as an awesome spectacle to behold and as a domain of political discourse that dominates contemporary world-making. Raymond Williams's "dramatized society" – the infusion of everyday life with more and more forms of theatrical, cinematic, and televised productions of "reality" – has now explicitly, almost breathlessly, become fixated on transnational political exchanges whose central terms are the lives and deaths of marginalized citizens.[1] Indeed, the very notion of "citizenship" has become analogous to the visual field, where mass surveillance becomes an accepted, even embraced, component of the everyday; where hidden "pocket cells" of militarized resistance are marked as outside the logic of the subject of "democracy"; where the demand for vengeful "justice" is so great that it can only be satisfied with endlessly reproduced spectacles of suffering both in the United States (and its fragile "coalition" of "allies") and abroad.

At the same time, such techniques of producing and reproducing violence have expanded the possibilities for resistance to the direst forms of state power. Beginning most infamously with the videotape documenting the beating of Rodney King, and continuing full-force through the public "leak" of images depicting prison torture at Abu Ghraib, representations of state-sanctioned violence have also taken center stage in the most colloquial forms of contemporary political discourse in many international locations and transnational networks (including the mass media and the internet). Just as the state has rallied behind what Dwight Conquergood called its "monopoly on violence," political resistors have strategically organized around their own substantial power to stage the image and the effects of

1

violence in photographs, low-tech video, street theatre, protests, documentaries, experimental films, pirate television shows, and other performance genres.[2] Consider, for example, the December 2001 transfer of political prisoners on hunger strike in Turkey from their dormitory cells to isolation cells in the new American-designed prisons. Knowing the prisoners would use any moment of exposure to document the torture inside the prisons, state agents planned for the transfer to take place secretly in a late-night ambush. But, as Metin Yegin's underground films document, the prisoners refused to allow such access to visibility to elude them: covered in scars, open wounds, and burns, and being moved quickly from a dark building to the police vans waiting to transport them, the inmates seized the chance to put themselves on stage and shout their testimonials, knowing the cameras were rolling: "Dozens were killed! [...] Look at my burns! [...] Six of my friends are dead!"[3]

Consider, too, the devastating images of New Orleans after hurricane Katrina made landfall in late 2005. Drawing comparisons between the ravaged landscape of the lower ninth ward and low-lying areas of "third world" South and South East Asian countries after a 2004 tsunami, media outlets infused the Mississippi delta with cameras and reporters. Videos and photographs, projected and relayed around the world, reproduced the spectacle of the newly homeless citizens of New Orleans desperate for assistance. As the American government slowly stumbled to organize relief, these citizens took advantage of their newfound (if still anonymous) hyper-visibility to document their suffering and to demand food and rescue. Stuck in flooded homes, many organized their few belongings into makeshift texts on their roofs, legible only when viewed from the perspective of cameras circling in helicopters overhead; congregating in overcrowded municipal and sports facilities, many others staged ad hoc demonstrations in efforts to capture the reporters' attention. Co-opting these techniques of representation and surveillance, New Orleans residents set the terms for a national debate on the intersections between the effects of a national disaster and discrete histories of political and infrastructural neglect.

The networks through which contemporary world politics are increasingly produced and performed are similarly intensified by the practice and the spectacle of violence. Just as American mourning became the dominant *tone* in immediate and ongoing responses to 11 September 2001, the threat and promise of violence has dominated the *practice* of staging new forms of nationalist discourse shaping global alliance-building in the last several years. Whether calling their project "jihad," "democracy," "freedom," or "terrorism," and however fragile their consensus, various global alliances have been radically forged and relentlessly facilitated through the continuous staging of violence as "resistance," "response," "coercion," and "liberation." Moreover, the moral qualities attributed to such acts – what we see most vividly and dangerously in language such as "if you're not with us, you're against us" – depends first and foremost upon the sanctioning of violence when performed by agents of the state. That is, to delineate between righteous brutality and insurgent aggression is simultaneously to endorse the productivity of violence with one side of the mouth while condemning that same productivity with the other. If, as Allen Feldman argues in the epigraph that opens this introduction, social representation has come to *depend* upon violence, then the question of who owns and consents to specific violent acts becomes critical.

As scholars in the field of performance studies, we are both uniquely qualified and ethically obligated to explore specific sites of violence as well as larger questions about how violence is specifically *performed*. Thus, the aim of the present volume is to consider the significance of performance in a world where political violence variously destabilizes and reinforces national boundaries and other articulations of belonging and abjection. As we can see in the examples cited above, our historical moment has witnessed the increasing mediatization of violence saturating the public sphere and creating new spectacular forms of non-state terror while rendering banal institutional forms of state terror. The authors represented here explore the ways in which alternative performance genres intervene into or reproduce dominant, mainstream accounts of sub-national, national, and transnational violence.

Using the analytical categories offered by the burgeoning field of performance studies, these authors explore the relationship between performance and political violence. This collection reflects recent developments over the last decade, which has witnessed a shift from studying violence within plays to understanding the performative role of violence in socio-cultural contexts. This theoretical shift from the paradigm of theatre to performance has crucially altered the field's contours and objects of study. Most notably, performance studies has ventured outside of the traditional theatrical space to explore various other forms of representation – film, television, photography, and the written word, among others – and has expanded, by re-imagining the very concept of performance and offering sustained political critiques of our modern world. Further, the expansive model of the "performative" initiated by J. L. Austin – and, in Jill Dolan's words, its "promiscuous" citation across the disciplines – has altered our understanding of how language produces and shapes the notion and the experience of subjectivity.[4] The chapters included in this volume convene under these terms, and argue for a re-consideration of what performance and performativity can teach us about local and global enactments of violence.

Not only has the growth of performance studies offered productive analytical tools for the study of violence, but the dramatic changes occurring in global politics also require a fresh interrogation of the relationship between violence and representation; 11 September 2001 was a pivotal historic moment that vividly demonstrated the spectacular power of transnational networks of globalized violence. Subsequently, the world witnessed the globalization of resistance against the wars in Afghanistan and Iraq in de-territorialized enactments of political communities. While attentive to the cataclysmic upheavals 11 September produced within America, this volume also decentralizes discussions of American suffering by locating it within global geographies of violence. In this way, the chapters collected here consider the internal dynamics and struggles specific to various countries in the global North and South.

Although we do not attempt to propose a specific theory of violence *per se*, we assemble within a theoretical terrain that bears some explanation. In what follows of this introductory chapter, we lay out several suggestions for how we may rethink the nature of violence as it is specifically *performed*, before turning, finally, to describe the collected chapters. These proposals are tentative, offered as starting points for a reconfiguration of the relationship between violence and performance rather than as definitive answers.

First, we argue that enactments of violence are both *spectacular* in their cultural impact and *embodied* in their transaction and effect. We begin with the acknowledgement that violence has become foundational to contemporary visual culture and thus has acquired, at breakneck pace, a profound power to command attention. The visual power of violent acts begins with the horror – often experienced with a perverse delight – they initiate in those who have access to the technologies that purport to "capture" them, and accelerates through the accumulation of mass audiences who are alternatively moved, sickened, inspired, or repelled by the terrifying images of their effects. Indeed, the spectacle of violence has so dramatically acquired momentum that it has spawned a cognate form of viewership: the seemingly reflexive picture of individual and collective witnessing. But the sheer range of images in this secondary category – ranging, for example, from photographs of onlookers in shock after the United States military dropped a bomb on a wedding party in Afghanistan to video coverage of Palestinians described as "cheering" as news of 11 September spread around the world – speaks not to the self-awareness of those who produce and display footage of violent acts, but rather to the overwhelming desire to reconcile viewers' discomfort with their own inability to look away.

In other words, the perpetuated practice of witnessing – perhaps most compellingly critiqued as "impossible" in Giorgio Agamben's *Remnants of Auschwitz* – is neither accidental nor innocent in the context of contemporary politics. At the same time that violence dominates our visual field, it remains intensely localized in its enactment. That is to say, while the spectacle of violence has operated similarly to Anthony Giddens's famous image of the planet Earth in producing a globalized modernity, the effects of violence are most acutely felt at the level of the inexorably mortal human body. Despite the potential for empathy facilitated by the photograph or the videotape – the sense that we, after looking, have "really been there" – it is the experience of suffering that is most considerably lost when images of violence inundate the visual realm, acting as surrogates for productive transnational discourse. Violence, then, acquires its immense significance in a delicate pivot between the spectacular and the embodied; it is precisely this quality that demands consideration by scholars in performance studies.

Second, we argue that violence is a *binding, affective* experience that cross-cuts the domains traditionally registered and distinguished as the physical, the psychic, and the social. The most banal and ultimately dangerous apology for violence is that it is the "last resort" in situations where diplomacy (or, in carceral systems, "rehabilitation") has failed to produce desired results. Such an argument implies that violence is merely a symptom of failed negotiations or, as we often see in state-produced propaganda, the only "civilized" response available in confrontations with "primitive" or "barbaric" communities. It is not difficult to see through such claims to the underlying motives of perpetrators of political violence; what is forgotten, however, is the manner in which acts of violence initiate networks of global alliance and discord that define contemporary politics. In other words, the global significance of violence is not simply in its creation of a distinction between aggressor and casualty, but in its pseudo-contractual function to bind involved parties in a mode of engagement defined by domination and facilitated in suffering.

Further, the suffering that extends from violent acts is illegible on any single, isolated register – it is not simply physical, psychic, or social. Rather, the effects of violence oscillate between the individual and the collective, and are unconsciously as well as somatically significant. The binding power of violence operates on each of these levels, initiating a future-history of trauma that does not merely *describe*, but performatively *produces* power relations. In other words, the interlacing physical, psychic, and social effects of violence performed extend forward into a political structure in which the promise of power is answerable to, and fully dependent upon, an ongoing relationship of coercion.

Third, we argue that conventional distinctions between "victim" and "aggressor" are often ill-suited to fully explain the effects of violence. This is not to deny that in discrete scenarios certain parties take on an active role in the perpetuation of violence; nor is it to ignore the destructive, often deadly intentions behind such actions. Rather, we mean to situate violence within a network of conflict whose complexities are forgotten in the binary language of domination and resistance. Conceiving of violence purely in terms of cause and effect – and organizing against violence around the theme of victimization – dumbs down the intricate problem of violence's productivity in the contemporary political sphere.

Arguments against such so-called "victim cultures" have been staged across the disciplines, but are perhaps best articulated in political theorist Wendy Brown's landmark *States of Injury*. Brown compellingly contends that the "wounded attachments" at the core of many contemporary political claims radically de-historicize the experience of suffering and of harm, and in the end reproduce the spectacle of various communities as defined by the aggression that has affected them. She writes: "History becomes that which has weight but no trajectory, mass but no coherence, force but no direction: it is war without ends or end. [...] The past cannot be redeemed *unless* the identity ceases to be invested in it, and it cannot cease to be invested in it without giving up its identity as such, thus giving up the economy of avenging and at the same time perpetuating its hurt."[5] In reconsidering violence not as that which will define forever the individuals and communities affected, but as an intensely significant moment in the context of historicity, we divest violence of its power to perpetuate its own effects on its victims (*as victims*) and we thus empower those who have experienced the horror of violent acts by refusing to see them as forever classified by victimization.

Fourth, we argue that representations of violence are not innocently mimetic, and risk extending and perpetuating the very trauma they aim to expose. This suggests, of course, that representations of violence are both descriptive and performative: not merely involved in staging and framing specific acts of violence, but also of *producing* the context in which violence is rationalized and excused as a symptom of inter-cultural encounter. In granting prominence, for example, to images documenting the events of 11 September 2001, violence performed in response is justified by the context (and in the interests) of American mourning; similarly, in continuously reproducing the image of ravaged communities in Afghanistan and Iraq before the American invasions, we perpetuate the myth of "democracy" as a justification for centuries of territorial occupation, marginalization, and neglect. On the other hand, in exploring representations of violence as intricately involved in ongoing cultural formations, we remain attentive to the significance of representation in the "disjunctural" cultural flows that characterize globalization and transnational exchange.

Further, imagining violence as both performance and performative radically reconfigures the "Cartesian problem" that underwrites distinctions between the discursive and the material. Often imagined as the difference between representation and "reality" – or, in other spheres, between "theory" and "practice," interpretation and authenticity, "talking the talk" and "walking the walk" – such distinctions promote a generalized amnesia around the politics of visibility or, in the language of semiotics, the power of the sign. This amnesia is nowhere more dangerous than in response to representations of violence, where it threatens to further impugn the victims of violent effects while, at best, redeeming (or at worst, glorifying) violence itself. The authors collected in this volume explore representations of violence not simply as residual artifacts of, in Saidiya Hartman's words, "scenes of subjection," but as actively engaged in the promotion of violence as a cultural force.

Finally, as suggested above, we argue that scholars in performance studies are ethically obligated to explore specific sites of violent acts as well as larger questions about the performative ontology of violence. Already colloquially accused of working outside the "real world," we bear full responsibility for how and where we direct our attention and to what ends we use our proficiency as cultural critics. That is to say, as scholars we bear witness to the subjects we choose to study, and witnessing is a political act. In bringing into dialogue the subjects explored by the authors represented here, we stage a new interrogation of the performance and the performativity of violence in the context of a world circumscribed by suffering, by marginalization, and by differential access to the economy of cultural capital. We do not enter the conversation about violence lightly; we know that the stakes in doing so are precisely life and death. We offer this collection as an urgent summons to scholars working in performance studies, and as a step towards developing an expanded vocabulary on violence with scholars from other fields.

Given these suggestions for how the field of performance studies may offer original contributions to research on violence, we should now trace the logic of inclusion for the collected chapters. At its core, *Violence Performed* is intended to reflect upon the multiple ways in which political violence mediates conceptions of self, community, and nation; the chapters included here thus explore the role of physical, psychic, and institutional violence in the constitution of race, ethnicity, gender, and nationality as these categories of human experience provide the context for political subjectivity. While paying close attention to specific locations, these chapters are also attentive to the global economies of identification and difference that define the conditions of contemporary world-making. *Violence Performed*, then, spotlights the structured, asymmetrical relations of global power and the epistemic and material violence of cross-cultural and inter-racial encounters. By concentrating on rather than sanitizing the asymmetrical relations of power in what Mary Louise Pratt refers to as the "contact zone" of racial, cultural and national alterity, these chapters redress the de-politicization that often characterizes scholarly discourses of inter-culturalism.[6]

The collection opens with Leslie A. Wade's discussion of an "ethics of inter-subjectivity" that lays out the discursive framework within which the subsequent chapters unfold. Bringing Levinas, Lyotard, Derrida, and Nancy into conversation, Wade suggests in Chapter 1 that inter-subjectivity – what he calls "being with or alongside" rather than being *as* – is initiated when a subject, faced physically or psychically with radical otherness, reforms his or her sense of self. Concluding his discussion with David Hare's *Stuff Happens* Wade contends that staged representations of such encounters – and their production of a social scenario in which an audience is similarly faced with alterity – potentially refashions the terms of the self-other binary and "initiates an attitude of hospitality for the stranger, the refugee, the abject."

What Wade calls the "traumatic" encounter between self and other emerges as the grounds on which the subsequent chapters unfold as interrogations of violence performed. In Susan Haedicke's, Laura Edmondson's, and Eddy Kent's contributions, the scene of that unfolding centers on the dialectic of nation and migration. In these three chapters, the nation (and its base unit, the normalized citizen) becomes defined by the racialized production of the other in the heart of the neocolonial metropole. Haedicke explores the interface of self and other by interrogating the various ways in which the national "interloper" – the refugee or the exile – is resisted, assimilated, or otherwise marked as "outsider" in France. This exploration is founded on a re-consideration of both the discursive distinctions between "home" and "street" and the ethical dimensions of forging and maintaining social space; the street thus emerges in Haedicke's chapter as both a physical and symbolic arena in which the boundaries between the private self and the public social are thrown into crisis. Concentrating specifically on immigration as a political problem negotiated in public space and contested in theatrical representation, Haedicke turns in Chapter 2 to French street theatre and argues that such public performances offer social actors the opportunity to stage new possibilities for "post-national" citizenship.

While Haedicke considers the critique of institutional and colloquial perceptions of the grotesque "outsider" offered by street theatre, our next chapter explores the romanticization of the refugee in mainstream Western media. In Chapter 3 Laura Edmondson analyzes discourses of displacement in three British plays that address the 1994 Rwandan genocide. Examining how the "displaced" Rwandan body becomes an object of fascination for the normative Western citizen while the "placed" Rwandan is ignored, Edmondson calls our attention to the violent erasure of Rwandans confined within the nation-state. The play between Rwandans "at home" and those forced to live "abroad" appears in this chapter as both a crisis for, and the condition of, the production of contemporary nationalisms.

Leaving one's "home," especially under conditions of violent duress, radically disorders customary notions of belonging and territorial allegiance. The real or potential severance from territorial affiliations offers an occasion to consider the emotional valences within the notion of "home" as a site of stability and refuge but also as one fraught with repressions and cruelties. While Edmondson considers the ways in which the displaced Rwandan refugee becomes an object of Western fascination, Eddy Kent is interested precisely in the unhomeliness experienced by the "partial citizen" within the racist and exclusionary social structures of the United Kingdom. Kent offers in Chapter 4 a reading of Kureishi's *My Son the Fanatic* by exploring the abjection of racialized Muslim minorities in a mythical, working-class English town. Assuming that the goal of a liberal society is to make its members feel at home in their world, Kent asks, what can be done when the institutions of one's world are being transformed by forces over which individuals have very little power? Working with Derrida's understanding of hospitality, Kent argues that if the goal of liberalism is to create the conditions in which each citizen positively affiliates with the social and political institutions of the state, then Kureishi's *My Son the Fanatic* is an exemplary – and deeply troubled – expression of the ideology of postcolonial liberalism.

Haedicke, Edmondson and Kent point to the complexities that racialized "outsiders" experience as peripheral subjects, evoking their otherness through the troubling specter of non-European ethnic roots and transnational routes. The double temporality that these "outsiders" negotiate positions them at once as local and global, with affective transnational affiliations. In the two subsequent chapters, Barbara Lewis and Mary Karen Dahl consider the everyday, institutionalized violence by which minorities are produced as outsiders within their own nations. By exploring the violent racialization experienced by minority subjects, these two chapters offer a critique of the exclusions upon which the fraternity of the nation is imagined. As Ella Shohat has reminded us, "representation of an underrepresented group is necessarily within the hermeneutics of domination, overcharged with allegorical significance."[7] Lewis and Dahl draw on Shohat's caution in their investigations of the structural forms of violence that, taking spectacular and/or more everyday forms, dehumanize, socially exclude, and depersonalize racialized minorities.

Barbara Lewis explores how lynching photography has contributed to the formation of a white American citizenry by making spectacular the racial abjection of African American bodies. Using historical photographs, tourist artifacts, and museum exhibitions, Lewis examines the inter-articulation of minstrelsy and lynching rituals in the production of an objectified, aberrant, and grotesque black body. But rather than casting the lynch-victim as a ghostly reminder of democratic "progress," in Chapter 5 Lewis demonstrates that the spectacular abjection of racialized bodies cannot be mourned simply as what *was*. Indeed, the recent repression of minorities in democratic and liberalized nations ironically corresponds to the "free" market and the increasing mobility of global capital. The question of race and minority citizenship is taken up again in Dahl's analysis in Chapter 6 of the recent murder of Stephen Lawrence in the United Kingdom; in a harrowing echo of both Barbara Lewis and Eddy Kent, Dahl maps the violent ways in which minority bodies are produced as incomplete citizens. In considering documentary film and drama based on the Stephen Lawrence murder, Dahl asks, "How are we to assess representational practices that effectively violate the subject of violence a second time, subsuming their deaths in another agenda? Does the act of representation inevitably violate its object?"

Here, Dahl echoes concerns raised by Leslie Wade in his discussion of *Stuff Happens*. While criticizing the erasure of Iraqi voices from a play that ostensibly represents the Iraq war, Wade remarks, "their otherness may have been subordinated to the ends of a message, to the advancement of a moral or political vision." Dahl and Wade render visible the politics of the mediating artists and are careful to disentangle the conflation of representations that Gayatri Spivak has identified as "proxy" or the political act of speaking for and "portrait" or aesthetic re-presentation of reality.[8]

Maya Dodd's chapter on violence and political belonging in post-Independence India joins Lewis and Dahl to consider the centrality of violence in the construction of political community in the global South. Through a taxonomy of infamous, famous and anonymous bodies, Dodd delineates in Chapter 7 how mortal bodies serve the logics of communitarian identity politics in India. Here, individual acts of violence are re-narrativized through the collective logic of communities. The significance of the physical body to a national project reveals that "literacies of corporeality" are as much a response to the daily violence of the state in fixing social meaning as they are to conflict resolution by other means. According to Dodd, sacrificial violence performed in the name of community, seeks not only to affirm bounded community identity, but also embodies responses to the Indian Constitution's avowal of secularism and socialism. Historicizing and contextualizing the liberal construction of the "private sphere" within postcolonial communities reveals the material contingencies that underpin recurrent violence in post-Independence India.

If, as Dodd argues, justice in India is reclaimed and re-staged by violent communitarian assertions in the spectacular public sphere, then how is performance deployed in the larger field of transitional justice in South Africa? Attending to the embodied, public, and performed dimensions of the South African Truth and Reconciliation Committee, Catherine M. Cole argues in Chapter 8 that it is in the ruptures of performance that the "completeness" of the vision of the apartheid past emerges. According to Cole, detailed analyses of moments of performed testimony deliver the "completeness" that the TRC was mandated to achieve in ways that the macro-narratives and quantitative analyses of the commission have been unable to provide.

While Dodd and Cole explore the co-implication of retributive and restorative justice and performance in the competing narratives of community and state within postcolonial societies, Patrick Anderson's chapter turns to the prison, an exemplary site of institutional power, and evaluates the hunger strikes that began in prisons across Turkey in 2000. Considering the hunger strike as an instance that dramatically fuses the subject and object of state violence into a single body, Anderson asks in Chapter 9 what kind of political subject is produced, and how her/his relationship to the state is redefined.

Dodd, Cole and Anderson explore the ways in which institutional state power is interrupted, and called into question, albeit incompletely, by spectacular and embodied enactments of resistance. While Dodd, Cole and Anderson deploy "performance" to study a range of embodied practices within communitarian, juridical, and carceral sites within the territorial and jurisdictional precincts of the Indian, South African, and Turkish nation-states, our next chapters query if the political imagination is necessarily bound to national membership.

Can we imagine forms of political communities outside the nation? Our next two chapters take up this question to explore the ways in which political art can mediate bounded and exclusive notions of culture and community.

Ketu H. Katrak demonstrates in Chapter 10 that the creative and political work of two "ethno-global," artist-activists, Denise Uyehara and Arundhati Roy, with political investments in spaces beyond their resident nation-states, offer a corrective to the rhetorics of democracy, patriotism, and development in the United States and India. Uyehara's critique of hate-crimes directed against racialized immigrants in the wake of the September 11 attacks and Arundhati Roy's denunciation of American military imperialism, and the rhetorics of development that undergird the Narmada Valley Dam projects in India position them as artists committed to speak to social inequities within and beyond national borders.

Sonja Arsham Kuftinec's activist and academic work takes up the potential for theatre to intervene into and positively affect the violent public sphere by offering alternative ways to imagine the relationship between self and society. Kuftinec contends that participatory theatre can interrupt violent assertions of ethnic community and steer towards a way to imagine ethical communities instead. Focusing on three projects in Kabul, Macedonia and Jerusalem that utilized Image Theater techniques, in Chapter 11 Kuftinec explores the efficacy of the theories and practices of Boal and Freire to raise the critical consciousness of youths in conflict or post-conflict situations.

Freddie Rokem, like Kuftinec, turns to theatre to explore the ability of art to mediate, reflect, and comment upon terror. Moving from racialized violence within nations to conflict between nations, in Chapter 12 Rokem considers the critical conflict between Israel and Palestine through the lens of dramatic representation. Focusing our attention on the potential for art production to effect "real world" social change, Rokem asks, "Can theatre bring about a change of attitudes towards such phenomena within the Israeli political and ideological contexts?" Here Rokem echoes Jill Dolan's influential attempts to explore the potential for theatre to revise public narratives of political history. As Dolan puts it, "Just the act of going to the theatre, of demonstrating a willingness to see and hear stories that might not otherwise be accessible, models a hopeful openness to the diverse possibilities of democracy. [...] Theatre and performance and the academic departments in which they're studied are ideal places to rehearse for participatory democracy."[9]

Mark Phelan's chapter also rehearses the unfinished negotiations of belonging and displacement that are the legacy of partitioned nations. Turning to Northern Ireland, in Chapter 13 Phelan considers the ghosting of the Disappeared, interred in the womb of the motherland. Here too the landscape consumes bodies; the Irish Republican Army secretly abducted, "executed," and buried eight people in unmarked graves in the 1970s and early 1980s. Phelan contends that David Farrell's photographic exhibition that captures the desecration of "Innocent Landscapes" offers a profoundly ethical medium for representing the Disappeared because of the "explicit dialectic between the material presence of photographic artifact and the profound sense of a doubly absent photographic subject in the form of the 'Disappeared', as the innate ambiguity of photography, given its paradoxical sense of materiality and loss and the simultaneity of presence and absence."

Suk-Young Kim, like Phelan, demonstrates in Chapter 14 that the pristine beauty of idyllic landscapes is inscribed and continually haunted by the public memories of violence. Here, again, landscapes are not innocent as the North Korean regime cannily reconfigures and stages "nature" for tourist consumption. Kim discusses the disjunctive modes of national performance manifested via tourism and hunger that reveal the government's desire to display or conceal its people in specific modes of behavior on an international stage. Violence becomes the North Korean regime's major mode of regulating its citizen's bodies, both in its brutal punishment of those who attempt to escape the country and in its crafty display of stage performers and tourist guides. Kim's analysis takes "drama" out of the conventional theatre space and into North Korean governmental and touristic venues, examining the directorial strategies used in the regime's violent regulation of the bodies of its people. The representations of partitioned nations, whether in India-Pakistan, Israel-Palestine, North-South Korea or North-South Ireland, engender profound private and public memories that surface not only as sinister hauntings of secreted violence but also function as an enduring, contentious struggle over belonging and place.

The final two chapters (and the closing Afterword) take us to the scandalous drama of prison torture that unfolded at Abu Ghraib. While Phelan and Kim elucidate the ways in which nationalist narratives are inscribed upon abjected bodies, in Jon McKenzie's analysis, the US "torture machine" both inscribes and reads the prisoner's body, punishing and gathering intelligence simultaneously. Using the concepts of executive performativity, constituting "speech acts by state and non-state actors," and theatrical performativity, "live and mediated forms of embodied action," in Chapter 15 McKenzie illustrates the anachronistic return of the scaffold and the sovereign, their power and scale now multiplied and dispersed across the globe. The current of the US torture machine flows both ways: the executive performativity of decision-makers enacted upon abject, imprisoned Iraqi bodies flows back as the theatrical performativity of interrogation yields "actionable intelligence."

Tony Perucci's chapter meditates upon the covert operation of neoliberal corporate power that drives the American Military Industrial Complex and

Prison Industrial Complex, institutional networks that, in turn, have eventuated the enactment of torture in Abu Ghraib. Perucci insists upon the
need to properly contextualize the spectacular display of torture in terms
not of individual aberrance, but rather of the structural and institutional
conditions of liberal democracy's nation-building. He explains, "Abu Ghraib
represents the staging of America's new penal culture on the world stage,
which celebrates a racialized brutal violence – a culture that is produced
both by the economic structures and the cultural logic of neoliberalism."
In Chapter 16 Perucci explores how the spectacular display of torture represented in the photographs of Abu Ghraib conceal the covert operations of
neoliberal corporate America.

Our closing chapters return us to the question of the ethical imperative
that structures the self/other relation that opens this volume. The centrifu-
gal organization of the chapters moves from the local to the global, but is
careful to complicate this binarism by looking at the co-implication of roots
and routes. From telescoping the powerful racialization of "outsiders" and
minorities within nation-states, the volume expands its geographical focus
by considering violent conflict between and within nation-states. While
several of the included chapters consider the repressions inherent in the
mechanics of national identity-making, the volume concludes by magnifying the scope of enquiry outside the frame of the nation and explores the
ways in which global performativity and neoliberal capital are implicated
within multiple geographies of violence.

Collectively, the chapters in this volume offer multiple and heterogeneous analyses of political violence that circumvent the monologism of
national and international accounts of political conflict. As an ensemble,
these essays explore the ways in which violence insidiously infiltrates the
borders between the self and society, and initiates a dissolution of the boundaries separating the intimate and the public. The nostalgia of unfinished
territorial affiliations, the trauma of corporeal violence, and the haunting
remorse of perpetrating violence all signify the collapse of a moral universe,
a world where the unthinkable is rendered banal. These experiences are
located in far more recalcitrant archives and it is to these incommensurable
narratives that this volume bears witness.

Notes

1. Raymond Williams, *Drama in a Dramatised Society: An Inaugural Lecture* (New York: Cambridge University Press, 1975).
2. Dwight Conquergood, "Lethal Theatre: Performance, Punishment, and the Death Penalty," *Theatre Journal*, 54:3 (2002): 342.
3. Cf. Patrick Anderson, "'To Lie Down to Death for Days': The Turkish Hunger Strike, 2000–2003," Chapter 9 in this volume.
4. J. L. Austin, *How to Do Things with Words* (New York: Oxford University Press, 1965).

5. Wendy Brown, *States of Injury: Power and Freedom in Late Modernity* (Princeton: Princeton University Press, 1995).
6. Mary Louise Pratt, "Arts of the Contact Zone," in Gail Stygall (ed.), 2000, 573–87.
7. Ella Shohat, "The Struggle over Representation: Casting, Coalitions, and the Politics of Identification," in De La Campa and Kaplan, 1995, 166–78.
8. Spivak, Gayatri, "Can the Subaltern Speak?" in Nelson and Grossberg, 1988, 271–313.
9. Jill Dolan, "Rehearsing Democracy: Advocacy, Public Intellectuals, and Civic Engagement in Theatre and Performance Studies," *Theatre Topics*, 11:1 (2002): 1–17.

Works Cited

Agamben, Giorgio. *Remnants of Auschwitz: The Witness and the Archive*. New York: Zone, 2000.

Austin, J. L. *How to Do Things with Words*. New York: Oxford University Press, 1965.

Brown, Wendy. *States of Injury: Power and Freedom in Late Modernity*. Princeton: Princeton University Press, 1995.

De La Campa, Román and E. Ann Kaplan (eds). *Late Imperial Culture*. London: Verso, 1995.

Conquergood, Dwight. "Lethal Theatre: Performance, Punishment, and the Death Penalty." *Theatre Journal*, 54:3 (2002): 339–67.

Dolan, Jill. *Geographies of Learning: Theory and Practice, Activism and Performance*. Middelton, CT: Wesleyan University Press, 2001.

Dolan, Jill. "Rehearsing Democracy: Advocacy, Public Intellectuals, and Civic Engagement in Theatre and Performance Studies." *Theatre Topics*, 11:1 (2001): 1–17.

Feldman, Allen. *Formations of Violence: The Narrative of the Body and Political Terror in Northern Ireland*. Chicago: University of Chicago Press, 1991.

Giddens, Anthony. *Runaway World: How Globalization is Reshaping Our Lives*. New York: Routledge, 2000.

Hartman, Saidiya. *Scenes of Subjection: Terror, Slavery, and Self-Making in Nineteenth-Century America*. New York: Oxford University Press, 1997.

Kaufman, Moises. *The Laramie Project*. New York: Vintage, 2001.

Kureishi, Hanif. *My Son the Fanatic*. New York: Faber & Faber, 1999.

Nelson, Cary and Larry Grossberg (eds). *Marxism and the Interpretation of Culture*. Chicago: University of Illinois Press, 1988.

Pratt, Mary Louise. "Arts of the Contact Zone." In Stygall (ed.), 2000, 573–87.

Shohat, Ella. "The Struggle over Representation: Casting, Coalitions, and the Politics of Identification." In De La Campa and Kaplan 1995, 166–78.

Spivak, Gayatri. "Can the Subaltern Speak?" In Nelson and Grossberg (eds), 1988, 271–313.

Stygall, Gail (ed.). *Academic Discourse: Readings for Argument and Analysis*. Fort Worth: Harcourt College Publishers, 2000.

Williams, Raymond. *Drama in a Dramatised Society: An Inaugural Lecture*. New York: Cambridge University Press, 1975.

Yegin, Metin. *After...* (Film). Distributed by the International Action Committee, Boston, MA.

Yegin, Metin. *F: To Lie Down to Death for Days* (Film). Distributed by the International Action Committee, Boston, MA.

1
Sublime Trauma: The Violence of Ethical Encounter

Leslie A. Wade

Performance events – from the brutalism of Sarah Kane's *Blasted* to the documentary style of Moisés Kaufman's *The Laramie Project* – have attempted to address the phenomenon of violence through various presentational strategies. Such recent works, while timely and innovative in effect, reiterate a persistent fascination of the Western theatrical tradition with the agonistic encounter, *sparagmos*, and the flow of blood. This chapter will not directly explore staging approaches per se but will primarily investigate the epistemic modalities of violence: how violence may effect a fundamental rupture of representation and a dissolution of identity relations under the law of the Same. I am chiefly interested in how violence functions as a critical touchstone in postmodern ethical theory and a guiding trope in an ethics of otherness, what Emmanuel Levinas identifies as a "first philosophy," and how this way of thinking may invite a re-imagining of the theatrical event as an experience that values interruption and generosity over domination and closure.

The fundamental focus of this chapter and its speculations concerns the rudiments of self-other relations. Much has been written about the violence that modernism and Western presumption have enacted, in totalizing attempts to eradicate alterity, to subsume and arrogate the Other. I rather wish to focus on a reverse dynamic – how recent ethical writing has asserted the violence that the Other brings to bear upon the self. Violence emerges as a key motif in the works of these writers (I here include Lyotard, Levinas, Derrida, and Nancy), as the phenomenon of violence, which defies representation and destabilizes assurances of the ego, parallels the sort of interior collapse experienced by the subject in an ethical encounter with the Other. In sum, violence operates as an ethical analogue; the encounter with otherness is on a basic level an experience of trauma (which ironically brings salutary effects).

The ethics of otherness provide a unique and illuminating conceptual lens for understanding violence and its theatrical evocation. Given the Western dramatic tradition's fascination with violence, and the continued obsession with violent events in contemporary performance, we can profitably question when and how theatre artists should approach the phenomenon of violence. On a basic level, we might consider how best to stage a violent moment and how to choreograph the victim and the aggressor. Outside of the practical matters of blocking and stage images, we can explore and evaluate how theatre productions materialize ethical relations and whether their stagings demonstrate impulses of ownership or domination. These concerns move the discussion from the staging of specific events or moments of violent action to the basic orientation of the theatrical form and approach. Attention to the theatrical figuring of ethical relations poses serious problems – both aesthetic and ethical – for the performative act and theatrical artists. Such concerns challenge conventional modes of representation and complicate the very possibilities of significative utterance.

In this light, if we view normative representation as a kind of appropriation, then traditional stage representation may enact a violence upon otherness, by "fitting" the Other to the theatrical frame. The staging process itself may thus enact a dynamic of violence. How, then, can the contemporary theatre proceed? How might it renounce such staging practices and explore modes that avoid the assertion of the self/same (at the expense of the Other)? How can the theatre draw from the insights of postmodern ethical writing and realize an approach that turns violence from the Other to the self, bringing an interrogation and challenge to the assumptions and complacency of the author/creators? Is it possible for a non-autonomous, non-Cartesian self to engage in an aesthetic enterprise whereby the event involves no "legislation," but rather works to open up a space for the Other?

I want to suggest in this chapter that the dissolution of the self – the violence that the Other brings upon the self – does not finally lead to solipsism or complete passivity; it rather calls the self to obligation. As Levinas asserts, the face of the Other demands responsibility. Explaining this motif in the thought of Jean-Luc Nancy, E. Jeffrey Popke affirms that living ethically involves acknowledgment of a "shared Being" and participation in a "collective spatial politics in which a commitment to the Other is our abiding concern."[1] This way of understanding the self–other relation and the onstage materialization of this relation challenges the theatre practice that commonly promotes the sacrifice or ritualistic obliteration of the Other for the assurance of communal bonds and identity confidences. Rather, this model envisions a way of "being with" or "alongside" the Other in a manner that highlights a reciprocity and obligation incumbent upon the self.

The disapprobation that recent ethical theorists have exhibited toward the Western philosophic tradition and its understanding of knowledge cannot be overestimated. Indeed, the limitations, strictures and erasures of this tradition have in large measure fueled the desire for new models and articulations of ethical relations. In sum, it is the aim of postmodern ethical writing to challenge the rationalist egoism of Western thinking, which understands the world (and others) only through modes of self-identification and a logic of the Same. According to both Levinas and Derrida, the history of Western philosophy has correlated thought with being and has struggled to system-atize existence under the laws of knowledge, operations that territorialize alterity under the rubric of likeness. In short, the Other is brought into the knowledge of the subject under the terms of the subject. This kind of knowledge inaugurates a power relation that demands a subjugation of the Other, as something to be overcome, and the subject/other relation in the Western tradition has consequently taken a profoundly agonistic if not vio-lent aspect. Hegel represents the epitome of this line of thinking; his history authorizes the dialectical overcoming of the slave/other in the advancement of Ideality.

Discussions of the Holocaust appear prominently in recent ethical writing, as its events have fundamentally challenged notions of morality, mem-ory, community, and representation. Much of postmodern ethical theory has been read as a postscript to the Holocaust, and indeed for Levinas, a Lithuanian-born Jew who was himself incarcerated in a German prison camp, this atrocity informed and consolidated his conception of otherness. For these thinkers, the Holocaust exposed the presumption and the limits of the Western tradition. The rationalist enterprise of the modernist project – which understands difference under the law of the Same – revealed itself as ineffectual and mute. Giorgio Agamben has explored the impossibility of rationally assessing the experience of the death camps in *Remnants of Auschwitz*.[2] Edith Wyschogrod has written that the violence of the Holocaust has acted as a kind of "nihil," a black-hole in Western history that defied all categories of reason and understanding.[3] Not only did the rationalist tradi-tion reveal itself as inadequate to comprehend the event, but the Western will to knowledge has been cited by some as complicit in the atrocity, evidencing a desire to suppress otherness: the gas chamber as solution to absolute alterity. Levinas's writing thus may be viewed as a profound defense of difference, a rebuttal to modernist assurance, and an insistence on the sanctity of the Other.

In this recent tradition of ethical investigation, the Holocaust has served as a profound limit point. Lyotard asserts that the Holocaust exploded the very possibility of thinking and speaking as a "we," effecting what one critic has described as a "suspension of the monopoly of the cognitive regime of phrases," that is, the confounding of all authority of knowledge and catego-rizations of experience.[4] In this light, the extreme violence of the Holocaust and its attendant trauma defy the very possibility of knowing and expose the limits of signification, demanding a moral and cognitive agnosticism. This kind of agnostic effect was the aim of Claude Lanzmann in his famous documentary film *Shoah*, which utilized extensive interviews with Holocaust survivors. For Lanzmann, "not understanding" was "the only ethical way to approach a representation of the Shoah"; his film did not attempt to mediate the event in critical or expository fashion, but pointedly worked to obviate what the film-maker identified as the "obscenity of understanding."[5]

The terror of genocide precipitated a profound distrust of "knowledge" in the conceptions of such figures as Levinas and Lyotard. Indeed, for these thinkers, the opposition to rationalism provoked a refashioning of knowledge and ethical relations, leading to the valuation of "not knowing" or "being otherwise." Their work thus undertook a reassessment of the ego and its will to knowledge. The film-maker Lanzmann's attitude towards "understanding" points toward what Lyotard would describe as the postmodern sublime.[6] Not to be confused or associated with notions of beauty or the ideal, the sublime here equals an expression of the *differend*, or that which cannot be apprehended or translated. Rather than bringing an exaltation or transcendent affirmation, the sublime exposes a disconnect between idea and feeling that shatters belief. For Lyotard, the subject in this experience does not have an adequate capacity to access or assess the event. The effect is to limit and humble. The self here is revealed in its rational poverty; the self is chastened. While the experience of the sublime can be devastating, the negative aspect of the experience can bring positive results: it can trigger an awakening to cognitive and representational limits. This disclosure of limits, according to Lyotard, is something that one should accept gratefully.[7]

In this line of thinking, attempts to address violence and its trauma are not so much concerned with the bodily mutilation of victims per se as with the psychic repercussions upon the survivor/viewer. At the risk of inviting cynicism, this outlook finds the violent event, in its horror, to bear a positive aspect. Violence's impact productively illuminates the extent of possible knowledge. Nameless victims thus promote a revelatory re-evaluation of the self and its prerogatives of power. And it is perhaps the salutary shattering of the subject – and its dynamics of sovereignty – that inclines writers of postmodern ethical theory to utilize tropes and images of violence.

Though it is certainly possible to read violence onstage (and in actuality) as the instantiation of dominance and the assertion of ego, an ethics of otherness invites an alternate reading, one that locates the "value" of violence not as overcoming, but as self-renunciation. Levinas clearly seeks to overthrow the structural models of Western rationalism; his portrayal of human relations reverses the poles of power, whereby the self bows prostrate before the Other, acknowledging and honoring the face of the unknown, the irreducible. This encounter has disturbing and disorienting effects upon the self, consonant with Lyotard's expression of the postmodern sublime. In this model, violence is not directed toward the Other but is experienced by the self, as a byproduct of ethical encounter. Images and metaphors of violence consequently pervade writings in recent ethical theory. Lyotard writes of this experience as disruptive, blazing, and lightning-like.[8] The Other is oftentimes described as having the effect of shock upon the self, inflicting a laceration in the language of the Same. Derrida, in fact, speaks of this otherness as a sort of circumcision, a wound that opens the self to the Other.[9] Levinas himself writes of the primacy of the wounded and hurting existent.[10]

Focusing only on violence and its radical effect upon the self, however, overlooks a central element in postmodern ethical relations. This decentering of the subject is certainly a recurring theme in the writings of Levinas, Derrida, and Nancy, one that serves as a counterbalance to Western rationalism and its presumptions. While this de-emphasis of the self may be viewed as a needed redress to modernist overweening, attention to this diminished aspect alone can invite charges of nihilism, foregrounding vertigo and foundational evacuation – the self in collapse and solipsistic withdrawal. These thinkers, importantly, point beyond any negative or injurious implications. Rather, they conceive a subjectivity that seeks to honor the distant stranger, the nameless victims, all absent others. Despite the inefficacy of signification and the self that is called into question, the sublime trauma of otherness engenders a "responsibility [...] in the face of something that exceeds symbolic guarantee."[11] In this context, trauma involves a beneficent pain, inaugurating a realigned relation between self and Other, an exploration of relational modes outside of reason and domination. The self is not left immobilized, shocked, and impaired, but called to responsibility for the Other. In this regard, the violence of encounter does not eradicate the Other nor does it immobilize the self; rather, it initiates an attitude of hospitality for the stranger, the refugee, the abject.

The insights of thinkers such as Levinas and Derrida encourage a questioning of normative moral imperatives and expand the ethical conceptualizations of philosophical discourse. Their writings can also activate speculation about the use and deployment of the stage – how to ethically engage the Other in a theatrical context and how to address otherness without resorting to moral maxims or notions of community. We are, in effect, challenged to imagine how we might be "in relation" with one another at the deepest and most profound level. What insight these thinkers might offer – in regard to both theatre art and social praxis – ultimately involves the matter of responsibility, or how to ethically honor one's obligation to the singular Other.

The definition and operation of responsibility can be illuminated by the ways in which recent theorists have reframed the contexts of ethical decision-making, highlighting a sensitive regard for singularity (as opposed to universal law) and an understanding of the self that is not independent or autonomous but always "in relation." Traditional moral conceptions have privileged the "rights" of the ego and principles of rationalist determination that act as a calculus in assessing ethical relations between self and other. Kant, for example, argues that actions can be determined to be responsible according to a principle of universality, that an action is just if it can be followed logically in all situations and cases of that kind. He argues that a just act, therefore, must accord with a maxim that operates independently of any given situation – attention should focus on the principle, not any individuals in question. Indeed, the tradition of Western philosophy has sought to intellectualize ethics. From Aristotle to Spinoza to Mill, the Western tradition has attempted to conceptualize ethical behavior in systems of justice based upon rational orders of operation and justification.

Described by one scholar as a sort of "radical altruism," the ethics of Levinas challenge the traditional moral appeal to law or principles of justice.[12] According to Levinas, the appeal to universal law can work to shroud alterity, in effect, to disconnect the self from responsibility with a mediating legalism – one in which otherness is elided. In this type of appeal, a moral code or set of maxims serves to separate the self from the Other. Levinas insists that no mediating element should intervene, that the self's obligation to the Other is not informed by law, duty, precept, or religious coercion. As Richard Cohen explains, "it is not the law in the other person that a moral agent respects, but the very otherness of the other person [...] in the very flesh of the Other, the Other's mortality, aging, degradation, suffering."[13]

The writings of Levinas – like those of Nietzsche and Heidegger – seek the demolition of the absolutism that is evident, for example, in Hegel's systematic philosophy. However, Levinas believes that Nietzsche and Heidegger point only to an ultimate egoism, the self alone in the world operating under the will to power (rather than the modernist will to truth). Levinas rejects the *uber mensch* of Nietzsche; in place of the sovereign ego, Levinas posits the call of the Other, which positions the self in a relation of responsibility toward the Other's alterity. The ultimate aim of Levinas's work is thus to reclaim the Other in inviolable relation to the self. In this view, the difference of the Other can never be subsumed (and therefore should not be elided via appeal to principles of justice or morality).

For Levinas, and for this chapter's rumination on both actual and theoretical violence, conceiving of human relations structurally is paramount. Levinas holds that the rudiment of human existence is not solitude but sociality. He rejects the independent ego of Western rationalism and offers an ethics of intersubjectivity, what one critic has described as a "preontological relation to alterity."[14] In his two most prominent works, *Infinity and Time* and *Otherwise than Being*, Levinas argues that ethics precedes knowledge, that ethics exists as a "first philosophy." For Levinas, the Other is anterior to the self, prior to discourse (and knowledge), and the self comes into being only in relation to the Other. The ethical relation is thus coincident with becoming, alterity its precondition. In this manner Levinas undermines the traditional subject/other relation of philosophy in which the subject, as author of his world, precedes and maps the Other. Moreover, the ethical relation is defined as a given, a constituent of existence, rather than a consequence of a social contract, universal maxim, or human-constructed code of rights and obligations. Levinas reverses the axis, so that, rather than deciphering or appropriating the Other, the self must "justify" its existence before the face of the Other, which commands as though from on high. The relationship is asymmetrical, not based on knowledge or recognition. One does not discern threat in the face of the Other but weakness, vulnerability, the bereft (the widow, the orphan, the homeless, etc.), and is thus called to responsibility. The self experiences a sort of structural guilt, a debt that never can be repaid. For Levinas, ethical obligation precedes the self's particular entry into the social order:

> The ethical exigency to be responsible for the other undermines the ontological primacy of the meaning of being; it unsettles the natural and political positions we have taken up in the world and predisposes us to a meaning that is other than being, this is otherwise than being.[15]

Levinas contends that the face of the Other – which is no specific Other – calls one to ethical responsibility. This does not mean that ethics is purely abstract; human encounters are material and particular, but situated in a relationship that is prior to any significance ascribed or mediated by philosophy, culture, politics, or place.[16] Ethical relation for Levinas is a constitutive, phenomenological structure that orients the self in its becoming toward otherness. Obligation is not something taught but pre-given. As Dorota Glowacka summarizes Levinas, the I is made possible so it can "reach out and aid."[17] Levinas's writing is notoriously difficult, and his views find no easy translation into political program or specific injunction (or prohibition). In fact any moral codification would countervail the thrust of his assertions. Yet his passionate insistence on respecting the singularity of the Other extends a clarion call to responsibility, one that reveals the ethical aspect of every human relation.

On the occasion of Levinas's death, Derrida spoke at Levinas's funeral and credited the deceased with awakening him to an understanding of responsibility.[18] Indeed, Derrida's later writing owes a great deal to Levinas's ethical formulations. With varying degrees of approval, scholars have noted a shift in Derrida's work in the last decade, a movement from textual criticism to a messianic politics. Nina Pelikan Straus has described this turn as one from Nietzschean resistance to Dostoevskian self-submission.[19] This assessment highlights the Levinasian element, of the constitutive obligation between self and Other in Derrida's later work. It also sheds a new light on the aim and object of deconstruction. In Derrida's late writings, intertextual critiques of logocentrism morphed into messianic expressions of the ethical. Derrida has in fact claimed that deconstruction is justice. One finds that Derrida espouses the Levinasian notion of ethical singularity, a view that holds imposition of creed, law, or ideology as a violence upon the other.

Richard Rorty has criticized Derrida's late work as overly obscure and private, failing to demonstrate an applicable ethical principle that could be considered public.[20] Rodolphe Gashe argues in Derrida's defense that these late writings intend to avoid advocating moral precept or dictum. It is indeed the disavowal of legislation that informs Derrida's conception of justice, which views justice not as an appeal to general principle but as an occasion of singularity. Justice, for Derrida, is in sum a moment that dismisses moral or juridical mediation in favor of a unique encounter with the Other. In keeping with Levinas before him, Derrida holds that moral codes occlude and systematize relations, actually rendering the Other invisible through an appeal to law or principle. By deconstructing the normative ethical relation, the Other is granted honor, her singularity. Justice, according to Derrida:

> depends, *at every instant*, on new assessments of what is urgent in, first and foremost *singular* situations, and of their structural implications. For such assessment, there is, by definition, no pre-existing criterion or absolute calculability; analysis *must begin* anew every day everywhere, without ever being guaranteed by prior knowledge.[21]

This assertion is not meant to defer engagement or to invite anarchy so much as to respect and honor the eventfulness of the encounter, the occasion of the decision, and the unique being of the Other.

This insistence upon the singularity of the moment, the undecidable nature of the moment, informs Derrida's concern for the messianic – for that which interrupts the economy of the Same – and brings the self into an attendant posture, awaiting the call of the other. Christopher Wise has discussed this feature of Derrida's thought as a "universal messianic structure,"[22] a kind of "hopeful anticipation, or a waiting for the truthful word of the other."[23] The messianic does not involve any identifiable messiah, or any supernatural or transcendent intervention; rather, Derrida focuses on the messiah "effect" as a particular attitude of the self that keeps it open to interruption and difference. In this light, deconstructive gestures or encounters function to perform justice, opening the event "to the incalculable, non-programmitizable response by an Other for whose arrival they have opened the way."[24]

In examining the ethical writings of both Levinas and Derrida, one may identify two different kinds of violence conveyed in imagistic and conceptual form. For both theorists, the egoistic inclination of the self always threatens violence upon the Other, always threatens to deform or appropriate the difference of the Other, and to bring the Other under its own knowledge. In this light, a recognition or representation of the Other is tantamount to subjugation. Both writers denounce rationalist presumption and decry the violence of systematization and legalism. The other type of violence described by these authors, however, concerns the rupture of the self and its egological prerogatives. In this type of violence, for Levinas, the Other wounds the self, breaking self-containment and composure. For Derrida, this dynamic is given messianic implications; the opening of the self allows for the "coming" of something other. In the work of both figures, this kind of wounding or laceration proves hopeful, revealing the intersubjective nature of the self and its primary formation.

This sort of ethical modeling found in Levinas and Derrida may spur fervor and a passionate regard for otherness, but their insights can prove frustrating when brought to bear upon particular, material moments of cultural exchange. And while their writing prompts valued speculation as to the nature (and possibility) of ethical social actions and utterances, they give no clear directive as to how an ethics of otherness might inform aesthetic acts or performative gestures, especially those pertaining to violence and trauma. One can at best only surmise how a performance event might evoke, represent or respond to an engagement with otherness.

The insights of Levinas and Derrida urge both caution and imagination for a theatre artist. Often these writers cite the Old Testament injunction against "graven images," a reference that highlights a dubious regard for representational practices. Levinas is profoundly distrustful of any representational or mimetic mode. Seen as a grand egological enterprise, he argues, representation mutilates, denies, and erases. The high degree of emphasis that Levinas extends to the Other suggests that the self should (and can) never speak for the Other and should never attempt to embody or represent the Other (which would translate into a form of idolatry). Indeed, such a view cuts to the root of theatrical representation. For Levinas, the Other can never be brought under the self's knowledge, and any representation would thus prove a deformation. As Jill Robbins observes, in "objecting to the 'theatricalization' of ethical rapports in figural interpretation, Levinas seems also to object to the possibility of theatrical representation itself."[25]

This profoundly anti-representational bias in Levinas invites consideration of alternate modes or tactics. Robbins notes that Levinas often mentions novels and dramas in his writing in favorable terms. Levinas cites Dostoevsky, and often the tragedies of Shakespeare. Robbins suggests that Levinas as a rule disapproves of mimetic forms, but that he seems fascinated with works that approach the experience of trauma, that come to the limits of representation. Levinas indeed extends favor to artworks that grapple with the Shoah. According to Glowacka, Levinas does leave open the possibility that certain poetic language may function outside the economy of representation and thus may work in a non-mimetic fashion, effecting an experience of shock and disturbance; Levinas claims this effect in the language of Blanchot's *The Writing of the Disaster*.[26]

A possible alternative to representation is suggested by Lyotard, who asserts that representation cannot hold incommensurabilities. He thus argues for non-mimetic art forms, negative and nonfigural, that "witness" before the impossibility of containment.[27] Lyotard's argument suggests that witnessing may prove an alternative to representation, and certainly the "call to witness" runs throughout contemporary ethical writing. What witnessing involves and how it can be conveyed in a performance event bears close scrutiny and considerable theorizing. Vivian Patraka's work on Holocaust museums, which explores the complexities of representing or commemorating genocidal violence, asks the sort of questions that theatre practitioners may profitably confront.[28] How can actors witness onstage, honoring the singularity of the Other? How can dramatic texts avoid the representational violence of surrogation or appropriation? How can performance provide the violent laceration that opens the self to the Other? How can theatre be justice?

Claude Lanzmann's vigorous assertion that his film *Shoah* should evidence "no moralizing gesture" points toward a kind of aesthetic posture that may honor the ethics of otherness.[29] What may prove a central and crucial element in an aesthetic form of witnessing is its conversational aspect. Both Levinas and Derrida renounce monologic assertion and suggest a type of aesthetic encounter in which the self attempts to remain open and attentive to the Other, as a listener or a witness. For Levinas, this relation may be understood in the differentiation between the Saying and the Said, the latter indicating a fixity of assertions and relations, and the former highlighting a continuous interaction that remains open to change and difference.[30] The Saying and the Said, in essence, embody different attitudes to the Other. Saying, which Levinas privileges, should be understood as a mode of address or a type of greeting. It is, in short, a verbal acknowledgment of the Other that functions as a sort of invitation. Conversely, the Said pertains to content, to the constative impact of the utterance, which implies thematizing and knowledge, that is, the Said operates in a declarative mode, conveying a content that has already been determined or thematized by the outlook of the speaker. For Levinas, an emphasis on the Said seeks to bring the Other under the subject's knowledge, and this translates to an act of subjugation and violence. As Glowacka explains, ethical language enacts a kind of dispossession, where the I is evicted by the Other from its home in language.[31]

While an ethics of otherness can both challenge and inspire, its implications for theatrical practice warrant increased investigation and scrutiny. Both theatre scholars and practitioners may benefit from further exploration, from experiments in staging the Saying as opposed to the Said. Before concluding this chapter I wish to engage in a brief assessment of the recent David Hare play *Stuff Happens*, a work that foregrounds Western politics, ideological imperatives, and the violence enacted upon nameless others. Hare's play serves well as an ethical test-case, as the work generated very heated and often polarized reactions, drawing attention to the delicate and complicated nature of political commentary and the docudrama approach. In identifying and addressing key elements of the play and its varied critical responses,

I hope to forward a few compelling questions and speculations concerning this work and its particular manifestation of ethical relations, questions that might invite broader consideration of the stage and its efficacy as a vehicle for witnessing and justice.

Premiering in the fall of 2004 at the Royal National Theatre, *Stuff Happens* dramatizes the political maneuverings of the Bush administration in the run-up to the Iraq War, representing onstage a wide array of contemporary figures, from Tony Blair to Donald Rumsfeld to Jacques Chirac. Focusing upon a recent historical flashpoint, Hare's play exhibits curious parallels with Moisés Kaufman's much-discussed piece *The Laramie Project*. While the latter explores the local context of Laramie, Wyoming in response to the murder of a gay college student, both works highlight and emphasize their use of factual transcribed material; both convey multiple personal and political viewpoints; and both investigate an instance of violence that reveals wider cultural tensions and ideological commitments. On a basic level these plays function as indictments of aggression, as pleas for respect and moral responsibility. Jill Dolan characterizes Kaufman's intentions as "sympathetic and benign,"[32] and Hare's motives too – presumably enlightened and compassionate – seem in keeping with a Levinasian outlook, exhibiting a concern for difference and a wariness of political expediency.

In *The Laramie Project* Father Roger Schmit, the local Catholic priest, challenges the Tectonic company with this cautionary note: "I will trust you people that if you write a play of this that you [...] say it right, say it correct. I think you have a responsibility to do that."[33] The burden of such responsibility does not escape David Hare, who acknowledges the obligation that comes with staging historical events. In the author's note to *Stuff Happens* he conscientiously accounts for his methods and material. Quoting sources verbatim, he assures that his work has been "authenticated from multiple sources, both private and public" (Hare enlisted the services of a professional researcher from Columbia University).[34] While Hare does not identify the piece as a documentary and forthrightly acknowledges that many private scenes derived from his own imagination, as speculations, he affirms: "Nothing in the narrative is knowingly untrue."[35]

This quasi-ethnographic approach gives platform to various voices and perspectives; it gives the impression of Saying, of bringing disparate utterances into a kind of open dialogical debate. Many critics reported some astonishment that Hare did not assume a more polemical, censorious authorial stance, given the malfeasance of the Bush administration. The Reviewer for *The Daily Telegraph*, for instance, found the piece "admirably fair and even-handed."[36] Most viewers did not encounter the expected, facile caricature of George Bush, but a sphinx-like figure of complicated resolve (if anything Blair comes off worse in Hare's dramatization). One of the most highly charged speeches in the play proceeds from a war supporter, the "Angry British Journalist," who decries the luxury and excess of the West, its disputation over the "style" of the invasion, and its devaluation of "the splendid thing done." He honors the Iraqi liberation – "freedom given to a people who were not free."[37]

Stuff Happens, nonetheless, may invite the criticism that has targeted other practitioners of docudrama, such as Kaufman and Anna Deveare Smith, that the factuality of the interviews feeds the rhetorical design of the author's intent. Stephen Bottoms emphasizes this point in his analysis of Hare's work, noting that "such plays can too easily become disingenuous exercises" obscuring the manipulation of authorial motive and opinion.[38] In this view, the multiplicity of voices (gained through transcripts) produces the semblance or effect of polyphony, while a prevailing aesthetic structure controls and shapes. In Hare's work an overt theatricalism frames the play; like *The Laramie Project*, *Stuff Happens* opens with a sequence that foregrounds the "actors" who will appear in the subsequent drama. Serving as something of a narrator, the figure identified as "An Actor" introduces the historical figures and gives commentary through the course of the play. On several occasions this figure is less than impartial. When Cheney declares that he had "other priorities" that prevented him from enlisting in the military during the Vietnam War, An Actor retorts: "Cheney proves himself willing to take on responsibilities others shirk."[39] The play then cites a 1974 memo in which Cheney admits he cannot correct a "drainage problem" in the first-floor White House bathroom. Other instances reveal a stronger authorial hand, such as the scene involving the phone conversation between Blair and Bush that suggests the Bush administration called a halt to the British Army's advance upon Osama bin Laden at Tora Bora so that US troops could intervene and gain the headlines. And numerous critics have noted the "ennobling" characterization given to Colin Powell, who is made to serve as a dramatic foil to the president, emerging as a sympathetic Brutus figure.

While *Via Dolorosa*, Hare's short play on the Israeli/Palestinian conflict, reveals the moral incertitude of the playwright and his reluctance to take sides, *Stuff Happens* may demonstrate a different authorial posture. In interviews Hare has spoken of the Iraq War with clarity and conviction; he characterizes the war as a "crime," the "biggest mistake in foreign policy since Suez."[40] Detractors of *Stuff Happens* consequently view the cultivated ambiguity of the play as a ruse, regarding the work less as a play than a political lecture. Such criticism draws attention to matters of intent and audience – for whom is the play intended? For what end? Whose position is affirmed and empowered?

Hare's harsher critics regard the play as self-serving and self-congratulatory. Alastair Macaulay observes: "There's a narcissism with Hare's theatre-as-journalism [...] elegantly making a spectacle out of his even-handed humanity."[41] Such an assessment sees *Stuff Happens* performing a kind of openness that the text in fact undercuts. Many note Hare playing to a receptive leftist audience, offering images and perceptions that only corroborate already-held assumptions. For Rosie Millard, "Nothing was challenged, no surprises were sprung"; she continues: "Weedy jokes about how jejeune the Bush rhetoric is got a riotous reception, as if left-wing London had never heard anything so right-on."[42] The reviewer for *The Independent* also recognized the sympathetic alliance of author and audience and opined: "Those who described the National's staging of David Hare's *Stuff Happens* as 'brave' should perhaps have paused for a moment to think about how much nerve would have been required to put on a play which argued that the Prime Minister was right to go into Iraq."[43] Such criticism points to Hare's surety of viewpoint and finds that the play's conclusion preceded its investigation and execution.

American critic John Lahr expresses admiration for Hare's work and views the drama as a kind of redress, a counter to Rumsfeld's obfuscation – that "stuff happens." If we, as Lahr, see the play as "rectifying that omission," then Hare's piece on some level promises understanding, revelation of the motives and machinations that brought the United States to war.[44] This promise, however, points to a fundamental dilemma in the theatrical enterprise, that is, how to "fill" an "omission" in an ethically responsible manner. Such theatrical surrogation brings up crucial concerns of authority and assertion. *Stuff Happens* functions as an in-depth investigation of power, the limits of knowledge in the service of power; yet, we can question the play in terms of its own authority (and limits). To what degree does the play's "knowledge" accurately or ethically compensate for the lacuna attending the war (as we may question whether *The Laramie Project* fills the "loss" of Matthew Shepard with its own politics of tolerance and inclusivity)?

Jon Erickson is one of the first theatre scholars to have explored the implications of Levinas's thought on theatre practice, and he warns against "mere self-serving attitudinizing" in performance; he writes that the stage "is there not to 'teach' the audience a lesson."[45] Erickson's injunction points to an element of *Stuff Happens* that is both powerful yet suspect, that is, its implicit condemnation of aggression and abusive political prerogative. Few would argue for the glorification of war, the indiscriminate slaughter of innocent civilians, or any unilateral colonialist enterprise; yet, Levinas draws caution to moral certitudes (how to take sides), to the imposition of ideals on the singularity of others. Despite the play's best efforts to operate in a dialogical fashion, to convey a laudable message of conscience in face of Bush's militarism, *Stuff Happens*, like *The Laramie Project*, may promote a heart-felt call to ethical responsibility while reducing the Other in so doing. The play's didactic affirmation, an assertion of the Said, may overshadow and pre-empt any interrogation of the Saying.

While *Stuff Happens* implicitly condemns the arrogance and presumption of the Bush administration, which has acted unilaterally in its invasion, revealing itself as deaf to the call of the Other, Hare exhibits his own presumption. In this light, Bush and company emerge as the Other for Hare, and the playwright exhibits little reluctance in assigning motive and intent to these individuals. The focus on Western political figures leads to a striking feature of Hare's work, that is, the absence of Iraqi voices in a play about the Iraq War. And when such an Iraqi figure does appear, he comes at the closing of the drama, arguing that nothing will change until Iraqis take up the banner of freedom for themselves. Such a closing may introduce an Iraqi viewpoint (though this pronouncement may affirm a Western individualism). Still, the ending seems somewhat strained; the play concludes with an unsettling sense of anonymity, that bombs will fall, and nameless innocents will die.

What may be at stake is the very singularity of the Iraqi, of any or all Iraqis, that their otherness may be subordinated to the advancement of a moral or political vision (Hare has been quite outspoken in his critique of the Israeli/Palestinian conflict and has linked it to the Iraq war). Viewed in this way, the aggression of violence implicit in *Stuff Happens* does not bring a destabilization of the self, a rending of assurances, but an opportunity to "represent" or "replace" alterity with a surrogate moral imperative. As Jill Robbins argues, "to approach the Other armed with a concept such as community [...] (or any other humanistic platitude) would destroy the alterity of the Other in the very guise of respecting him or her."[46] To explore the injustice of the Iraq War – to attempt to "understand" the violence directed at nameless victims – as an occasion to theatrically advocate for tolerance and community would, for Levinas, itself prove an occasion of violence and injustice. Jill Dolan has criticized the New York production of *The Laramie Project* with representing "the citizens in the play as a means to an end, rather than ends in themselves."[47] Dolan's concern points to the over-valuation of ideas or politics at the expense of individuals. In its surety of purpose, *Stuff Happens* may assert its conviction as a declarative act and in so doing prove deaf to the call of the Other.

That plays such as *The Laramie Project* and *Stuff Happens*, so profoundly political (and moral) in content and orientation, can engender ethical complexities indicates the depth and difficulty of claiming and embodying ethical relations on stage. Conventional representation enacts violence upon otherness, reforming alterity according to the knowledge of the self. Moralizing enacts violence, bringing the singular under universal law. For theorists such as Levinas and Derrida, salutary violence is that experienced by the subject when confronted with the irreducibility of the Other, when the self is torn and wounded. One can perhaps only imagine a theatre that achieves this effect, that operates as an invitation, an opening for the arrival of the Other, and a witness to a momentary justice. What intrigues me is the possibility of performing the Saying, the address to the Other that invites, that demands no closure or thematizing, that serves less to interpret than to interrupt. Such a theatre might liken itself to the dwelling described by Adriaan Theodoor Peperzak, a place disordered by the coming of the unknown:

You surprise me by coming to me. Even if I invited you, your coming disturbs my world. Indeed, your entering into my dwelling place interrupts the coherence of my economy; you disarrange my order in which all things familiar to me have their proper place, function, and time. Your emergence makes holes in the walls of my house.[48]

Notes

1. Jeffrey E. Popke, "Poststructuralist Ethics: Subjectivity, Responsibility and the Space of Community," *Progress in Human Geography*, 27:3 (2003): 312.
2. Giorgio Agamben, *Remnants of Auschwitz: The Witness and the Archive*, trans. Daniel Heller-Roazen (New York: Zone, 2002).
3. Edith Wyschogrod, *An Ethics of Remembering: History, Heterology and the Nameless Others* (Chicago: University of Chicago Press, 1998), 14.
4. Dawne McCance, *Posts: Re Addressing the Ethical* (Albany: State University of New York Press, 1996), 53.
5. Quoted in Linda Belau, "Trauma and the Material Signifier," *Postmodern Culture*, 11:2 (2002): 15.
6. See Stuart Simm, *Jean François Lyotard* (London: Prentice Hall, 1996), 98–105.
7. Simm, 105.
8. Simm, 105.
9. John D. Caputo, *The Prayers and Tears of Jacques Derrida: Religion without Religion* (Bloomington: Indiana University Press, 1997), xx.
10. Dorota Glowaka, "Ethical Figures of Otherness: Jean-Luc Nancy's Sublime Offering and Emmanuel Levinas's Gift of the Other," in Krzysztof Ziarek and Seamus Deane (eds), *Future Crossings* (Evanston, IL: Northwestern University Press, 2000), 183.
11. Belau, 1.
12. George Kunz, *The Paradox of Power and Weakness: Levinas and an Alternative Paradigm for Psychology* (Albany: State University of New York Press, 1998), 3.
13. Richard A. Cohen, *Ethics, Exegesis and Philosophy: Interpretation After Levinas* (Cambridge: Cambridge University Press, 2001), 6.
14. Popke, 303.
15. Emmanual Levinas and Richard Kearney, "Dialogue with Emmanuel Levinas." *Face to Face with Levinas*, ed. Richard Cohen (Albany: State University of New York Press, 1986), 13–33.
16. See Richie Howitt, "Frontiers, Borders, Edges: Liminal Challenges to the Hegemony of Exclusion," *Australian Geographical Studies*, 39:2 (2001): 233–45.
17. Glowacka, 187.
18. See Jacques Derrida, *Adieu to Emmanuel Levinas*, trans. Pascale-Anne Brault and Michael Naas (Stanford, CA: Stanford University Press, 1999).
19. Nina Pelikan Straus, "Dostoevsky's Derrida," *Common Knowledge*, 8:3 (2002): 559.
20. Rodolphe Gasché, *Inventions of Difference: On Jacques Derrida* (Cambridge, MA: Harvard University Press, 1994), 4.
21. Quoted in Popke, 307.
22. Christopher Wise, "Deconstruction and Zionism," *Diacritics*, 31:1 (2002): 59.

23. Wise, 58.
24. Gasché, 11.
25. Jill Robbins, *Altered Reading: Levinas and Literature* (Chicago: University of Chicago Press, 1999), 50.
26. Glowacka, 184.
27. Simm, 104.
28. Vivian Patraka, *Spectacular Suffering: Theatre, Fascism, and the Holocaust* (Bloomington: Indiana University Press, 1999).
29. Belau, 17.
30. Fabio Ciaramelli, "Levinas's Ethical Discourse Between Individuation and Universality," in Robert Bernasconi and Simon Critchley (eds), *Re-Reading Levinas* (Bloomington: Indiana University Press, 1991), 83–105.
31. Glowacka, 178.
32. Jill Dolan, *Utopia in Performance: Finding Hope at the Theater* (Ann Arbor: University of Michigan Press, 2005), 117.
33. Moisés Kaufman, *The Laramie Project* (New York: Vintage, 2001), 66.
34. David Hare, *Stuff Happens* (London: Faber, 2004), author's note.
35. Hare, author's note.
36. Charles Spencer, "Hare Victory as Theatre becomes War Forum," *Daily Telegraph* 11 Sept. 2004: news, 6.
37. Hare, 15.
38. Stephen Bottoms, "Putting the Document into Documentary: An Unwelcome Corrective?" *The Drama Review*, 50:3 (2006): 57.
39. Hare, 5.
40. Quoted in John Kampfner, "Strange Stuff Happens," *New Statesman*, 5 June 2006: 13.
41. Alistair Macaulay, "US History Lesson Lacks Surprise," *Financial Times*, 13 Sept. 2004: arts and style, 17.
42. Rosie Millard, "Notebook: Review of *Stuff Happens*," *New Statesman*, 9 Sept. 2004: 41.
43. Thomas Sutcliffe, "Civilizations Built on Sand," *Independent*, 17 Sept. 2004: features, 5.
44. John Lahr, "Collateral Damage," *New Yorker*, 27 Sept. 2004: 155.
45. Jon Erickson, "The Face and the Possibility of an Ethics of Performance," *Journal of Dramatic Theory and Criticism*, 13:2 (1999): 10.
46. Robbins, 5.
47. Dolan, 126.
48. Adriaan Theodoor Peperzak, *Beyond: The Philosophy of Emmanuel Levinas* (Evanston, IL: Northwestern University Press, 1997), 60.

2

The "Outsider" Outside: Performing Immigration in French Street Theatre

Susan Haedicke

Immigration is a fraught and potentially destabilizing challenge facing the European Union (EU) today, yet it constitutes one of the cornerstone issues confronting the building of a common Europe. The scale of migration movement has increased significantly in the last few decades with an annual growth rate now of 2.9 percent according to the International Organization for Migration (World Migration Report). In spite of the hardships and dangers experienced by migrants, many thousands of people enter the EU both legally and illegally each year to establish residence, and movement between member states of the expanded EU continues to rise. Accurate statistics on migration flows are difficult to obtain, and many official papers begin with a disclaimer as to the availability and quality of the data, but it is estimated that around 25 percent of the world's migrants now live in Europe.[1] Those entering the EU come for a variety of reasons, from armed conflict, natural disasters, economic distress, and political or religious repression in their homelands to a chance to participate in a mobile work force, family reunification, and better health care in their adopted country. As a result, daily contact with diverse cultures, religions, and languages is a reality for most European citizens. With the increase in diversity comes not only an escalating politicization of immigration at national and international levels, but also an increased sense of the potential threat that immigration poses. Immigrants, regardless of their legal status, are often blamed for economic instability and unemployment as well as for fears about local and national security issues ranging from increased crime to potential terrorist acts. In addition, immigration is seen as a serious threat to national identity as native publics see their familiar way of life eroding. As a result, many immigrants in Europe, especially those from countries in Africa, the Middle East, the Caribbean, and Asia, suffer ostracization, discrimination in housing, education and employment, and emotional and even physical harm in their adopted countries.

France is not unique among EU nations in seeing immigration take center stage as a contentious and pressing issue. This has resulted in repressive and often exclusionary legislation and the espousal of an anti-immigration rhetoric in the national discourse on immigration. The "guest worker" program that began as a practical solution to reconstruction needs after World War II has transformed into a volatile problem that threatens national, social, and political cohesion. In late October and early November 2005, France's ambivalent and often belligerent position on immigration became front-page news as civil unrest in the *banlieues* (outskirts or suburbs) of Paris spread rapidly from one city to the next. President Jacques Chirac, in a gross understatement, belatedly called the car-burning, rock-throwing, and looting that occurred every night for over two weeks and that resulted in a national state of emergency a "profound malaise" revealing "a crisis of identity" in the country. The rioting, triggered by the deaths of two youths of North African descent allegedly fleeing from the police, exposed a long-simmering rage felt by the segregated and impoverished immigrant community, referred to as "excluded" in the official language.[2]

The way immigration has traditionally been viewed in France grows out of a concept of the unified French Republic and a secular ideal that insists on assimilation of immigrants into the French way of life and rejects the viability of a multicultural society that recognizes ethnic and religious difference. The failure of that approach became painfully obvious during the 2005 uprisings, which revealed the extent of racism and the hopelessness, rejection, and anger felt by immigrants and by French citizens who were either born in former French colonies or born to immigrant parents in France. France's restrictive, often repressive, immigration policies, which Chirac promised to tighten as a response to the rioting, further intensify this outsider status. While the official designation of "immigrant" (*l'immigré*) is limited to those born abroad without French nationality and who now live in France, on the street those who look different or practice different cultural traditions are *seen* as *immigrés* regardless of their citizenship. As French citizen Semou Diouf, born in Senegal when it was a colony, explained in the wake of the rioting, the problem is "the French don't think I'm French."[3]

The "us" versus "them" tension on the French socio-political stage makes immigration issues a fertile source of material for the theatrical stage, as several recent productions testify.[4] Janelle Reinelt claims that "theatre is especially well-suited to influence as well as reflect the course of history by providing imaginative mimesis, transformative models, and observant critique" and that performance in tandem with other cultural practices can play a significant role in reinforcing or changing social attitudes and behavior.[5] The sheer number of theatrical productions dealing openly or metaphorically with immigrants, refugees, and "outsiders" implies an assumption about their potential to have an impact on public opinion. In the thriving street theatre (*théâtre de rue*) phenomenon, the theatrical scene in France has an additional vibrant performance venue through which to thrust immigration issues into the public eye. With its goal of democratizing "high culture," *théâtre de rue* is particularly well-suited to intervene in the official and popular discourses on immigration. Street theatre at its best responds quickly and vividly to societal attitudes, prejudices, and concerns and to official actions and edicts, so it is not at all surprising that many artists would create performances that focus on such a pressing issue for Europe.[6]

Theatre on the street is certainly not a new phenomenon, but what is today considered *théâtre de rue* exploded on the public stage in Europe in the 1970s. Inspired by the same anti-establishment impulses that led to the May 1968 student/worker uprising on the social stage in France, street artists insist that their guiding principle is to create art in public spaces, outside of traditional theatres. *Le Goliath 2005–2006*, the annual guide to arts of the street published by HorsLesMurs, lists close to 1000 street theatre companies or artists practicing today.[7] Every year over 500,000 spectators flock to approximately 300 street theatre festivals to experience a wide range of performances (busking, puppetry, stilt-walking, acrobatics, story-telling, installations, musical ensembles parading through the streets or suspended from the trees, *nouveau cirque*, site-specific events, theatre of fire, and elaborate *spectacles en déambulation*) in locations outside the predefined spaces of conventional theatre.[8]

French *théâtre de rue* offers sometimes startling, sometimes mundane, interventions into the immigration discourses currently going on in Europe. Experimenting with form and content, street theatre artists seek to create an experiential performance that is not bound by institutional structures or paying audiences, that transforms a public space into a performance space, that breaks down the barriers between the imaginative (or aesthetic) world of the performance and the everyday world of the spectator, and that engraves theatrical moments into "the imaginary of the everyday life of the people."[9] These alternative discourses "from below" expose or counteract the myths in the popular imagination and in official proclamations that identify the migrant as the key source of the nation's problems. As such, it is particularly well-suited to have an impact on audience attitudes.

After giving a brief overview of immigration policies in France, this essay will explore three strategies that certain street theatre companies have used to "perform immigration" and thus to interrupt the immigration discourse and potentially to sway public opinion. One strategy, inspired by the republican paradigm of assimilation in France, forces the public into the role of citizen, of "insider," as the performer becomes the "outsider." In Friches Théâtre Urbain's *Melgut*, the spectators stare at, reach out to, shun, laugh at, mock, or even confront the character, Melgut.[10] The character then exaggerates the outsider's response to the actions of the insider and thus puts a spotlight on the audience by drawing attention to the spectators' intolerant, even racist, behavior. The second strategy again casts the public in the role of those who belong, but here, the performance offers a possible vision of post-national citizenship, a utopian spectacle of acceptance of difference as created by Royal de Luxe's *Le Géant Tombé du Ciel*.[11] Compagnie Dakar's production of *Braakland* offers just one example of a third possible strategy, one that thrusts the spectators into the world of the immigrant or refugee and encourages them to experience, physically and emotionally, the hardships faced by the outsider.

While these three performances differ greatly, they all intervene into the discourses on immigration through metaphor and image rather than through a documentary or literal approach. They also all share the goal of establishing "a direct rapport, without barriers, with the public."[12] This special actor/audience relationship represents a key characteristic of street theatre practice: "the rapport with the public is the motor itself of the shows" and is an important source of its potential to have an impact on attitudes toward immigration.[13] The spectator plays a key role not only in the intellectual process of the construction of meanings, but also in the practical process of quasi-role-playing in the actual show. Denied the position of passive observer and the anonymity of a darkened auditorium, the spectator enters the performance in a variety of ways and thus inhabits both imaginary and actual worlds simultaneously. It is the public that penetrates the barrier between everyday and aesthetic experiences and, as a result, that *experiences* both positive and negative possibilities in relation to the issues of immigration.

Immigration and France

On 24 April 2002, Jean-Marie Le Pen, the leader of the Front National, an extreme right-wing party espousing anti-immigration policy as a core issue since its founding in 1972, announced: "I will stop and reverse the flow of immigration into France."[14] He promised to withdraw France from the European Union, to re-establish French borders to control the "flow of foreigners" into France, to expel illegal immigrants, to tighten French nationality laws, and to promote the doctrine of "national preference" that gives preference in housing, education, employment, and benefits to French and EU nationals over others. His anti-immigration platform helped place him second, ahead of Lionel Jospin, the left-wing candidate, in the first round of voting in the 2002 presidential election, and thus forced voters to choose between a right-wing conservative (Jacques Chirac) and an extremist. Tensions over issues of immigration and national identity continue today and arguably contributed to Nicolas Sarkozy's presidential victory in 2007. For at least the last four decades, a significant percentage of the French populace has been active in the push to close the borders to "outsiders," especially as immigration is defined more and more in terms of a threat to security, both on the national and local levels.

After a period of relatively open immigration policy in the form of temporary or "guest worker" programs in the two decades following World War II, opinions about "foreigners" began to change as the face of the immigrant started to shift from a European to North and sub-Saharan African one and as "temporary" workers stayed in France, married, and raised children born on French soil. By and large, these families were ghettoized and housed in low-income projects in the *banlieues* of major cities. The oil embargo of 1973 and the economic slowdown and consequent job losses hardened negative attitudes towards migrants. In 1974, France banned labor immigration, although professional and other highly skilled personnel, most often those from other European or Western countries, were exempted. Since these skilled professionals were referred to as *étrangers*, the more pejorative term *immigré* came to be associated with unskilled workers from the Third World or non-European French citizens thus attesting to the racialization of immigration discourse.

This misrepresentation of citizenship has its roots in the French Republic that promoted the idea of the transformation of *individual* into *citizen* through the erasure of cultural and religious identities and the adoption of a common French identity.[15] "*Only* individuals are citizens, citizens are *equal*, therefore *all* individuals are equal citizens" became the foundational assumption of French citizenship.[16] That concept has fostered an institutionalized resistance to the acknowledgement of an ethnically diverse society. In spite of France's position as both "a defender of individual rights and a country of immigration, willing to grant citizenship and rights to those who wished to become French," being accepted as French is not so easy.[17]

The goal of complete assimilation into the French way of life or *habitus* (to use Bourdieu's term) depended on a socialization process in the schools, work places, military, and social organizations with the intended result being the erasure, as much as possible, of foreign origins. But as many immigrant groups not only would never "look French," but also began to seek cultural pluralism over assimilation, the lines of demarcation between *insider* and *outsider*, between those perceived to be French and those perceived to be *immigré*, hardened. Not surprisingly, the result was increasingly repressive legislation, worsening living conditions, and numerous incidents of discrimination and violence against those perceived to be "foreign." In fact, the mid-1980s saw the introduction of the word *clandestin* (clandestine) to describe immigrants, a word implying illegality and "invoking the idea of the immigrant as a threat to the political community, a scapegoat for the imagined collapse of a social order."[18]

In 1986, newly elected right-wing government leaders attempted to reform French nationality laws so that children of immigrants born on French soil would not be granted automatic citizenship. Only the loud outcry by the opposition forced the government to retreat from the proposed reform, but anti-immigration rhetoric continued to escalate. Old republican notions of *assimilation* morphed into support for a policy of national *integration*. Although seemingly pluralistic, integration was no more tolerant of difference than assimilation since it advocated the absorption of minorities into the socio-cultural "mainstream" not by erasing religious and cultural difference, but by relegating it to the private sphere so that it did not disrupt national cohesion in the public sphere. These attitudes came to a breaking point in the 1989 *affaire des foulards*, when the headmaster of a Paris suburban school prohibited three Maghrebi girls from attending if they wore headscarves, which he argued were religious symbols and thus opposed Republican secularism. That year, 75 percent of the respondents to an opinion poll conducted by *Paris-Match*[19] expressed concern that France was in danger of losing her national identity if the influx of foreigners was not limited.[20] This incident eventually resulted in a 2004 law banning the wearing of all "ostentatious" religious symbols in public schools. Failure to comply results in expulsion, which a number of young Muslim women have experienced for refusing to remove their headscarves.

François Mitterrand declared in 1990, "France is no longer and can no longer be a country of immigration. We cannot welcome all the world's poor." In 1991, Jacques Chirac, then Mayor of Paris, complained of the "overdose of immigrants" and, in a now infamous statement, sympathized with the "decent" French citizen "who sees his next door neighbor – a family where there is one father, three or four wives and twenty-odd kids, getting fifty thousand francs in social security payments without going to work: add to the noise and smell and it drives the French worker crazy."[21] In 1993, French Interior Minister Charles Pasqua initiated what were popularly called "Pasqua Laws" restricting entry and residence rights, permitting police identity checks, and limiting access to French nationality and announced that the prime goal of the center-right government that had just taken office (Chirac's party, RPR) was "zero immigration." Although he later revised that to "zero illegal immigration," the original comment reflects "an impossible but hugely appealing objective in the eyes of many ordinary French men and women, for it carries the promise of somehow ridding the country of all problems linked in the public mind with people of immigrant origin. In a word, it encapsulates the notion that immigrants and their descendents are fundamentally out of place in French society.[22]

When he was still Interior Minister, Nicolas Sarkozy echoed Pasqua's absolutism in his response to the civil disorder in the *banlieues* in 2005. His labeling of the rioters as "scum" and his apparent dismissal of the real issues of second-class status and lack of educational and employment opportunities earned him the reputation of a hard-liner, but his tough approach appealed to the many who worry about the impact of immigration on French national identity and influenced his 2007 election as president. Violence against those perceived to be immigrants continues in institutional blindness to discrimination in employment, housing, and education, in widening gaps in rights between nationals and non-nationals, and in the label of *immigré* for those considered immigrant regardless of citizenship in the public discourse. As immigration has gained importance as a key social issue focusing on national identity, the depiction of outsiders and the exploration of hostile attitudes toward them began to appear in street theatre.

Performing immigration on the street

Street theatre establishes an artistic and social intervention into the actual life of the city: streets must be closed and traffic rerouted, unsuspecting pedestrians find their way blocked by a crowd, and noise and litter increase. These interruptions in the flow of daily activities are the key source of intervention into the discourses on immigration. Street theatre is a visceral performance practice in which the artists forge an often intense experiential relationship with the audience – those who come to see the show and those who are just passing by. In street theatre, the discomfort that immigration issues often arouse is transposed to a real physical discomfort and imposed upon the audience. The outdoor performance site and often the dramatic form itself increase the participatory aspect of the performance by creating an experience in which the audience members cannot ignore the discomfort of their bodies as they sit on the pavement under the hot sun or a cold rain, stand packed tightly among others struggling to see the show, or walk, sometimes up to a mile, to follow the actors on a *déambulation*. Their physical engagement contributes to an experiential understanding of the aesthetic event that, in turn, allows for a reassessment of attitudes toward and understanding of the "outsider."

As the actors and spectators share the same un-self-conscious public place, transforming it into a self-conscious aesthetic space by occupying it and playing there, the imaginative world of the performance is superimposed on the actual world of the street.[23] These two worlds, visible simultaneously, rub against each other, unsettling both. Street theatre inserts a performance imaginary into the social life of the street as performers bridge the gap between the daily activity of the spectator and the artistic activity of the performance, inhabiting the world of the street and the world of "theatre" simultaneously. They play with the cityscape turning it into scenic décor by, for example, scaling walls or climbing trees, hiding in doorways, leaning on parked vehicles or out of open windows, or walking into traffic, forcing cars to stop, swerve, or honk. This aestheticization of a familiar, everyday place defamiliarizes both the space and the activities going on there and thus encourages the spectators to see them differently. It is the many ways that this art form wedges itself into the social life of the street that audiences find so engaging. Sarah Harper, Artistic Director of Friches Théâtre Urbain, a professional Paris-based street theatre company, explains that street theatre strives "to get art inside life; to get theatrical sensations inside the tiniest folds of the real world; to express those feelings and ideas in a living space, in the public's space, and not a space that belongs to cultural institutions; [...] and to give an ordinary space in a town public memories for those who saw the show there."[24]

The transformation of public place into aesthetic space forces the spectators to be witnesses to, and celebrants of, that shift of reality and shakes them from the role of passive observer. Gwénola David, in *Scènes Urbaines*, a publication of HorsLesMurs, pushes the active participation of the audience even further when she argues that "the street, a symbol of opposition-forces transformed into an agora that is as open to aesthetic experiences as to political demands, becomes a place where a direct link with the public can be recreated. The artists of the street establish an egalitarian relational exchange with the public. The spectator is called upon to become a partner in the action, or even an actor in the performance."[25]

This special actor/audience relationship represents a key characteristic of street theatre practice and is an important source of its potential to have an impact on attitudes toward immigration. The spectator plays a key role not only in the intellectual process of construction of meanings, but in the practical process of quasi-role-playing in the actual show. The audience member penetrates the barrier between everyday and aesthetic experiences and, as a result, *experiences* a varied range of possibilities in relation to the issue of immigration. Thus street theatre alters, both actually and metaphorically, the way the public "sees" a place, a performance, and often an issue. And it is this "re-vision" that makes the issue of immigration and public attitudes toward "foreigners" so engaging for street theatre artists. Few of the street theatre performances about immigration offer facts or background information on migration movements or even workable solutions to real-life problems; rather, they explore the life of the individual who does not fit in, the "foreigner" who lives outside the context of the established norm, through images so vivid and provocative that we cannot remain detached and indifferent. The significance is not *what* the performances say about immigration, but *how* they say it since the form is so often the content in street performance. Not a text-based theatrical form, but one in which the words, if there are any at all, support the visual or spectacle text, *théâtre de rue* launches an attack not only on conventional theatrical forms, but also on the hegemonic culture and discourse that these forms represent.

Friches Théâtre Urbain's *Melgut* represents *théâtre d'intervention*, an improvisational form of street theatre that depends on inserting the imaginary world into the actual one by placing unusual images in an ordinary setting or confronting the public with fiction in order to alter our experience of the world. The *outsider*, the character who stands out as different from and incapable of fitting into his surroundings, is a natural choice to juxtapose reality and fiction. *Melgut* is a one-man show in which actor Pascal Laurent presents Melgut (see Figure 2.1), a spider-like and hunched character on one-meter stilts, dressed head-to-toe in red flowing robes.[26]

Figure 2.1   Pascal Laurent in *Melgut*, Friches Théâtre Urbain
Source: Photograph courtesy Friches Théâtre Urbain.

Living outside the city, maybe in a dump or in the tunnels of the sewage system, the old and limping man arrives in the center of town to find something important to his survival like food, news, or warmth. Melgut appears out of nowhere among unsuspecting pedestrians who suddenly find a foreigner – disoriented and scared – in their midst. This "foreign" character forces us to recognize our responses to his difference; we cannot just passively watch the outsider being shunned, mocked, or feared. Instead, we are compelled to engage in that racist behavior ourselves by staring at him, laughing at his behavior, sometimes grabbing a stilt or the robes, or just walking away and dismissing the outsider. While some spectators try to connect with Melgut by, for example, offering a cigarette, he gets the audience to laugh at him by his surprise and over-reaction to the gift and thus turns the act of kindness into another way to mock the outsider. The problematic power of this performance arises from the embodiment and reinforcement of the stereotype of the outsider that, in turn, pushes the spectators into negative stereotypical responses.

Melgut hopes to blend into the crowd, to assimilate, but his size (about three meters tall), his vivid red attire, and his erratic and unsocialized behavior, so different from that of the pedestrians he meets, make that impossible (see Figure 2.2).

Although he would like to be as invisible as the normal inhabitants of the town, he discovers that he is not as we surround him, stare, and point at him even when he crouches motionless against a wall for several minutes, wait to hear what he has to say, follow him, or force him to engage with us. He cowers against buildings, frightened of the world he finds himself inhabiting; he stops and leans against a wall, curling up to reduce his oversized

Figure 2.2   Pascal Laurent in *Melgut*, Friches Théâtre Urbain
Source: Photograph courtesy Friches Théâtre Urbain.

body. He nearly falls asleep and trembles from his nightmares; he moves only his fingers or his head, constantly aware of danger. Only his muttering allows the audience to know Melgut is not asleep. A circle of curious passers-by gathers around him. Suddenly a noise attracts his attention and little by little he discovers our feet, then lifting his eyes he discovers our hands, then our eyes. Melgut suspects that since he is so different, so visible, the "insiders" are going to chase him, to try to catch him, to lynch him if they can, so he hides along the walls, keeping an eye on his back. He is clearly out of place in the ordinary landscape of the street, and he will never be able to assimilate.

Melgut is always the outsider. He has entered the public space, and his precisely enacted awareness of that intrusion firmly places the crowds that surround him in the role of insiders. The impossibility of assimilation presented in this fictional situation exposes the paradox of the republican model by showing that Melgut will never fit in, never be an insider, and by revealing how easily the spectators can become intolerant and prevent the assimilation of someone who looks and acts different. However, Melgut is so blatant a stereotype that it is not hard to lose sight of his metaphoric significance.

There are some moments of connection with the insiders as Melgut touches someone's head, leans on someone's shoulder, or allows people to hold his hand, but he often pushes too far and so he begins to turn the crowd against him. In one show, an unannounced performance intervention in June 2002 near Saint-Germain-des-Prés in Paris, a crowd of children surround him, pulling at his robes and grabbing his hands. Suddenly, he lifts one child up and swings her around three or four times. The others are stunned; some want a turn, others back away, but Melgut takes off across the street, stopping traffic, trapped among the cars. His actions dramatically convey the image of being hemmed in and afraid, of being a rat caught in a corner. He hides in the folds of his robes and stares at those who have begun to surround him again. A child points and laughs, and Melgut stands and stares. He starts talking, arguing grumpily with the public about current issues, accusing us of ignorance, indifference, blindness or worse. He attacks us with words and rushes into our midst, making us scatter. "I know you want my head, but you will not have it so easily. You may kill me, but before you do I will tell you what you are [...]" The text that follows can vary depending on the composition of the crowd; the subject may be environmental abuse of the earth, the death penalty, racism, violence, nuclear weapons, or globalization. The text becomes more tense, moving from submission to power to compassion to tenderness, and back to power. Melgut then escapes in laughter, crossing the road in the middle of traffic. Moving quickly, but not so fast that the spectators cannot follow, he establishes a new performance space and starts again.

Melgut lashes out against his audience one minute and tenderly engages with a child the next. He will not let his audience connect with him or accept him for long; he is too wary of the consequences. He knows that those who seem so friendly can turn on him in an instant when he steps out of line. Trapped in a world that makes him uncomfortable and seems to reject him, Melgut brings the outsider's feelings of isolation, fear, anger, and longing into sharper focus as he highlights the impossibility of assimilation for "outsiders." He exposes the fallacy of the republican paradigm that assumes the success of assimilation, of a voluntary erasure of cultural specificity in order to become a citizen, and of the concurrent acceptance into the mainstream of the reformed foreigner. He challenges the myth that the desire to fit in, to become French, insures success as his too big and too red body sets him apart quite vividly. But he also forces the audience to experience how easy it is to discriminate against someone who looks and acts different.

Melgut's actions do not really allow "us" to accept him: he harangues us too fiercely, he invades our space too aggressively; his mercurial temperament is too frightening. Spectators leap out of the way, often bumping into or tripping over one another; some giggle nervously, others swear. Melgut most often forces us into a fearful or a hostile posture and so compels us to be the ones who reject or hassle him. He magnifies and highlights the public's racist behavior by exaggerating his reactions to it, wincing and cowering, speeding away, or spinning around and attacking like a cornered animal. But the performance intervention goes one step further. While Melgut understands he is an object of scorn and a potential victim of violence, he does not really understand this alien world; in fact, he is somewhat blind, unable or unwilling to open his eyes. Like those who torment him, he is sort of racist himself, ready to lash out at others when he feels at risk. So not only do we experience being the one excluding the "Other," but we also experience the receiving end of abuse.

Exploring this symbiotic relationship between space, spectator, and show, Michel Siminot writes, "The peculiarity of a street performance is that it transforms any place, open or closed, into a performance space by the single fact of its irruption. It creates both a socio-political intervention and an artistic intervention simultaneously."[27] The forced opening of our eyes caused by this intervention is often uncomfortable, but it can be transformative. Because it is so ordinary and taken-for-granted, it demands a "re-vision" of what we have overlooked. Thus, in a Frieirean sense, this street performance parallels "problem-posing" education that turns passive students into "co-investigators" of social and political structures.[28] Here, passive spectators become active collaborators in the artistic process.

The experience of having Melgut as "outsider" enter and focus our "insider" space and status dramatizes the rise of French racism based on cultural difference. This performance intervention highlights the unassimilability of the "identified" outsider, and it intensifies the ease with which we can reject or mock difference, making it more clearly discernible. Melgut offers a hyperbolic and metaphoric characterization of the *immigré*, and while that characterization is offensive as it reinforces negative stereotypes, it also exploits extreme exaggeration to dispel any possibility of being construed as an accurate portrayal and thus shifts the focus onto the experience of the spectators as we react to this obviously alien being. Melgut balances precariously on the line between being so outrageous that he is just entertaining and being the source of biting social critique.

Like *Melgut*, Pno Pneu's *Les Quiétils Réfugiés politique* (2004) highlights the "us" versus "them" conflict, although much more gently (see Figure 2.3). The Quiétils (written as a made-up word, but when spoken sounds like "qui est-il" or "who is it") are refugees from somewhere foreign, unknown. The spectators are the insiders who represent the norm, whereas the characters are clearly outsiders working hard to understand and appreciate this new world in which they find themselves, simultaneously fascinated and frightened by what they see and touch and by how the people who surround them react to them and interact with them. The Quiétils cannot

Figure 2.3   Les Quiétils Réfugiés politique
Source: Photograph by Susan Haedicke.

blend in with the crowd and become invisible because their bizarre appearance and unusual behavior embody otherness. The heads of the creatures' bumpy, stone-like "naked" bodies are attached to the chest or belly of the actor whose actual head is hidden in the large pack the refugee carries on his hunched back.

These refugees (sometimes only two, sometimes several) can easily amuse themselves for several minutes with a bit of litter found on the ground or with a passer-by who has stopped to take a photograph. They never speak. Since the Quiétils are not aggressive, the crowd can easily embrace them and always begins to "teach" them how to "fit in." The republican model of assimilation is played out time and time again as the Quiétils are instructed to throw litter in the trash bin, to sit properly, or to use a camera. The audience rewards them as they learn the proper etiquette. The performance, sometimes tongue-in-cheek but sometimes simply nostalgic, expresses the joys and the promise of the assimilation paradigm, at the same time as it shows that assimilation's price is cultural amnesia.

The performers who intervene into the daily activities of the street draw attention to themselves by highlighting the incongruity of their presence with the space around them for spectators looking on in amusement, staring, engaging with, or even taunting them. The living interchange between actor and audience transforms a neutral and static place into a dynamic public space, but it also reveals how exposed and potentially vulnerable one can be in a public space. As de Lage explains, "A public space is a space that finds itself between private spaces [...]. It is a space for everyone and at the same time, a space for no one."[29] These performers draw attention to the emptiness of public spaces even when they are filled with people and thus dramatize the "aloneness" of the outsider. As the characters move back and forth between their own fear of, and fascination with, the crowd and between the hostility and friendliness they receive from the spectators, they performatively embody the experience of the outsider, the immigrant who is struggling to survive in an alien world. But the performance form goes beyond offering a peek into the hardships of the "foreigner" as the performances cast the audience in an active role in the experience. It is the audience that accepts or rejects the newcomer; it is the audience that can begin to recognize the impossibility of the republican paradigm.

Royal de Luxe, one of France's pioneering and most famous street theatre companies, also presents an outsider who cannot fit into the ordinary world of the city in *Le Géant Tombé du Ciel*. This performance, lasting several days and nights, offers audiences the opportunity to live with a foreigner in their midst – a foreigner who demands attention. The outsider in *Le Géant Tombé du Ciel* is a giant marionette, more than 10 meters tall, attached to a scaffolding (his "cage") and manipulated by approximately 20 men dressed in long red coats. For several days before the theatrical event begins, leaflets appear, knocking off balance the everyday life of the city by inserting a fairy tale into the urban landscape as a news item:

> Once upon a time, a giant lived in the clouds. One day he fell from the sky onto the boulevard. When he awoke, he found that the people had tied him to the ground. Over the next few days, they paraded him in a huge cage to amuse the city. But every night, the giant dreamed, and his dreams terrified the people. So they built a huge wall of light to prevent him from sleeping. That night the giant had such a powerful dream that he broke from his cage and disappeared in the night [...].[30]

Then, one morning, the townspeople discover a huge fork stabbing a car as though it were a sausage and find one large shoe, measuring one and one-half meters and weighing 40 kilos, dangling from a street light, and another from a balcony of the bank. The next morning, Léonard, the giant, is found tied, Gulliver-like, to the ground in front of the church.[31]

The audience is clearly cast as the Lilliputians. The giant breathes and opens his eyes and mouth. An opera singer entertains him with arias. The men who attend him shower and dress him, feed him tons of sausages and gallons of water, and brush his teeth. They help him stand up and encase him in a large cage so that he can be paraded through the city. With incredibly life-like movements in each part of his body, from his eyes, eyelids, and mouth to the tilt of his head, his hands with jointed fingers, and his feet and toes, he walks through the city, looking through second- and third-story windows into people's private lives in their apartments and visiting the church, the port, and the town hall – the three sources of institutional or public power representing religion, commerce, and government. Each day he charms the crowds that follow him with his clumsy innocence and child-like wonder, disarming the people's tendencies toward cruelty that often plague outsiders. But each night, Léonard snores in his sleep, his eyelids flutter, and he dreams. The result of his dreams, or nightmares, are visible the next day in cars crushed and tangled or buses cut in half by a huge knife still poking from the roof. To prevent his destructive nightmares, his "minders" build a wall of powerful spotlights to keep him awake, but their strategy fails and the next morning the city awakes to find the giant gone and the cage and nearby cars broken and crushed. In the documentary video of the theatrical event performed at Le Havre, audience members mourned his "escape": "He's left a hole in my heart" and "I have to say I fell in love."

During his stay in the city, Léonard's routine activities of eating, sleeping, washing, and even relieving himself, openly performed on such a grand scale, magnify the plight of the "outsider," the homeless person or refugee forced to "live" in public, and the visibility of his intimate acts blurs the distinction between public and private space. The whole city becomes a gigantic residence as Léonard takes over the city with his size and his slow procession through its streets. Normal life must come to a halt, but it is "thanks to his clumsiness and ignorance of how to use public spaces [that] he [and, by extension, the audience] interrogates time and place. With him, we ask about the geography of the cities and the regions, about the times and history."[32] As the spectators watch Léonard "see" the city, they are encouraged to see it through his eyes, much as an adult rediscovers the mystery of nature on a walk with a child.

Like *Melgut*, *Le Géant Tombé du Ciel* focuses on the quest of a solitary individual for *la vie sociale*, a life in the company of other human beings, but unlike *Melgut*, this performance offers the possibility of a post-national identity (or transnational citizenship), one that does not depend on assimilation into an established society, but rather on acceptance of and working with difference. Zillah Eisenstein, in a chapter entitled "Humanizing Humanity: Secrets of the Universal" in *Against Empire: Feminisms, Racism, and the West*, writes:

> Thinking without our skins and with our bodily desires and needs promises a possibility for recognizing human connectedness. Once one recognizes the human claims of one's bodily needs for food, shelter, love, privacy, and sexual autonomy they subvert the isolated self. These are shared meanings of what it is to be human.[33]

And it is precisely through Leonard's "bodily desires and needs" that we come to know and love him. He highlights a common humanity with "the possibility of resemblance and equivalence between people," and on a very human level (although on a supra-human scale), he points the way toward "humanizing humanity," toward a transnational citizenship that promotes inclusivity and unity but not uniformity.[34] "Differences, of any kind," writes Eisenstein:

> are translatable because human differences are also connections. Only when the self is visioned as completely autonomous and individual do differences become totally distinct and separate. Humanity transcends and articulates polyversality simultaneously because no individual is ever completely different or totally the same as another. This is why I can know differences that are not my own; I can push through to a connection that allows me to see variety – even if in translated form through my own experience which is never identical with any other site.[35]

But an embrace of difference is not so easily accomplished as Fiorella Dell'Olio shows in *The Europeanization of Citizenship*. She argues that the creation of European citizenship has failed to take the next step of the development of a European identity. Instead European citizenship is "often seen as a threat to cultural specificity because of its own peculiar supra-national nature," and has strengthened the public's ties to their own national identities.[36] *Le Géant Tombé du Ciel* dramatizes this notion of difference as a threat and yet, simultaneously, plays with the possibility of a post-national identity by enabling the audience to experience love for the outsider. Léonard's enormous size magnifies an existence "on display" as he is stared at and pointed at like a freak in the circus, his promenade through the city signifies the charged confrontation between "us" and "them," between insider and outsider, and his violent nightmares cause real damage. While the inhabitants of the city seem to welcome him, they do confine him to his cage and they try to prevent him from dreaming. His route from church to port to town hall represents the "impossible encounter between an individual and the group and vice versa between the group and an outsider."[37] Léonard's size, his ponderous gait, and his cage magnify his solitude – he can never "fit in" with a crowd that is so different and that "knows how to welcome him only by blocking his dreams and obliging him to adopt their customs, only by making him a fairground puppet."[38] He embodies the dysfunctional assimilationist model of citizenship.

Although Léonard forces the audience to recognize an unforgiving and intolerant world (portrayed in his nightmares that acquire a physical presence as the townspeople find forks and knives stuck into cars and buses), he also offers the hope of a utopian dream of unification across difference as he and the huge crowds that follow him and come to love him rediscover the spatial and temporal urban landscape with the wonder of a child. The joyful faces and loving comments of the spectators recorded in the documentary on the Le Havre event clearly testify to the profound acceptance the people feel toward the stranger in their midst. The affection that the people express toward the huge marionette comes, of course, partially from the skill of his many manipulators, but it is more than that. It is Léonard who is appreciated, and his difference, so vividly represented in his size, does not make us fear him as a huge giant, but rather reassures us by magnifying his child-like innocence. "Diversity and not uniformity underlies life's 'complex oneness,'" writes Eisenstein. "The self must be present if an earnest commitment to community is embraced. The love of humanity – as an inclusive polyversal – must therefore start close to home, and our home is the body."[39] The size of Léonard's body and the visibility of his desires and needs underscores this inseparable connection between body and community. And it is the acceptance of Léonard's difference (as contrasted with the enforced rejection of Melgut) that offers the possibility of a post-national identity across borders and in spite of difference. *Le Géant* creates an imaginary world of solidarity with diversity that the spectators enter "to deepen understanding by sharing and decentering the self with a newly fulfilling complexity."[40] We, the audience, our city, and our future become the performance that we are watching.

The hurdles faced by transnational citizenship in a New Europe are vividly enacted in Compagnie Dakar's *Braakland*.[41] Rather than open borders and "humanized humanity" pointing the way toward a global democracy, the audience experiences a *braakland* – a *wasteland* – "a world without a name and without a soul, where all reason to live has disappeared," where all are outsiders who abuse and kill each other – the weaker die first, but even the strongest cannot survive.[42] Unlike the companies discussed thus far, Compagnie Dakar is a Dutch street theatre company founded in 2001 by Guido Keene. I saw *Braakland* at the street theatre festival in Chalon-sur-Saône in 2005 in a space well outside the festival limits.

When I get to the rendezvous point where a bus is supposed to pick up the audience and take us to the performance, the doors of the bus are closed, and about 30 people stand nearby waiting. As the hour for the start of the show approaches, the audience begins to get restless and wonders aloud when the driver will arrive. Finally, with just 15 minutes to go, a company member announces that the bus will leave in about 15 minutes, but if we want to get to the show on time, we had better walk [...] and quickly. Thus begins the trek of the spectators. The promotional booklet for Chalon-dans-la-rue describes the adventure: "On a wasteland, nine actors lead us into an enigmatic and mysterious world 'Braakland,' they are the men who approach what they don't like to see. They embrace exactly what they don't want to know. Because they can't do otherwise, because they don't know how to flee, they yield to violence, hatred, indifference, and despair. A show that has to be tried."[43] Teasing the spectators with the possible thrill of an uncomfortable and frightening participatory performance, *Braakland* actually goes beyond the advertising as we soon learn.

I, along with the other audience members, briskly march out of town along the river, past a large industrial complex and warehouses, most of which look abandoned. After about 20 minutes of walking, we, the compliant audience members, arrive at a chain-link fence with the gate slightly open. Our leader seems to have disappeared, and as we approach the gate, we are told to stop by several people standing on the other side. Our explanations that we are there to see a show are met with disdain, but finally they let us in one-by-one and point vaguely off to the right through a scruffy looking field. I am very glad I am wearing long pants and closed shoes as I make our way through burrs and thistles. I have no idea how the person at the front knows where to go, but everyone follows without question. After another five minutes or so, I arrive at a dark building, but rather than entering, I make my way around it. On the opposite side, I find myself in front of a very large field (probably six to eight football fields in size), a wasteland similar to the one I just traversed, but much larger and more hilly. In the distance I can see a working factory with smoke rising from the many chimneys. As the sun begins to set, lights on the factory make it look quite sinister. The audience is told to sit on hard backless benches staring into this empty field for what seems like a very long time.

If I had been in this space with these 30 or so strangers outside the festival context, I would actually have been quite scared. Even knowing I am attending one of the shows of an "in-company" (invited company), I can't ignore my feeling of anxiety and discomfort as I wait and wait. What Compagnie Dakar has accomplished by this long and uncomfortable "gathering phase" experientially approximates the refugee journey. The audience members actually take a "voyage" that physically hurts to a place that is unfamiliar and uncomfortable and that seems threatening. And they take this journey as themselves, not as a character role-playing in a dramatic enactment. The feeling of danger is intensified by this route, down narrow roads outside the city limits, past abandoned buildings in the gathering darkness (not the kind of route one would choose to walk on one's own), so there is a sense of entrapment: one would hesitate to leave the group and walk away from the performance to return to the town alone.

The strategy of "living the experience" shifts once I arrive at the performance site as I am now a spectator of the drama, not one of its actors (even though I did not realize that had been my role until much later). Now, rather than telling a story of exile and of the despised outsider, Compagnie Dakar surrounds the audience with a vivid image of this wasteland devoid of compassion, decency, and humanity. Rather than the reality of the immigrant experience, the actors offer a vivid metaphor of exile and otherness at its worst. After several minutes of waiting in the gathering darkness and increasing chill in the air, I see a man appear in the distance and slowly make his way toward the audience. There is no sound, no music to signal the start of the show, and in the gradually approaching darkness, the only lights are from the factory in the distance and a street lamp on the road off to the right. When the man gets to about 30 feet from the spectators, he sits on the ground staring in front of him. Then nothing happens. After several minutes of watching the motionless figure and waiting, I see another man carrying a pitchfork appear in a far corner of the field. He begins to cross painstakingly slowly, but as he gets two-thirds of the way across, he collapses.

Again, nothing happens for a long time, and again, eventually I begin to make out the arrival of more of these silent, zombie-like people. Suddenly, the man sitting close to the audience whom I have forgotten, jumps up and runs full speed back to the collapsed man. As he does so, others run toward the fallen figure as well and within seconds, the corpse is stripped bare. Each returns slowly to his spot carrying his prize. One builds a fire; another starts to dig a hole. Across the field, a woman is attacked and raped quite brutally by a fellow exile. He throws her to the ground, drops his pants and stands naked over her momentarily until he throws himself on her. In the tall grass, his backside rhythmically rises and plunges. Then all motion stops, and after a little while, he gets up, wipes himself off, raises his pants and walks off, leaving the woman on the ground. She lies hidden in the tall grass a long time – long enough for the audience to look for something else and forget her – until she finally gets slowly to her feet and then rushes to a bucket of water and washes and washes. At one point, a woman dressed all in yellow arrives; she really stands out from the others and is soon robbed and stripped.

Over the 90 minutes in which I experience *Braakland*, I see someone strangled, another beaten to death; another whose neck is snapped. All are dragged to a large hole that some of the characters began to dig at the start of the show – the grave. By the end, only the woman in yellow is left alive and in the final moments of the piece, she dives head first into what seems like a large rabbit hole some distance from the audience. I watch as she buries herself alive almost as though she is being sucked into the hole. Although I have not moved from my bench, I feel bruised and drained from my "embodied" experience that has encouraged me to "live" the event by placing me within its frame. That location, in turn, alters the way I understand and incorporate the information. The intensity, prepared for in the long walk and long wait at the beginning, increases relentlessly throughout the performance. The size of the field and the dimness of the available light allow the characters to perform unbelievably violent acts and, given the performance site and the very real enactment of brutality, my imagination fills in whatever is not really depicted. The actors choose not to provide an outlet during the performance for the building tension, but rather strive to keep increasing the intensity, and, not surprisingly, many in the audience just shut down, burying their faces in their hands or staring blankly at the distant factory. Others laugh or mock the events.

Although for the final ten minutes or so of the show, most audience members had distanced themselves emotionally from the action as a form of self-protection, I do not think that the impact of the experience was dissipated. On the contrary, I believe that being pushed to that point, remaining unmoving on the hard bench, struggling to retain separation from the violence unfolding before me, and then making the long walk back to civilization, this time in the dark, engraved the experience in my body more deeply than had many other participatory performances. I attended the show with others and, while the main comment was "what was that?," *Braakland* became the performance we discussed over the next week more than any other. While the specific details fade, the physical experience of the excessive violence caused by displacement and otherness remains.

Street theatre offers its audiences multiple perspectives on immigration and places the public at center stage as it dismantles official policies and hegemonic discourse. Through its cultural interventions, it forces a "re-vision" of the immigrant and his/her experience into the public sphere. Since street theatre plays with the notion of taking over and transforming public space, it can draw attention in a vivid and visceral way to the contrast between open and restricted access to a space. It is "around the notion of restricted access that a rapport between space (territorial) and identity is thinkable," whereas, the idea of an open space "breaks with the territorial dimension and is envisaged as a space of universality, of indifference to difference [...] of democracy."[44] The spatial experience of the insider/outsider conflict dramatizes the immigration debate, probably unconsciously, on a visceral and aesthetic level. Thus, it offers a different experience of the issue, an aesthetic one to be sure, but also an embodied one. These street theatre performances expose official "myth-making" and strategies of "socialization" that harden anti-immigration attitudes. They lure spectators into experiencing with their bodies the hardships of the immigrant or the prejudices of the native, making them uncomfortable with these attitudes and responses. Street theatre artists thrust the debate into public spaces and, through the use of metaphor, place the spectators in the role of decoding the images.

Notes

1. Gallya Lahav, *Immigration and Politics in the New Europe: Reinventing Borders* (Cambridge: Cambridge University Press, 2004), 1. The World Migration report was published by the IOM in Geneva in 2000.
2. Riva Kastoryano, *Negotiating Identities: States and Immigrants in France and Germany* (Princeton: Princeton University Press, 2002), 73.
3. Craig S. Smith, "What Makes Someone French?" *New York Times*, 11 Nov. 2005: A1.
4. These productions include, but are not limited to: Ariane Mnouchkine's *Le Dernier Caravansérail*; Michel Deutsch's *Skinner*; Mohamed Kacimi's *1962*; Leila Sebbar's *Les Yeux de ma Mère*; Fatima Gallaire's *Rimm, La Gazelle* and *Molly des Sables*; Noureddine Aba's *Une Si Grande Espérance, ou Le Chant Retrouvé au Pays Perdu*; and a participatory performance on refugees sponsored by Amnesty International and other human rights organizations entitled *Un Voyage Pas Comme les Autres sur les Chemins de l'Exil*. See my article, "Politics of Participation: *Un Voyage Pas Comme les Autres sur les Chemins de l'Exil*," *Theatre Topics*, 12:2 (2002): 99–118.
5. Janelle Reinelt, "Performing Europe: Identity Formation for a 'New' Europe," *Theatre Journal*, 53:3 (2001): 366.
6. There are close to 50 street theatre performances created in the last decade that in some way touch on issues of immigration. It is very difficult to collate this information since there are no electronic databases that index street theatre performances or companies. The best source of information is the librarians at HorsLesMurs (http://www.horslesmurs.asso.fr). HorsLesMurs is the only archive devoted solely to street theatre and circus, located in Paris, France. Special thanks to Anne-Laure Mantel, who provided me with the names of these productions.
7. *Le Goliath: Guide Annuaire 2005–2006 des Arts de la Rue et des Arts de la Piste* (Paris: HorsLesMurs, 2005).
8. Very little scholarly work on French street theatre exists, especially in English. That situation is beginning to change with the establishment in February 2005 of "Le Temps des Arts de la Rue," a program initiated by the Federation of Street Arts in collaboration with the Ministry of Culture and Communication and HorsLesMurs, to develop street arts in the areas of production, training, documentation, and scholarship. Two million euros were allocated for 2005 with additional

financing available through 2007 ("Le Temps": 3). For additional information, go to http://www.tempsrue.org.
9. Michel Siminot, "L'Art de la Rue. Scène Urbaine – Scène Commune?" Spec. issue of *Rue de la Folie*, 3:1 (1999): 6. All translations of the texts originally in French are my own.
10. Friches Théâtre Urbain. Box A01059. HorsLesMurs. Paris, France.
11. Royal de Luxe's *Le Géant Tombé du Ciel* and *Le Dernier Voyage* were produced in 1995 in Paris, France.
12. Siminot, 6.
13. Siminot, 6.
14. France 2 television.
15. For further explanation of the republican paradigm, see Kastoryano. Also see: Christophe Bertossi, "Politics and Policies of French Citizenship, Ethnic Minorities, and the European Agenda," in Agata Górny and Paolo Ruspini (eds), *Migration in the New Europe: East-West Revisited* (New York: Palgrave Macmillan, 2004); Fiorella Dell'Olia, *The Europeanization of Citizenship: Between the Ideology of Nationality, Immigration, and European Identity* (Aldershot: Ashgate, 2005); Jane Freedman, *Immigration and Insecurity in France* (Aldershot: Ashgate, 2004); Agata Górny and Paolo Ruspini (eds), *Migration in the New Europe: East-West Revisited* (New York: Palgrave Macmillan 2004). For background information on immigration policies in France, see: Lauren M. McLaren, "Anti-Immigrant Prejudice in Europe: Contact, Threat Perception, and Preferences for the Exclusion of Migrants," *Social Forces*, 81:3 (2003): 909–36; Robert Miles and Dietrich Thränhardt, *Migration and European Integration: The Dynamics of Inclusion and Exclusion* (Madison, NJ: Fairleigh Dickinson University Press, 1995); Demetrios G. Papademetriou and Kimberly A. Hamilton, *Converging Paths to Restriction: French, Italian, and British Responses to Immigration* (Washington, DC: Carnegie Endowment for International Peace, 1996); Thomas F. Pettigrew, "Reactions Toward the New Minorities of Western Europe," *Annual Review of Sociology*, 24 (1998): 77–103; Maxim Silverman, *Deconstructing the Nation: Immigration, Racism, and Citizenship in Modern France* (London: Routledge, 1992); Rita J. Simon and James P. Lynch, "A Comparative Assessment of Public Opinion Towards Immigrants and Immigration Policies," *International Migration Review*, 33:2 (1999): 455–67; Mehmet Ugur, "Freedom ofMovement vs. Exclusion: A Reinterpretation of the 'Insider'-'Outsider' Divide in the European Union," *International Migration Review*, 29:4 (1995): 964–99.
16. Bertossi, 110.
17. Freedman, 9.
18. Freedman, 16.
19. *Paris-Match*, 14 Dec. 1989.
20. Freedman, 127–41.
21. Quoted in Jonathan Marcus, *The National Front and French Politics: The Resistable Rise of Jean-Marie Le Pen* (Basingstoke: Macmillan, 1995), 93.
22. Alec G. Hargreaves, *Immigration, "Race," and Ethnicity in Contemporary France* (London: Routledge, 1995), 1.
23. I am drawing on the distinction between *place* and *space* in Michel de Certeau's, *The Practice of Everyday Life*, trans. Steven Rendall (Berkeley: University of California Press, 1984). In particular, I am referring to Chapter IX, "Spatial Stories," 115–30.
24. Sarah Harper, email to the author, 20 May 2005.

25. Gwénola David, "Public Chéri: Un Autre Rapport?" *Scènes Urbaines*, 1 (May 2002): 26.
26. Pascal Laurent, personal interview, 11 November 2005.
27. Siminot, 6.
28. See Paulo Freire, *Pedagogy of the Oppressed*, trans. Myra Bergman Ramos (New York: Continuum, 1997); and Freire, *Pedagogy of Freedom: Ethics Democracy and Civic Freedom*, trans. Patrick Clarke (Lanham, MD: Rowman & Littlefield, 1998).
29. Christophe Raynaud de Lage, *Intérieur Rue: 10 Ans de Théâtre de Rue (1989-1999)* (Montreuil–Sous-Bois, Fr.: Éditions Théâtrales, 2000), 58.
30. Royal de Luxe. Box A00316. HorsLesMurs. Paris, France.
31. *Royal de Luxe: 1993–2001* (Arles: Actes Sud, 2001) includes many excellent pictures of the production. There are also some pictures of a slightly different production with Léonard at http://www.chez.com/cinetoile/Royal/Ballade.htm.
32. Sylvia Ostrowetsky, "Port-Royal de Luxe," in Sylvia Ostrowetsky (ed.), *Pour une Sociologie de la forme* (Paris: L'Harmattan, 1999), 22.
33. Zilla Eisenstein, *Against Empire: Feminisms, Racism, and the West* (London: Zed, 2004), 53–4.
34. Eisenstein, 54.
35. Eisenstein, 61
36. Dell'Olio, 82.
37. Ostrowetsky, "Port-Royal," 31.
38. Ostrowetsky, "Port-Royal," 144.
39. Eisenstein, 65.
40. Eisenstein, 6.
41. For pictures of *Braakland*, go to Compagnie Dakar's website at: http://www.compagniedakar.nl/.
42. Advertisement flyer from Chalon-dans-la-rue.
43. *Chalon-dans-la-rue. Festival Transnational des artistes de la rue* (Chalond-sur Saône, 2005), 21–4.
44. Jean-Samule Bordreuil, quoted in Philippe Chaudoir, "L'Interpellation Dans les Arts de la Rue," *Les Langages de la Rue* (Paris: L'Harmattan, 1997), 171.

3
The Poetics of Displacement and the Politics of Genocide in Three Plays about Rwanda

Laura Edmondson

During and immediately following the 1994 genocide in Rwanda, in which approximately 800,000 Tutsi and Hutu moderates were slaughtered between April and June,*Time* and *Newsweek* devoted a total of three cover stories to the event.[1] Each cover bore a quotation or headline that strained to convey the magnitude of the horror: "There are no devils left in Hell [...]. They are all in Rwanda," "Hell on Earth: Racing Against Death in Rwanda," and "This is the beginning of the final days. This is the apocalypse."[2] These headlines do not, as one might expect, accompany graphic images of the carnage in Rwanda, despite the availability of such photographs.[3] Instead, each cover documents the flight of approximately two million Rwandans into Zaire and Tanzania in one of the most rapid and largest exoduses in recorded history.[4] The first quotation is superimposed upon a photo of a living refugee woman holding her baby in Tanzania; the second cover displays piles of corpses felled by disease in Goma, Zaire (now DR-Congo), as a small child stands in the forefront. The third cover bears a particularly shocking image of hundreds of trampled refugees at the Zaire/Rwanda border. The accompanying stories vividly conveyed the plight of these millions of refugees through detailed descriptions of the stampedes, cholera, dysentery, and emotional trauma that daily added thousands to the death toll.

The intensity of violence and disease that the refugees experienced was, to use a theoretical catchphrase, unspeakable.[5] Unimaginable. Incomprehensible. But, to return to the actual genocide itself, what of the unspeakable *in* Rwanda? Although the quotations refer to the apocalypse and hell "in Rwanda," the visual images and accompanying articles defer the reader's focus to the transnational humanitarian catastrophe rather than the bloodshed that occurred within its borders. Indeed, the intensity of the refugee stories began to blot out the genocide itself, which received relatively scant coverage in comparison, not only in the US mainstream media but also in France and the United Kingdom.[6] In condemnation of this disproportionate coverage, Alain Destexhe, who was the Secretary-General of Médecins San Frontières/Doctors Without Borders at the time, writes:

> Yesterday the genocide of the Tutsi by the Hutu militia, today the geno-
> cide of the Hutu refugees by the cholera? This comparison, which one
> can see widely used in the press, puts on the same plane things which
> have nothing to do with each other. Through this confusion the original,
> singular and exemplary nature of the genocide is denied and the guilt of
> the perpetrators becomes diluted in the general misery.[7]

Destexhe's conflation of Hutu refugees with perpetrators is extremely troubling, especially in light of evidence that indicates a relatively small percentage of the Hutu population participated in the killings.[8] Still, the dilution
of the genocide to which Destexhe refers produced considerable political
and material consequences. After weeks of steadfast refusals to invest a cent
to stop the killings, the international community was finally galvanized
into action and contributed millions of dollars of aid for the camps, many
of which provided shelter for the local officials and militias who orchestrated the slaughter.[9] This contradictory response raises the question, why
would the fleeing, presumably Hutu body be favored over the murdered,
presumably Tutsi one trapped "at home"?

Media scholars Jo Ellen Fair and Lisa Parks provide a straightforward
answer. In their analysis of the coverage in the United States of the 1994
genocide, they attribute this preoccupation with refugees to a basic journalist
preference for clear-cut victims and perpetrators: "Covering genocide meant
having to understand politics to assess both domestic and international
accountability. By comparison, covering refugees was simple."[10] Since 1994,
explanations of why the genocide occurred have also become fairly "simple." Issues of domestic and international accountability have been boiled
down to a standard narrative that places most of the blame on Belgian colonial policy that not only hardened ethnic divisions but also favored Tutsi
as the "superior" group.[11] A preference for clear-cut categories has also prevailed, as the genocide is usually understood in terms of Tutsi victims versus
Hutu perpetrators. At the time, however, the bloodshed in 1994 was apparently inextricable from the cycles of pogroms and genocides against both
Hutu and Tutsi that had occurred in the Great Lakes region since the late
1950s.[12] In contrast, refugees could be understood as "good people to whom
bad things were happening,"[13] and a media frenzy rapidly ensued around
this seemingly comprehensible crisis.[14]

But to return again to the magazine covers, which simultaneously invoke
and evade the genocide, I am interested in the tension produced by this gesture of deferral. I believe that this tension resonates with deeper issues of
displacement versus confinement, which in turn helps to determine crucial
matters of visibility on the global stage. This tension pervades three British

plays that address the genocide, all of which have premiered in London since 2002: *I Have Before Me a Remarkable Document Written by a Young Lady from Rwanda* by Sonja Linden, *Sanctuary* by Tanika Gupta, and *World Music* by Steve Waters. *A Young Lady from Rwanda* has also become the most produced play about the Rwandan genocide in the United States, with productions in at least eight cities at the time of this writing.[15] These compelling plays use the Rwandan refugee as an entry point into the genocide and thus comment upon the peculiar Western penchant to view a *displaced* Rwandan as an object of fascination, whereas a *placed* Rwandan is something to be ignored.

Anthropologist Liisa Malkki helps to clarify the fascination that the refugee commands. In a landmark study of Burundian Hutu refugees living in Tanzania, she discusses the relative palatability of refugees, noting that the refugee is easily "stripped of the specificity of culture, place, and history" and thus serves as a "tabula rasa" (p. 13) upon which multiple agendas and ideologies can be projected. Images of refugees are generic, universalized, and more easily detached from the political quagmire of the Great Lakes region.[16] The fluidity of this figure helps to explain how the refugee can be pathologized on the one hand as a "threat to 'national security'" (p. 7) but also contained and made "knowable" on the other. Writing just a few weeks after the 1994 genocide began, Malkki predicts that a "curtain of silence" will eventually envelop Rwanda, which will be left invisible and *incomprehensible* by the press (pp. 294–5; her emphasis). In contrast, those Rwandans who make it across the national border "will emerge knowable again, on the other side, to international wire services and international relief organizations and development agencies and scholars as 'African refugees,' objects of a special philanthropic mode of power" (p. 296). In contrast to the incomprehensible masses of mutilated corpses in the churches, schools, latrines, streets, and homesteads of Rwanda, the masses of refugees appeared more familiar, palatable, and malleable.

The three plays discussed in this chapter offer a new spin on these images – the refugee as heroic survivor. In *Staging Place: The Geography of Modern Drama,* Una Chaudhuri coins a phrase that helps to clarify the *dramatic* appeal of the refugee. She identifies "a victimage of location and a heroism of departure" as two main principles that pervade the plot and politics of modern drama or early modernist drama; in other words, place is defined as the primary problem, leading the protagonist "to a recognition of the need for (if not an actual enactment of) the latter."[17] Although she argues that this trope of geopathology unraveled mid-century as "the massive and agonizing dislocations of the modern age [left] the poetics of exile far behind,"[18] I believe that these three plays indicate that the massive and agonizing dislocations are all too easily domesticated and contained into a new kind of poetics – a poetics of displacement – one that pushes the victims "at home" off the radar of international concern. In the process, the genocide threatens to become a mere backdrop for the refugees' pain, which assumes center stage.

To use the term "center" to describe the plight of refugees is, of course, a problematic move. This argument threatens to privilege massacred Rwandans over fleeing Rwandans (who could also be massacred[19]) in a useless hierarchy of suffering. Although refugees command considerable attention on the world stage, they are invariably a target of anxiety and fear as manifested in increasingly restrictive state legislation.[20] As Gabriele Griffin finds in her analysis of Gupta's *Sanctuary*, these plays "problemat[ize] the political debates on asylum seeking the issue of refugee status"[21] and help draw the audience "into that experience of uncertainty, humiliation, invasion of privacy, and process of abjection that the refugee/asylum seeker undergoes."[22] Although Griffin's approach could easily be applied to all three plays, it is not my aim to consider how these plays speak to and/or challenge refugee policies in the United Kingdom. Instead, I focus on how the "eternally remote" African dead haunt and destabilize the "ever-so-photogenic" refugee.[23] Given the prevalence of Rwandan refugee characters in plays about the genocide,[24] it seems imperative to confront and unpack this preference for the displaced, transnational body over the incarcerated "native."[25] My hope is that this awareness will open up new approaches for representing the gut-wrenching destruction of genocide.

The poetics of exile in *A Young Lady from Rwanda*

The figure of exile haunts literary representations of refugees. As Edward Said ruminates in his classic essay, "Reflections on Exile," exiles are "cut off from their roots, their land, their past,"[26] leaving them ripe for theoretical imagining and romanticization. Although this *tabula rasa* image resonates with Malkki's analysis of refugee discourse previously discussed, the figure of the exile is made distinct through "a touch of solitude," whereas the word refugee suggests "large herds of innocent and bewildered people requiring urgent international assistance" (p. 181). Despite Said's insistence that "our time [...] is indeed the age of the refugee, the displaced person, mass immigration" (p. 174), he clearly finds the figure of the solitary exile more compelling, preferring to emphasize the "lyrics of loss" (p. 179) penned by solitary exile poets rather than delve into the sociopolitical complexities of the "uncountable masses for whom UN agencies have been created" (p. 175).[27] Particularly in the framework of conventional drama, which emphasizes the individual protagonist over the collective, the figure of the exile provides an evocative medium for exploring flight from the homeland.

In *I Have Before Me a Remarkable Document by a Young Lady from Rwanda* (2003), Sonja Linden refuses to romanticize the refugee through the poetics of exile, a phrase I use to describe what Caren Kaplan calls the "mystified universalism" of the individualized exile, whose "dislocation is expressed in singular rather than collective terms, as purely psychological or aesthetic situations rather than as a result of historical circumstances."[28] This intriguing play is grounded in playwright Sonja Linden's experiences as the writer in residence at the Medical Foundation for the Care of Victims of Torture, where she worked with Rwandan refugees to use writing in order to "discharge"[29] the trauma of their pasts. Juliette, the "young lady" of the title, is based on Lea Chantal, one of Linden's Rwandan clients at the foundation. Unlike Said, Linden carefully articulates the unique circumstances of political refugees through emphasizing the myriad complexities that make up their transient existence. Ultimately, however, the character's plight as a refugee detracts attention from the genocide – not simply because the play focuses on a lady "from" Rwanda rather than a lady "in" Rwanda, but also because the play domesticates the genocide as an internal trauma to be resolved through writing. As in Said's essay, the exile blots out the refugee.

The plot itself is deceptively simple. After conducting extensive library research, Juliette has produced a thick manuscript in the language of Kinyarwanda that explores the historical conditions of the genocide. With hopes of immediate publication, she brings the manuscript to Simon, a burned-out poet and failed novelist who helps clients with their writing at the Refugee Centre in London. Through a series of visits, Simon guides Juliette through a process of revisions that culminates in a book about her personal experiences instead of the dense socio-political analysis she had originally written. Because no other character appears in the play except Simon and Juliette, the play frequently relies upon monologues to fill in its gaps. The dramaturgy of the play itself reflects Juliette's growing awareness of the value of personal testimony and the concomitant emphasis upon individual subjectivity.

These early scenes closely adhere to the poetics of exile. As Juliette yields to Simon's guidance and begins to write a more personal tale, she increasingly indulges in nostalgia for her homeland. She recalls games she would play with her sister Dominique in the garden (pp. 63–4) and the chameleons that her brother Claude used to frighten Dominique (pp. 71–2). Simon encourages this flow of fond memories with enthusiastic praise. While reading about her family members in one of their meetings, Simon is "totally absorbed" (p. 70). He then commends her work: "These descriptions are wonderful! [...] It's marvellous" (p. 71). These effusive reactions stand in stark contrast to his obvious disinterest in Juliette's original treatise that wrestled with the complex historical factors that contributed to the genocide. In her revised document, the collective is marginalized in favor of the individual subject, producing the nostalgia of the exile but not the horror of the genocide. At this point, the narrative thread threatens to erase the moment of genocide through an emphasis upon tales of the "before" – the idealized, untroubled African life – and the "after" of exile.

Simon deliberately evades the subject of genocide. Juliette opens the play with an off-stage monologue in which she tells of the moment when her father opened the door the morning after President Juvénal Habyarimana had been killed to find their Hutu neighbors armed with machetes. Although this opening scene places the action squarely within the context of the genocide, the next scene shifts the emphasis to Simon and Juliette's relationship in London. During this first meeting, Simon is understandably reluctant to ask questions about her experiences in Rwanda. Instead, Simon focuses upon her trials as a refugee in London: "That must be difficult [...] to be here all alone, new country, new language, new customs, must be hard" (p. 5). In a later scene, Simon's interest in her London experiences reaches an unusual intensity, as he peppers Juliette with a series of questions about the complicated procedures for obtaining travel and food vouchers and phone cards from the Refugee Center. "God!" Simon explodes at the end of her tale of inconvenience and hassle (p. 33) – in contrast to his total lack of response to her earlier revelation that she is a survivor of the genocide (p. 7). Simon's fascination with her life as a refugee in England begins to obscure the cataclysmic event that made her into one.

At the play's climactic moment, however, Simon finally confronts the reality of Juliette's past. A frustrated Simon arrives at Juliette's room, demanding to know why she has repeatedly missed appointments. A despairing Juliette informs him that the Home Office had refused to grant asylum to her younger brother Claude, who had managed to escape to Uganda. In response to Simon's questions about her brother, she bursts forth with a horrific tale of the night when her family was murdered. In a subtle challenge to the modernist narrative of writing out the pain, a specific act of the British government, as opposed to an internal breakdown triggered through the writing of her book, generates her outpouring of emotion. As Simon listens, Juliette tells the story of the butchery of her father and brothers at the hands of their neighbors, followed by the raping and shooting of her mother, sisters, and herself. "We survived too much," she says of herself and her female relatives (p. 90) – an eloquent line that marks the impossibility of assigning hierarchies based on the magnitude of suffering. The sustained deferral set up in the opening monologue finally ends as the actual moment of genocide is inhabited and explored.

Simon's response to Juliette's testimony clarifies the significance of this scene. Instead of reacting with his usual expressions of misplaced outrage, he provides material assistance by buying her an airplane ticket so she can visit her brother in Uganda. This action marks the culmination of a journey in which Simon begins to understand Juliette in all of her complexity instead of simply romanticizing her as a beautiful "African princess" (p. 38) who provides him with the inspiration to start writing again. At the beginning of the play, Simon serves as a synecdoche of the Western world through his stubborn refusal to confront the reality of the genocide. Indeed, in the opening scenes with Juliette, he manages to avoid even using the word genocide, as if to echo a similar refusal on the part of the international community when the bloodshed erupted in 1994. Ultimately, though, he witnesses Juliette's pain and is moved to provide economic assistance.

Subsequent events further complicate the earlier tropes of exile. Juliette's visit with her brother helps to propel her on the road to healing: "Something seemed to have shifted after that [visit]," Simon observes. "Seeing him alive and well" (p. 96). Instead of a nostalgic return to the homeland, the visit initiates the forging of a post-genocide relationship with her brother, whose "terrible scar" (p. 93) from a machete serves as a stark reminder of what they have endured. Once Juliette finishes the book after returning from Uganda, she announces to Simon that she will relinquish writing and will instead study medicine like her father. Although Juliette's recovery is surprisingly sudden, it is grounded in the ties of kinship and therefore counteracts the romanticization of the solitary exilic writer perpetuated in earlier scenes. These latter scenes carve out an evocative space in which the complexities of Juliette's past and present begin to coexist.

The potential of this approach, however, is abruptly curtailed in the final scene when Juliette reads from her finished manuscript at a public reading. In a classic reification of Eurocentric notions of resolving anguish through writing, Juliette tells the audience: "It was very hard to write, very painful, but now I have finished it, I feel clean. I feel clean [...]. This book helped me so much. I can sleep. Only some nights I have bad dreams now. I can eat" (pp. 101–2). This scene not only reduces the genocide to an internal trauma to be purged, but also turns it into creative raw material for the soul-searching exile to domesticate and refine. When she begins to read aloud to the assembled spectators, the tropes of exile are again upheld through the invention of a mythical, idealized past. As Juliette reads from the manuscript in Kinyarwanda, Simon translates for the audience in the final lines of the play: "Once upon a time [...] in the heart of Africa [...] there was a small paradise, a beautiful country of forests and lakes and mountains [...] which we called the land of milk and honey [...] and the country of a thousand hills" (pp. 103–4). Simon colludes with Juliette in perpetuating a sanitized and romanticized notion of Africa – one in which genocide did not exist. Like Juliette herself, Rwanda has been purified and cleansed.

The minds of UK critics were also apparently cleansed. Although reviewers of the 2003 production in London often touched upon the horrors of the genocide, several conflated it with the difficulties of exile. *The Stage*, for example, categorized the play as "a real insight into the lives of exiles and their reasons for leaving and one which makes you wonder why the human race can be so cruel,"[30] and *Time Out* reduced it to "a stirring tale of human bravery in the face of adversity."[31] Even Simon's personal struggles with writer's block were placed on a par with Juliette's trauma: "If Juliette is in a state of stasis, alone in the world after seeing her entire family murdered, then so is Simon," Lyn Gardner writes in *The Guardian* "Together, they become unblocked."[32] The *Guardian* review also contained a particularly vivid example of obliviousness, as it describes Juliette's climactic telling of the slaughter in Rwanda – the one scene that unflinchingly takes on the genocide – as "a blistering good scene where Simon's resentments bubble to the surface." Although it is hardly surprising that British critics privileged Simon's point of view, the ease with which the play is described in the usual vein of culture clash suggests that the beautiful "African princess" was all too easily appropriated via the tropes of exile.[33]

Reviews of the several US productions are less easily summarized. Although some reviewers employ the "culture clash" rhetoric,[34] many are clearly compelled by Juliette's story. The *Los Angeles Times* emphasizes that "the most crucial component of the piece is Juliette,"[35] and *The Miami Herald* recalls that "[t]he picture [Juliette] paints with her words is devastating and unforgettable."[36] Indeed, the *Chicago Daily Herald* suggests that Simon should be removed completely: "[The play] would have been more intriguing as a one-woman performance [...] Having found her voice, Juliette deserves to be heard. Solo."[37] I suspect that this shift in focus has more to do with the timing of the productions rather than their geographical location. Given the startling number of documentaries and feature films that have appeared since the ten-year anniversary of the genocide in 2004, it is quite possible that the reviewers of the 2005 and 2006 productions in the United States were simply more prepared to *hear* Juliette and, like Simon, witness her suffering. Her tale has become more imaginable and comprehensible and therefore "unforgettable."

What the US and UK reviewers share is a collective acceptance of the play's premise: that writing can cure the extremities of crimes against humanity. Indeed, the problematic final scene is hardly mentioned at all. In the Indianapolis production that I attended,[38] Monet Butler, who portrayed Juliette with sensitivity and nuance throughout the performance, stood proudly on a bench in this scene as she read aloud portions of her manuscript. This moment stood in sharp contrast to an earlier scene in which Juliette crouched on the floor as she lit a candle for each of the ten members of her family who were killed. This contrast underscored her journey from devastation to unmitigated triumph. "I feel clean. I feel clean," Butler as Juliette repeated. Her literary endeavor had elevated her to a realm of purity, free of despair and pain.

Said gestures to this sense of purity when he speaks of the moral and creative superiority of the exile. Unsullied by the bane of provincialism and nationalism, the exile can "cross borders, break barriers of thought and experience."[39] Although the complicity of Said's image of exilic purity with the modernist romanticization of estrangement has already been addressed, its intersection with the representation of genocide sharply raises the political stakes. As Patraka asks, "how can we portray genocide, the mass noun, and signify the individual if fragmented subjectivities are subjected to these terrible events?"[40] In *A Young Lady from Rwanda*, the "modernist myths of authorship"[41] distill the mass noun of genocide into the figure of the solitary exile writing out her pain. Despite the specificity with which Linden articulates Juliette's trauma, in the end, these modernist myths cleanse Juliette of one of the most abhorrent events of the twentieth century. Her fragmented subjectivity is made whole.

Purity and diaspora in *Sanctuary*

Sanctuary, like most of Gupta's plays,[42] is a dense and fascinating exploration of diasporic and transnational characters living in the United Kingdom. What makes this particular play unique is that it also delves into the world of political refugees. Griffin, who calls the play a "problematized interrogation of identity, refugee status, survival, and the right to live,"[43] asserts that the play constructs a "diasporic space" in which "the displacements generated by political conflict create new and fragile micro-communities which remain haunted by their diverse pasts" (p. 228). I would go a step further and argue that the play uses the poetics of displacement to construct a diasporic utopia. By poetics of displacement, I refer to postmodern notions of transnational movements and communities that celebrate their ability to challenge the theoretical parochialism of the nation-state.[44] In Gupta's play, the concept of diasporic utopia holds considerable promise as a space of safety, if not empowerment, for these abject figures (p. 224) who have endured and/or witnessed severe and sustained trauma. This sense of safety, though, is sustained at a considerable cost.

Within the confines of a relatively conventional play that adheres to fourth-wall realism and linear action, Gupta has assembled a group of characters who exemplify the postmodern conditions of displacement and transnationalism. Rwandan refugee Michael Ruzindana works at a parish garden-cum-graveyard in London populated with a host of refugees and assorted border-crossing figures, including Sebastian from Trinidad, Kabir from India, and Ayesha, a mixed-race teenager. Their mother hen is Jenny Catchpole, a white bleeding-heart parish priest. The revelation that the church has been sold to a fitness club and the beloved garden will be transformed into a swimming pool sparks a series of events that threaten to overturn the very notion of sanctuary.

At first, the play refuses to romanticize these displaced characters. Kabir, who belonged to the Muslim minority in India, is consumed with guilt over his experience of watching his wife be raped and murdered by soldiers while he hid with their young daughter. As "an evil man saturated in my wife's blood,"[45] he is haunted by the figure of Satan, who watches and waits for him in the garden (p. 51). Meanwhile, Sebastian, a former international journalist, is frequently drunk and shabbily dressed. Michael and Kabir are united in their disdain for Sebastian, mocking his photography projects as "rubbish" (p. 40). These depictions stand in stark contrast to the noble, dignified figure of Juliette in *A Young Lady from Rwanda* and promise to complicate the poetics of exile.

Indeed, as the secrets of Michael's past are revealed, he is transformed into the antithesis of Juliette. In the first act, Michael comes across as a victim of the genocide, a meek and mild pastor who lost his son and, he hints, the rest of his family (p. 47). Michael exemplifies the steadfast refugee bent on carving out a new home in the same way he skillfully carves wooden spoons. But through the investigative efforts of Sebastian in the second act, Michael's disguise is stripped away to reveal his role in trapping thousands of Tutsi in his church to facilitate their slaughter. As if to mirror the UN, which unintentionally sheltered the Hutu extremists in the refugee camps, Jenny has inadvertently taken a ringleader in the 1994 genocide under her wing. Upon the revelation that a mass murderer lives within their midst, the characters' makeshift home in the parish garden is destroyed through internal turmoil in addition to the external force of commercialism.

The Rwandan refugee reveals the limits of their utopia. Among these border-crossing characters, he serves as the polluting element that must be eradicated in order to maintain the diasporic purity that the other characters possess. Upon discovering Michael's role in the genocide, Kabir and Sebastian assume the role of avenging heroes bent on punishing the wayward, evil African, who personifies Satan himself (p. 53). Sebastian, who worked as a journalist in Rwanda following the genocide, notifies the police when he identifies Michael as Charles Bagilishema, a pastor in Kibungo who ordered the massacre of 5000 Tutsi.[46] Once Sebastian has initiated the process, Kabir delivers the decisive blow. When Kabir apprehends Michael as he tries to escape the country, Michael confesses to the crimes in lurid detail: "Hobble the children first. Slash their Achilles tendons so that they could not move far. Then the parents would come running to protect them [. . .] those were my orders. It took four days and nights. When the dogs heard the cries of the people. They too began to howl" (p. 101). He also attempts to buy Kabir's protection with a diamond that he procured in Zaire. Kabir responds by aiming his machete at Michael's head: "I am seeking vengeance for all those people who were killed" (p. 105). As if to mimic the *gènocidaires*, Kabir uses his machete to cut off Michael's arm and drags him to the garden shed where he is burned alive. Michael's utter annihilation indicates the magnitude of his threat to the social order. As in the case of the *Time* and *Newsweek* headlines described earlier, the Rwandan is relegated to hell, to the devils, to apocalypse, consigning him to an otherworldly realm beyond the reach of international intervention.

Through the triumvirate of Michael, Kabir, and Sebastian, the play sets up a hierarchy of diasporic figures that positions Michael at the bottom. His fellow migrants occupy a more ambiguous, exilic space of borders within borders. Kabir signifies a "marriage of two margins"[47] as a minority within a minority – a member of the minority Muslim community in

India who is also part of the disenfranchised black British population. Similarly, Sebastian, who hails from Trinidad, represents the African diaspora and as such invokes the notorious history of slavery. Even Ayesha, the mixed-race teenaged friend of Kabir who darts in and out of the play, aims to become an airline attendant and see the world; as such, she exemplifies the border-crossing, transnational citizen par excellence. In contrast, Michael is more easily categorized as a black refugee from Africa and does not possess the evocative qualities of his comrades, who are far more fleshed out and complex. I suggest that in this particular sanctuary, Michael is not border-crossing "enough" and thus he must be eliminated so that the poetics of displacement and the purity of exile can be re-established.

Certainly, the atrocities that Michael committed play a significant role in his destruction. He is dismembered and burned alive as punishment for orchestrating the murder of 5000 people, not because of his lack of a certain transnational cachet. But the parallel dramatic structure that weaves together the unveiling of Michael's secret and the destruction of the diasporic paradise means that horror and anger at Michael's actions intertwine with a longing for the purity of exile and the nostalgia of home. Indeed, Sebastian's horror at the genocide intermingles with anger at how his cherished image of the African homeland was damaged: "My people – my ancestors – people I loved and respected – capable of such depravity. Children in the classroom, their arithmetic still chalked up on the board, mothers with their babies hacked to pieces in their arms, pregnant women with their wombs ripped out."[48] These moments clarify the ways in which the tropes of exile underpin the poetics of displacement.

Only with Michael's death can the diasporic paradise be reclaimed. When Kabir's murder of Michael is revealed, all of the characters – with the exception of Ayesha, who is kept ignorant of these events – agree to conceal the deed and bury Michael's remains in the parish graveyard. On the one hand, Michael's burial sends a message that the horror of genocide is not confined to Rwanda but also infects British soil. Moreover, it should be noted that Michael's eradication does not prevent the destruction of the parish garden. Once his remains are disposed of, Sebastian simply disappears, and Kabir and Jenny are resettled in a new parish. On the other hand, Michael's burial permits the play to move toward an oddly cheerful ending. In the final scene of the play, Margaret, Jenny's mother, acts on Kabir's behalf and gives Michael's diamond to Ayesha. The horror of genocide, so vividly depicted in Michael's confession, is blithely forgotten as a smiling Ayesha throws the diamond into the air (p. 116). As Margaret points out, Ayesha possesses the resources available to fulfill her dream of traveling the world. Unfettered by national boundaries, Ayesha can move on to realize a new version of diasporic utopia, untroubled by the 5000 victims who perished in Michael's "sanctuary."

An examination of critical response reveals, however, that these nuances and differences among the characters were largely overlooked. In a classic example of the Western penchant for classifying a refugee as a heroic victim, the reviewer of the *Daily Telegraph* carelessly describes Michael as a "survivor of the horrors of his native Rwanda."[49] In a similar vein, Michael Billington of The *Guardian* lumps Kabir, Sebastian, and Michael together, only adding the qualifying statement that Michael "questionably survived the Rwandan tragedy."[50] Gupta's distinctions among Michael, Kabir, and Sebastian are erased as these three are homogenized as heroic refugees. Instead of confronting the complexities of these characters, reviewers preferred to focus upon the symbolism of the garden's destruction even though the revelation of Michael's past is unquestionably the driving force of the play. Through this misplaced emphasis, the critics demonstrated a stubborn persistence to avert their collective gaze from the genocide. In what also could be perceived as a sign of resistance to Gupta's attempt to "rub our noses in global reality,"[51] critics dismissed the complexity of these characters and the intensity of its subject matter as manifestations of an "implausible,"[52] "preposterous,"[53] and "contrived"[54] plot.

And as a result, the genocide disappears from view. To return to Malkki's discussion of refugee discourse, the refugee becomes containable and knowable on the global stage, whereas those who are left behind remain unintelligible. In Gupta's play, Michael is contained and knowable to the point of becoming a cliché, and his victims do not possess the strength to haunt and destabilize the narrative. In a thoughtful critique of diasporic discourse, Aparna Dharwadker links the "current [academic] preoccupation with transnational movements"[55] with the tendency to overlook non-Western, non-Anglophone theatre, which "inhabit[s] a limbo outside the realms of canonicity and interpretability" (p. 94). In the case of Rwanda, I suggest that this limbo of unknowability contains far more than dramatic texts – it also contains the hundreds of thousands of dead Tutsi and Hutu moderates trapped within the nation-state.

Contamination and dissonance

Of the three plays discussed in this chapter, *World Music* contains the most powerful critique of the poetics of exile and displacement. Because of the play's emphasis on the character of Geoff, a white Member of the European Parliament (MEP) who is forced to confront the genocide, most critics interpreted the play as a critique of white liberal politics and post-imperialist guilt. An exploration of the play through the lens of exile and displacement, however, reveals its complex strategies for representing genocide. Through a nonlinear structure that juxtaposes images of "placed" Rwandans at home and displaced Rwandans abroad, the play calls attention to the deadly consequences of Western investment in the tropes of purity, exile, and displacement.

It should be noted that the word "Rwanda" does not appear in the play. Even though the political and historical events in the play are identical to those of Rwanda – indeed, the characters even use Kinyarwanda words – Waters created the fictional country of Irundi, in which the majority Muntu slaughtered the minority Kanga. In justifying his choice, Waters claims that "[i]n attempting to make visible the play of history around particular experience, to forgo hindsight and reductive interpretation, I had to have recourse to invention."[56] On the one hand, this strategy seems designed to deflect attention from Rwanda, allegorizing it as a source of Western misunderstandings instead of grappling with the material circumstances of the actual genocide. On the other hand, the superfluousness of his gesture, given the extreme resemblance between Rwanda and Irundi, seems more indicative of Waters's own anxiety over delving into such sensitive subject matter as a white European. As will be discussed later, this tactic might have made Waters vulnerable to attack from critics loath to confront the complexity of the subject matter itself.

The opening scenes, in which Geoff welcomes the Irundi refugee Jean Kiyabi to Belgium as an official guest of the European Parliament, quickly set forth the trope of heroic departure. Jean is carefully framed as a brave survivor; for example, when Geoff and a fellow MEP, Paulette, greet him at the airport, Geoff comments on the weight he has lost as a result of what is later revealed as typhoid. In a speech to the European Parliament, Jean passionately condemns the international community. "They have died in your fields to grow the tea you drink, the coffee you grind and roast, the beer in your bottles, they have died and you have built cities on their bones and you have forgotten this you have forgotten all of this."[57] Although subsequent scenes reveal that he is condemning the willingness of Europe to side with the Kanga minority in the aftermath of genocide, the immediate context suggests that he is a Kanga himself, excoriating the Parliament's lack of response during the slaughter of his people. In these opening scenes, Jean colludes with Western anxiety and desire to perceive him as the heroic survivor.

Much of this anxiety is embodied in the character of Geoff, who actively bolsters this image. Prior to Jean's arrival, Geoff is determined to challenge Parliament's aim to support the new Kanga government with "shitloads of Eurocash" (p. 20), claiming that the majority Muntu suffers under the new government. In a meeting with Paulette and Alan, a powerful MEP, Geoff clinches the argument with a fax he received from Jean describing the horrific conditions in a refugee camp on the Congo border: "Water is full of Typhus. No firewood as overcrowding has stripped land; food is powder and rationed; today three children died; cannot bury them so wrapped in plastic and taken in truck to be burned [...] We have lost everything" (p. 23). A final plea ends the message: "I pray the world will hear our prayers. We are like ghosts, and the world pays our tormentors" (p. 11). This passionate and strategic play upon European sympathies for the plight of the displaced has its desired effect, as Alan agrees to revisit plans to support the new Kanga government. Once Jean arrives in Belgium, he uses a similar strategy to deflect Geoff's questions about what happened in Irundi:

> *Geoff*: I saw footage, of the village.
> *Jean*: Please.
> *Geoff*: The church. The earth red beneath the church.
> *Pause.*
> It was all muddled up, what was said, the way they described it. But –
> heaps of bodies. I mean it was – what happened – can you – I mean
> where were you when –
> *Jean*: What do you think I was doing my friend?
> *Pause.*
> In one night the whole village was gone. In one night we fled. Like so
> many birds. Across the marshes, with our things in heavy bundles on
> our backs, our whole lives, all we could take, across the river –
>
> Waters, *World Music*: 30

Aside from Geoff's feeble attempt to glean more details of the genocide, he generally accepts Jean's performance as "the noble refugee" fleeing from persecution.

Geoff's determination to cast Irundi refugees in this light is clarified in alternating scenes in a bar with his son, Tim, throughout the first act. When he discovers that their waitress, Florence, is a refugee from Irundi, he immediately strikes up a flirtatious relationship with her and even invites her to stay at his flat. The arbitrariness of Geoff's invitation suggests the extent to which he has fallen under the spell of exile. "You don't even know who she is," Tim says incredulously (p. 28). Later in the play, when Florence herself expresses the same sentiment, Geoff responds that "I've got a pretty good notion" (p. 38), even though he still knows little about her beyond her refugee status. In his enchantment, he even experiences the nostalgia of exile himself, as demonstrated in two brief scenes in the first act in which Geoff lectures schoolchildren about his experiences in Irundi as a teacher. His presentations depict a peaceful community free of ethnic tension where the villagers danced together and swam in the lake: "The whole village swam in this lake," he tells the children, accompanied by the "faint sound of laughter, splashing" (p. 11). Like Simon in *A Young Lady from Rwanda*, Geoff deflects the reality of genocide by emphasizing romanticized notions of African communal society – the "before" – and the heroic state of exile – the "after."

But unlike Simon, Geoff has direct experience with the country that he insists upon romanticizing. In the second act, a clumsy seduction scene with Florence at his Brussels apartment alternates with flashback scenes of Geoff as a naive schoolboy spending his gap year teaching school in Jean's village in Irundi. Through the portrayal of a village rife with tension between the Muntu and Kanga, these scenes powerfully contradict the sanitized memories that he shared with the schoolchildren in the previous act. Jean, the village *bourgmestre* or head, tells Geoff misogynist and racist stories of Kanga women and vividly recalls years of mistreatment at the hands of Kanga. At school, one of Geoff's Muntu students calls the Kanga *inyenzi*, or cockroach, a notorious phrase used in the Rwandan genocide to dehumanize the Tutsi. These vivid illustrations of the complexity of Irundi society challenge the supposed unknowability of the African interior – a colonialist phrase I deliberately invoke here because of its connotations of mystery and "darkness."

The glaring contradictions between Irundi realities and Geoff's memory as a middle-aged MEP are further emphasized in alternating scenes with Florence in his Brussels apartment. In a scene strikingly similar to the one in *A Young Lady from Rwanda* in which Simon encourages Juliet to immerse herself in nostalgic memories of pre-genocide Rwanda, Geoff persuades Florence to talk about her memories of "[e]veryday things" such as her family (p. 52): "Not the violence," he says, "That we know about" (p. 51). Florence obliges through fond recollections of her mother, who taught traditional Irundi music and the "old ways." "Every word you say restores me," Geoff tells her blissfully (p. 51). The juxtaposition of these scenes with glimpses of pre-genocidal hatred powerfully conveys the extent of Geoff's carefully-constructed obliviousness.

Toward the end of the second act, it becomes evident that Florence confounds the image of the pure heroic exile: "What I have done I have done," a weeping Florence says to an increasingly uncomfortable Geoff (p. 54). This crack in Geoff's armor culminates in a series of revelations in the third act. Most of this act takes place in Alan's office in the European Parliament building, where Alan and Paulette confront Geoff with a report that tells in graphic detail how Jean ordered his fellow Muntu villagers to participate in the genocide. As an African-Caribbean MEP, Paulette can be likened to Sebastian and Kabir in *Sanctuary* as a kind of transnational hero, because she forces Geoff to confront the magnitude of Jean's participation in the genocide. In the end, though, she is excluded from this heroic role, since Jean slips away and migrates to Canada. In this play, the shadowy world of border-crossing provides a haven for criminals instead of a radical, transformative site.

The beautiful Florence also proves to be contaminated by genocide. The confrontation in Alan's office alternates with scenes at Geoff's flat the morning after his tryst with Florence. Despite Florence's powerful hints about her Muntu identity the night before, Geoff insists upon asking about her experiences during the genocide. In the course of questioning, he betrays his investment in seeing Florence as a victim when he mistakenly calls her

Odette, the name of a Kanga woman he knew in Irundi (p. 65). In a haunting echo of Juliette's attempts to purge her pain through writing in *A Young Lady from Rwanda*, Geoff abruptly urges Florence to write about her experiences: "I think it might be hugely – cathartic and – I'll write it down – have you written it down" (p. 72). Geoff's feeble attempt to hold on to the poetics of exile finally triggers Florence's outburst:

> *Florence*: It is written inside of me. No one must know.
> *Geoff*: There's no shame in this – hey – your reserve is admirable but it plays into the hands of the simplifiers –
> *Florence*: No one must ever know.
> *Geoff*: Your silence protects the real killers.
> *Florence*: Yes yes it must protect the killers yes –
> *Geoff*: What?
> *Florence*: I have killed and killed and watched killings and assisted killings and killers – I have – snatched babies from mothers and thrown them – I have drawn up lists of names of those to be exterminated – have ticked off lists of the dead, checked off against heaps of dead, tick, her, tick –
> Waters, *World Music*: 72

This speech introduces a torrent of confession designed to obliterate any final shreds of Geoff's romanticization: "once you begin, once you start, you must never ever stop, you must never cease to kill, daily, hourly" (p. 73).

Geoff's resistance finally disintegrates through the combination of Alan's report and Florence's confession. In a stark revision of his nostalgic memory of the lake filled with Muntu and Kanga villagers laughing and swimming together, he admits, "The lake where we swam [...] Filled with the dead" (p. 78). In his struggle to come to terms with Florence's revelation, he also clears a space for the complexity of Muntu/Hutu perpetrators that exceeds the stereotype of the African devil. He urges her to "tell the truth of it": "Some were led, some led, some were willing, some not so, some merely looked on, some waved it on from afar, some knew, some knew and chose to – or knew and couldn't – feel" (p. 78). He charges her with a crucial mission to challenge the tired dichotomy of evil perpetrator versus pure survivor – a dichotomy that cannot be sustained in the quagmire of Rwanda's and Burundi's history. In the final moments of the play, however, Geoff's promising, even brilliant, insight is abruptly cut short. He suddenly changes his mind and promises to get Florence to Canada: "maybe thereafter you are Kanga, yes, Kanga?" (p. 78). In a direct reversal of his earlier advice, he encourages her to follow Jean's lead and inhabit the role of the heroic survivor.

An understanding of the gendered codes of exile helps to clarify the significance of this ending. Throughout the play, a careful distinction is drawn between Jean and Florence. Jean comes to Belgium as an official guest of the European Parliament; even after his role in the genocide is revealed, the EU pays for his flight to Canada. Although Florence also flees to Belgium, she takes unofficial routes and lives in fear of the immigration authorities because of her lack of proper papers. Her clandestine status versus Jean's EU-sanctioned presence resonates with entrenched gender codes that consign the female body to a confined interior space, whereas the male is equated with public, exterior space and mobility. By transgressing these codes, the female transnational/exile/refugee is configured as unnatural, a woman without a home.[58] As the female migrant, Florence is carefully excluded from the privilege that her male counterpart enjoys.

The play's exploration of gender politics expands through a third Irundi character, a Kanga woman named Odette, who appears in the Irundi scenes in Act 2 as Jean's servant. As the "natural" African female who stays within national boundaries, Odette is firmly bound to Irundi. The play proceeds to critique these codes through the disturbing revelation of Odette's fate. In a brief scene toward the end of the play, a younger Geoff displays a photograph of Odette cooking a meal with the schoolchildren. He explains to the children: "It's hard for us perhaps not to see what we might describe as subjection as what it is, something quite natural" (p. 77). This "natural" status is equated with death, as this scene is interwoven with the older Geoff's discovery of Odette's name in a list of fatalities in Alan's report. Odette's confinement as the native African woman leads to her destruction. To be placed, the play suggests, is to be killed.

In contrast to the finality of Odette's fate, Florence's ambiguity offers new ways of thinking through the quagmire of genocide. At first, Florence is romanticized as the exotic (and thus erotic) foreign female. She is consistently associated with music throughout the play; for example, Geoff's interest in Florence is piqued in the opening scenes when she begins to play Irundi music in the bar (p. 9), and he becomes aroused when Florence, caught up in memories of her mother, begins to dance in his apartment (p. 53). This association of Florence with "world music," a phrase that itself invokes a homogenized concept of non-Western musical traditions, helps to reinforce Geoff's assimilation of Florence into his romanticized vision of exile. As the play continues, however, these musical connotations become increasingly complex. As Florence herself points out, Irundi music is "out of tune" to Western ears, as a "different music [in] a different scale" (p. 53). The final moments of the play emphasizes this quality of dissonance. The younger Geoff, played by his son Tim, plays a selection of Irundi music for schoolchildren: "I want to play you some music now, a tune that some of you will probably think is out of tune" (p. 78). As he plays the music, Florence, who has just delivered her horrific speech about her role in the genocide, begins to dance the same steps that she danced earlier in the play. In this context, the jarring tones of the music resonate with the disturbing quality of Florence herself, who has revealed herself to be utterly "contaminated." Even as Geoff urges her to play the role of a heroic survivor, Florence performs her discordant dance in a visual reminder of the excesses of truth.

"There are obviously no heroes here," one critic remarked in his review of the play.[59] Through a sustained critique of the purity of exile and the heroism of departure, the play challenges the Western tendency to divert its collective attention from the Rwandan genocide. But perhaps inevitably, critics were determined to be diverted. Some became fixated on Geoff and how "uninteresting"[60] and "noisily dull"[61] his character is – even though this characterization itself is a possible strategy to call attention to the more interesting Irundi characters who surround him. Reviewers also expressed considerable annoyance over Waters's choice to set the play in the fictional world of Irundi. They complained about "didactic, expository passages that might impact more if we could trust in their documentary truth"[62] or criticized it for creating a "woozy no-man's land between hard fact, allegory and creative fancy."[63] They were quick to dismiss the subject matter because of this so-called "fictionalizing,"[64] even though this approach was only a thin veneer for the realities of Rwanda and the international response to the genocide.

It is noteworthy that even in the midst of these various deflections, Florence struck a chord. Although she is occasionally ignored or dismissed, as in the example of one review that summarized her as a "pretty Muntu expatriate who herself turns out to be complicit in the Urundi [sic] atrocities,"[65] most reviews devote considerably more attention to her than to Jean. Her character inspires insightful comments about how she "hints at the moral mayhem of Rwanda at war; at how Africa makes Europe feel necessary, and how it provides us with vicarious drama."[66] Neither is Jean himself easily dismissed because of his charisma and passion, which was, as several critics noted, due partly to actor Ray Fearon's performance.[67] Still, Florence's combination of purity and contamination manages to penetrate Western obliviousness and position the play as "one that valuably dares us to look beyond the cosy confines of fortress Europe."[68]

This assertion stems from the considerable attention that Florence's testimony about the genocide commanded. In a feature article about *World Music* in *Time Out*, Jane Edwardes mentions that the play "does not, with good reason, attempt to put genocide onstage." Edwardes clearly takes it for granted that the reader will agree with Waters's sensible decision to avoid delving into the quagmires of representing African genocide. Her offhanded statement does, however, invoke Patraka's question, "Exactly what are the 'dramatic' or 'stageable' actions of genocide?"[69] None of the three plays discussed in this chapter attempts to depict the genocide itself; instead, each contains a monologue in which a Rwandan/Irundi refugee character bursts forth with a tale of her or his experience with genocide. Personal testimony has long played a key role in plays about violence since it can help evade what Patraka calls the "dangers of the visual," which include the possibility that "the anguish of suffering bodies will be conveyed on stage as 'real' and somehow comprehensible, manageable, able to convey what is actually an immeasurable absence" (p. 102).[70] But the choice to use African refugee characters as the medium for these narratives raises additional questions about the staging of violence. If the refugee is more "knowable" in the Western imagination than the incarcerated native and thus more easily appropriated, are these narratives more easily comprehensible, manageable, and subsequently ignored?

It is interesting, for example, that UK reviewers never mentioned Juliette's monologue about the genocide in *A Young Lady from Rwanda*, which is clearly the climactic moment of the play. As previously noted, one critic simply described it as a scene about Simon's resentment "bubbling to the surface,"[71] as opposed to a revelation of the horrors of genocide. Given the ease with which Juliette's character is assimilated and comprehended as the beautiful and heroic survivor, the subject matter of genocide was perhaps too easily forgotten. In a similar vein, the evil Michael was also easily comprehended and dismissed. Although two of the three reviews briefly mention Michael's "lurid"[72] and "graphic"[73] account of genocide, these references are scant in comparison to the attention devoted to the symbolic use of a garden as a sanctuary for the characters. In the critical reception of both plays, the genocide was perceived more as a backdrop than as a critical part of the dramatic action.

But to return to Florence, her combination of purity and contamination managed to penetrate these walls of disinterest. Writing for *The Financial Times*, Alastair Macaulay writes that Florence's testimony, "delivered in a near-hysterical fervour of guilt and horror and perplexity [...] must pierce any listener: an extraordinary first-person account of genocide from within."[74] The phrase "from within" touches upon the potential of this monologue to invoke the realities of genocide, albeit through a distanced, exilic figure. Florence's "extraordinary aria of self-immolation"[75] inspired Charles Spencer of *The Daily Telegraph* to find literary allusions, noting that "the climactic accounts of atrocities in Irundi bring home the resonant words in Conrad's *Heart of Darkness* – 'The horror! The horror!'"[76] Of all three plays discussed, Florence's speech clearly provoked the strongest reaction.

But, to return to the Conrad reference, what *is* the horror? Although the obvious response would be the horror of genocide, the specific context of Florence's monologue invites the possibility of a more complex understanding of this *particular* genocide, one that epitomizes Michael Taussig's notion of epistemic murk.[77] As the contaminated figure of Florence attests, the Rwandan genocide is a quagmire of horror in which evil perpetrators, innocent victims, and heroic survivors cannot be disentangled. While this is not the place to dissect the ambiguity of Conrad's novel, I do want to call attention to Said's analysis, which considers protagonist Marlow's attempts to "restor[e] Africa to European hegemony by historicizing and narrating its strangeness."[78] Regarding the death of colonialist adventurer Kurtz, Said explains, "Africa recedes in integral meaning [...] it had once again become the blankness his imperial will had sought to overcome"(p. 165). In the supposedly post-imperialist, postcolonial age, Africa has remained blank, unintelligible, and unknowable. Through their various attempts to grapple with the murk and horror of the Rwandan genocide, these three plays contain the promise of illumination. They help to defeat the blankness, invisibility, and incomprehensibility of the hundreds of thousands of victims, those who were unable to follow the example of Juliette, Michael, Jean, and Florence and escape to the knowable, intelligible world of Europe.

Notes

1. Although 800,000 is the most accepted statistic "for reasons which remain obscure" L. R. Melvern (in *A People Betrayed: The Role of the West in Rwanda's Genocide* [London: Zed, 2000: 223]), estimates range from 500,000 to 1,000,000.
2. Quoted in Nancy Gibbs, "Why? The Killing Fields of Rwanda," *Time*, 16 May 1994, 10 June 2007 <http://www.time.com/time/magazine/article/0,9171, 980750,00.html>; cover of *Newsweek*,1 Aug. 1994; cover of *Time*, 10 June 2007.
3. Graphic images of the slaughter in Rwanda were certainly available a few weeks into the genocide as demonstrated in two pages of photographs in a May issue of *Newsweek* (9 May 1994: 40–1). In contrast to the full-color photographs of the refugees, these particular photos were printed in black and white as if to make them more palatable to the reader. Other photos were relatively subdued, as in the example of a *Time* photo of a murdered mother and child. The lack of visible injury refutes the tale of bloodshed described in the accompanying text (16 May 1994: 59). See Edgar Roskis, "A Genocide Without Images: White Film Noirs," in Allan Thompson (ed.), *The Media and the Rwanda Genocide* (London: Pluto; Kampala: Fountain; Ottawa: IDRC, 2007), for a discussion of a similar pattern that occurred in the French press, which tended to ignore the genocide up until the "famous exodus" in July. All citations of the Thompson book refer to the on-line version at http://www.idrc.ca/rwandagenocide/#beginning and do not include page numbers.
4. Melvern, 218, and Alex de Waal, *Famine Crimes: Politics and the Disaster Relief Industry in Africa* (London: African Rights, 1997), 195.
5. See Marie Béatrice Umutesi's extraordinary autobiography, *Surviving the Slaughter: The Ordeal of a Rwandan Refugee in Zaire*, trans. Julie Emerson (Madison: University of Wisconsin Press, 2004), for a detailed (if controversial) description of the anguish, violence, and extreme deprivation that the refugees endured. It should be clarified that although these refugees were in the international spotlight in the period immediately following the genocide, those refugees who fled further into Zaire were resolutely ignored by the international community and humanitarian organizations, even though they were relentlessly attacked by various rebel forces in Zaire and the Rwanda Patriotic Front (RPF).
6. See Gérard Prunier, *The Rwandan Crisis, 1959–1994: History of a Genocide* (Kampala: Fountain, 1995), 303 n. 48, for specific citations of British and French media reactions that mirrored those of the United States. Several scholars have

called attention to the international media's preoccupation with the refugee crisis; see, for example, Melvern, 218–19, Larry Minear, Colin Scott and Thomas G. Weiss, *The News Media, Civil War, and Humanitarian Action* (Boulder, CO: Lynne Rienner, 1996), 62–7; and de Waal, 195–202. Johan Pottier, *Re-Imagining Rwanda: Conflict, Survival and Disinformation in the Late Twentieth Century* (Cambridge: Cambridge University Press, 2002), 59–67, provides a particularly thoughtful analysis of media coverage of the genocide, making a distinction between the French and anglophone (US and UK) press on the one hand and the rest of continental Europe on the other. Several essays in the anthology *The Media and the Rwandan Genocide* touch upon the intensity of the media response toward the exodus, see, for example, Roskis and Hilsum ("Reporting Rwanda: The Media and the Aid Agencies"). In that collection, Emmanuel C. Alozie ("What Did They Say? African Media Coverage of the First 100 Days of the Rwandan Crisis") provides a unique perspective through his analysis of the coverage of the event in the African press.

7. *Libération*, 27 July 1994, quoted in Prunier, 303. Lindsey Hilsum, who interviewed Anne-Marie Huby, the executive director of MSF-UK in the mid-1990s, provides another evocative quote: "In the general public's memory, the Rwanda crisis was people who die of cholera. I think people forgot the long-lens coverage of genocide. [In Goma,] I remember CNN saying 'This is genocide again.' We told the reporter that dying of diseases is not genocide."

8. René Lemarchand in, "U.S. Policy in the Great Lakes: A Critical Perspective," *Issue: A Journal of Opinion*, 26:1 (1998): 42, stresses that the *génocidaires* were a "tiny minority" of the population, and insists that the notion of "collective guilt must be scotched once and for all." See also John Mueller, "The Banality of 'Ethnic War,'" *International Security*, 25:1 (2000): 58–62.

9. See Samantha Powers, *"A Problem from Hell": America and the Age of Genocide* (New York: HarperCollins-Perennial, 2002), 328–89, and Melvern, 186–209 for an account of the US government's actions during and immediately after the genocide.

10. Jo Ellen Fair and Lisa Parks, "Africa on Camera: Television News Coverage and Aerial Imaging of Rwandan Refugees," *Africa Today*, 48:2 (2001): 47.

11. This narrative is pervasive in representations of the genocide. It can be discerned in feature films (*Sometimes in April, Hotel Rwanda*), documentaries (the 2004 PBS film *Ghosts of Rwanda*), and trade books (Philip Gourevitch's *We Wish to Inform You that Tomorrow We Will be Killed with Our Families: Stories from Rwanda* [New York: Farrar, 1998]). It is also easily discerned in stage productions, perhaps most emphatically in Groupov's *Rwanda 94*. (For a discussion of Groupov's production, see Josette Féral's "Every Transaction Conjures a New Boundary," in Janelle Reinelt and Joseph Roach (eds), *Critical Theory and Performance*, 2nd edn [Ann Arbor: University of Michigan Press, 2007]). For a thought-provoking and controversial counter-narrative, see Pottier.

12. I cannot emphasize enough that attempting to understand why the 1994 genocide occurred is a daunting task, given the extraordinarily complex history of the Great Lakes region (Rwanda, Burundi, Zaire/DR-Congo, Uganda, and part of Tanzania). This complex history is captured by the title of René Lemarchand's article, "Genocide in the Great Lakes: Which Genocide? Whose Genocide?", *African Studies Review*, 41:1 (1998): 3–16. Lemarchand identifies three separate genocides: of Hutu by Tutsi in Burundi (1972), Tutsi and Hutu by Hutu in Rwanda (1994), and of Hutu by Tutsi in what is now called the Democratic Republic

of Congo (1996–97). In particular, he calls attention to the 1972 genocide in Burundi, which he calls the "historical thread that enables us to make sense of subsequent developments" (p. 6).

13. Minear, Scott and Weiss, 64; qtd. in Fair and Parks, 48.

14. Hilsum provides a thorough and disturbing account of the various shenanigans pulled by journalists, aid agencies, and governments in the refugee camps.

15. *Sanctuary* opened in the Loft at the National Theatre in London on 29 July 2002 as part of the Transformation season, which was aimed at "younger than normal" audiences; the largest section of the audiences during the season consisted of 25 to 34-year olds (see Jeremy Austin, "No Transformation in Hytner's Tenure," *Stage*, 30 Jan. 2003: 2). *I Have Before Me* had its London premiere at the Finborough Theatre on 17 June 2003, which has a reputation for producing edgy, violent material. As artistic director Neil McPherson commented in 2004, "Somebody once claimed that the Finborough only does plays about genocide, war and disease – which I take as a compliment" (quoted in Aleks Sierz, "Theatre: Back in the Front Line,"*Independent*, 8 July 2004, 10 June 2007: http://www.artsindependent.co.uk/theatre/features/article46301.ece). After the US premiere at Union Station's City Stage in Kansas City in April 2005, Linden's play has been produced in Washington, DC, Milwaukee, Chicago, Los Angeles, Indianapolis, Atlanta, and Coral Gables. The London premiere of *World Music* took place in the Donmar Warehouse from 16 February 2004 to 13 March 2004. The play was the first new play to be produced since Michael Grandage took the helm as artistic director in 2002 (Jane Edwardes, "Out of Africa," *Time Out*, 11 Feb. 2004: 144).

16. Liisa H. Malkki, *Purity and Exile: Violence, Memory, and National Cosmology Among Hutu Refugees in Tanzania* (Chicago: University of Chicago Press, 1995), 12.

17. Una Chaudhuri, *Staging Place: The Geography of Modern Drama* (Ann Arbor: University of Michigan Press, 1995), xii.

18. Chaudhuri, 20.

19. See Pottier, 148–50 for a discussion of the massacres of Hutu refugees in eastern Zaire.

20. Although the numbers of world refugees have fallen to just over nine million, the lowest statistic in the past 25 years, a recent UNHCR report calls attention to an increase in xenophobia in the post-9/11 era as states pass legislation based on "the notion that asylum seekers are the agents of insecurity rather than its victims" (*The State of the World's Refugees: Human Displacement in the New Millennium* [Oxford: Oxford University Press, 2006], x. Available on-line at: http://www.unhcr.org.)

21. Gabriele Griffin, *Contemporary Black and Asian Women Playwrights in Britain* (Cambridge: Cambridge University Press, 2003), 225. I thank Mary Karen Dahl for calling Griffin's work to my attention.

22. Griffin, 224.

23. Roskis.

24. Film-makers seem less hesitant to set the action in Rwanda itself, as the films *Hotel Rwanda* (2004) and *Sometimes in April* (2005) indicate. The recent premiere of J. T. Rogers's *The Overwhelming* in May 2006 at the National Theatre should also be noted, as the entire action of the play is set in Rwanda shortly before the genocide occurs. In this instance a displaced American family in Rwanda provides the entry point for a Western audience; still, the success of this production raises the possibility that audiences and critics are more prepared to confront the 1994

crisis on stage as well as on screen. I return to this possibility in my discussion of the US reception of *A Young Lady from Rwanda* in the next section.

25. Here I am citing Arjun Appadurai's "Putting Hierarchy in Its Place," *Cultural Anthropology*, 3:1 (1988): 37, in which he called attention to the anthropological tendency to "incarcerate" the native, conceiving them "as confined to, and by, their places."

26. Edward Said, *Reflections on Exile and Other Essays* (Cambridge, MA: Harvard University Press, 2002), 177.

27. See also Caren Kaplan, *Questions of Travel: Postmodern Discourses of Displacement* (Durham, NC: Duke University Press, 1996), 117–20, for a discussion of Said's conflation of exile and refugee in this essay. Kaplan's book, which is devoted to a deconstruction of the poetics of exile, was instrumental in shaping my critique.

28. Kaplan, 4. As a special issue of *Modern Drama* on "Theatre and Exile" demonstrates, an exploration of exilic discourse in theatre opens up a variety of approaches that greatly exceed a simplistic tendency toward romanticization. Several of the articles contain thoughtful explorations of how exile is constructed and performed within particular historical and performative moments rather than imposing universalized theoretical frameworks (see, for example, Mary Trotter's "Re-Imagining the Emigrant/Exile in Contemporary Irish Drama," *Modern Drama*, 46:1 [2003]: 35–54). Nevertheless, the marginalized status of refugees in exilic discourse is largely unexamined.

29. Sonja Linden, *I Have Before Me a Remarkable Document Written by a Young Lady from Rwanda* (London: Iceandfire, 2003), viii.

30. Rob Speight, "I Have Before Me a Remarkable Document Given to Me by a Young Lady From Rwanda," *Stage*, 26 June 2003: 10.

31. Franzeiska Thomas, "I Have Before me a Remarkable Document Given to me by a Young Lady from Rwanda," *Time Out*, 25 June 2003: 154.

32. Lyn Gardner, "Review: Theatre: I Have Before Me: Finborough, London 2/5," *Guardian*, 23 June 2003: 20.

33. When the play premiered in Ireland at the 2005 Fringe Festival, Calypso Theatre company made an unusual choice to open the production with a short film of the carnage in Rwanda. As a result, it seems that neither Simon nor Juliette made much of an impact. Dismissed as "heavy-handed" by the *Independent* (Emer O'Kelley, "Devastating Truth Out of Africa is Tarnished," 25 Sept. 2005, 21 June 2007: http://www.arts.independent.co.uk.search; see also Susan Conley, rev. of "I Have Before me a Remarkable Document Given to me by a Young Lady from Rwanda," 20 Sept. 2005, 21 June 2007: http://irishtheatremagazine.ie/home/fringeArchive05.htm). In this particular instance, the graphic images of *placed* Rwandans diminished Juliette's exilic appeal. I thank Brian Singleton for bringing this production to my attention.

34. Damien Jaques, "Young Lady Finds Healing in Tragedy," *Milwaukee Journal Sentinel*, 20 Sept. 2005: B6.

35. F. Katleen Foley, "Touched by a 'Young Lady,'" *Los Angeles Times*, 24 Aug. 2006: E8.

36. Christine Dolen, "You'll Feel the Horror of Genocide," *Miami Herald*, 28 Feb. 2007: E3.

37. Barbara Vitello, "One Would Have Done Tale of Rwanda," *Chicago Daily Herald*, 3 Feb. 2006: 32.

38. I saw the production at the Indianapolis Repertory Theatre on 23 December 2006. My thanks to Barbara Edmondson for making it possible for me to see the production.

39. Said, 185. In a passage that bears a striking similarity to these discursive maneuvers, Malkki notes that the Burundian Hutu refugees in Tanzania had transformed the usually polluting image of the refugee "in a curious social alchemy into a state of purity" (p. 231).
40. Vivian M. Patraka, *Spectacular Suffering: Theatre, Fascism, and the Holocaust* (Bloomington: Indiana University Press, 1999), 99.
41. Kaplan, 120.
42. Some of Gupta's more recent plays include *Banglatown Banquet* (2006), *Gladiator Games* (2005), and *Fragile Land* (2004).
43. Griffin, 228.
44. Several critics have called attention to the "postmodern cult of the migrant," as Terry Eagleton, *After Theory* (New York: Basic, 2003), 21, categorizes the academic tendency to homogenize all border-crossing figures as transgressive bodies. Although the articulation of diaspora and transnationalisms has become more nuanced since the earlier celebratory studies of Homi Bhabha, James Clifford, and Ulf Hannerz (for an example, see Keila Diehl, *Echoes from Dharamsala: Music in the Life of a Tibetan Refugee Community* [Berkeley: University of California Press, 2002]), it is still common to find reflexive nods of approval to border-crossing figures in contemporary scholarship, indicating the compelling hold that they maintain on postmodern imaginations.
45. Tanika Gupta, *Sanctuary* (London: Oberon, 2002), 73.
46. Charles Bagilishema is based on the actual figure of Elizaphan Ntakirutimana, a Seventh Day Adventist pastor who orchestrated the killing of 5000 Tutsis. The connection is revealed when Sebastian finds a letter that Charles's victims sent to him that states, "We wish to inform you that we have heard that tomorrow we will be killed with our families." The playwright is quoting an actual letter sent to Ntakirutimana, which Gourevitch subsequently used as the title of his well-known book on the genocide.
47. Sara Suleri, "Woman Skin Deep: Feminism and the Postcolonial Condition," in Patrick Williams and Laura Chrisman (eds) *Colonial Discourse and Post-Colonial Theory: A Reader* (New York: Columbia University Press, 1994), 246.
48. Gupta, 93.
49. Charles Spencer, "Genocide and Jokes Make an Uneasy Mix," *Daily Telegraph*, 31 July 2002: 17.
50. Michael Billington, "Trouble Amongst the Herbaceous Borders," *Guardian*, 31 July 2002: 19.
51. Billington, 19.
52. Spencer, "Genocide and Jokes," 17.
53. Benedict Nightingale, "Sanctuary," *Times*, 31 July 2002: 17.
54. Billington, 19.
55. Aparna Dharwadker, "Diaspora, Nation, and the Failure of Home: Two Contemporary Indian Plays," *Theatre Journal*, 50 (1998): 74.
56. Steve Waters, "The Truth Behind the Facts: The Resurgence of Documentary-Style Theatre Underlines the Importance of Dramatic Imagination," *Guardian*, 11 Feb. 2004: 26.
57. Steve Waters, *World Music* (Sheffield: Sheffield Theatres, 2003).
58. Several feminist and/or postcolonial theorists have commented upon and deconstructed this trope. See, for example, Partha Chatterjee's *The Nation and Its Fragments: Colonial and Postcolonial Histories* (Princeton: Princeton University Press, 1993), and Anne McClintock's *Imperial Leather: Race, Gender and Sexuality in the Colonial Contest* (New York: Routledge, 1995).

78   *Violence Performed*

59. Sam Marlowe, "World Music," *Times*, 3 June 2003: 15.
60. Benedict Nightingale, "World Music," *Times*, 18 Feb. 2004: 15.
61. Rhoda Koenig, "World Music Donmar Warehouse London," *Independent*, 25 Feb. 2004.
62. Brian Logan, "'World Music' Donmar Warehouse," *Time Out*, 25 Feb. 2004: 141.
63. Nicolas De Jongh, "Lost in a Woozy World of Fiction," *Evening Standard*, 17 Feb. 2004: 43.
64. Michael Billington, "World Music: Donmar Warehouse, London 3/5," *Guardian*, 18 Feb. 2004: 26.
65. Nightingale, "World Music," 15.
66. Logan, 141.
67. De Jongh, 43; Nightingale, "World Music," 15.
68. Dominic Cavendish, "Flawed but Fresh," *Daily Telegraph*, 18 Feb. 2004: 18.
69. Patraka, 99
70. Although Patraka is specifically referring to the politics of staging the Holocaust, this assertion can also be applied more generally to the dramatic representation of state terror.
71. Gardner, 20.
72. Charles Spencer, "Confronting the Horror," *Daily Telegraph*, 4 June 2003: 20.
73. Billington, "Trouble," 19.
74. Alastair Macaulay, "World Music Donmar Warehouse, London," *Financial Times*, 18 Feb. 2004: 15.
75. Koenig.
76. Spencer, "Confronting the Horror," 20. A vivid exception to the critical praise of this monologue is found in Matt Wolf's review ("World Music," *Variety*, 8 Mar. 2004: 47). In addition to condemning the entire play as "emotional pornography," he singles out Florence's speech as "phonily written" and "brazenly manipulative." The intensity of his attack, however, is itself suggestive of the monologue's power; at the very least, he was unable to ignore it.
77. Michael Taussig, *Shamanism, Colonialism, and the Wild Man: A Study in Terror and Healing* (Chicago: University of Chicago Press, 1986).
78. Edward Said, *Culture and Imperialism* (New York: Knopf, 1993), 165.

4
The Case for Postcolonial Liberalism in Hanif Kureishi's *My Son the Fanatic*

Eddy Kent

According to one popular argument, made most recently and forcefully by Nancy Armstrong, the history of literature is the history of identity.[1] Literature is the place where communities are imagined, social bodies constructed, and sympathies disseminated throughout a reading public. It is also the site where difference is registered, alienation traced, and possible futures considered. Construed thus as a site of dialogue and engagement, literature assumes simultaneously a political and an ethical dimension. Foremost among contemporary practitioners of such ethical and controversial writing is British author Hanif Kureishi. Introducing his 2005 collection of essays *The Word and the Bomb*, for example, Kureishi considers the increased relevance of art and the artist at a time when politicians, sound-bite commentators, and media outlets speak uncritically in terms of civilizational clashes:

> In an age of propaganda, political simplicities and violence, our stories are crucial [. . .] [It] is in such stories – which are conversations with ourselves – that we can speak of, include and generate more complex and difficult selves. It is when the talking and writing stops, when the attempt is to suppress human inconsistency by virtue, that evil takes place in the silence.[2]

The evil Kureishi has in mind here encompasses not only the ongoing racialized discrimination that continues to taint human relations in postcolonial Britain, but also the periodic eruptions of race riots and, which occasions this chapter, the suicide attacks on the London Transportation Network in the summer of 2005. As the title of his collection makes clear, Kureishi sees the Word as inextricable from the Bomb. The conjunction in the title is key to understanding this relationship – this is not the Word *or* the Bomb – suggesting his commitment to exploring the ongoing relationship between the performative act of writing and the material effects of violence.

In negotiating this relationship, I read Kureishi's call to produce new stories and make new identities as a confirmation of his secular humanist allegiance. I say 'humanist' in the spirit of Giambattista Vico, who insists that the world is not given but made, a principle seemingly averred by Kureishi when he suggests that, by producing literature, we literally "generate" ourselves.[3] Reading Kureishi as a secular humanist is helpful when we come to approach the ethical position outlined in his fiction. Without it, the most we can claim is that his ethics are postmodern or anti-essentialist, positions which maintain that the only truth is that there is no truth. The Vichean principle, alternatively, ties the relativity of truth to human action and thus allows the critic and reader to historicize particular situations. In turn this provides the basis for a form of political action.

The consequence is clear for Kureishi's moral injunction that we must continue "conversations with ourselves" in order to ward off "evil": if truth, or the good, derives from our response to a given situation, then participation rather than deliberation or reflection becomes the measure of ethics. For the social institution called literature, truth is not found in the production of "good" (i.e., valuable or moral) literature, but rather in the act of production itself. Thus evil, usually understood as a corruption of the truth, now appears as silence, as the absence of agency or the failure to act. In the postcolonial reality of Kureishi's Great Britain, where once disparate human communities and cultures are challenged to coexist peacefully and for each other's mutual benefit, our attention must be concentrated on an ethics of action, on how literary characters participate in the maintenance, cultivation, and operation of the social institutions that constitute their worlds.

But this relationship between individuals and institutions is complicated, and Kureishi's fiction unsparingly stages the human effort to produce an individual identity amidst competing narratives of belonging. Consistently over the course of novels, screenplays, short stories, and essays, this writer has taken as his central subject those who do not feel at home in their worlds. His focus on the intersubjective nature of identity has resulted in characters that consistently refuse to be bound to the nominal categories of class, race, ethnicity, nationality, gender, or sexuality. Thus, it is not difficult to see why Kureishi's works have appealed to theories of the postcolonial and cosmopolitan conditions. Recently, though, the same critics who once celebrated Kureishi's radical identity politics have begun to wonder openly whether he has lost his touch or, worse, sold out, because of the way he depicts British Muslims in his later works.

Devout Muslims began to play central roles in Kureishi's work in the 1990s, in texts like *The Black Album* (1995)[4] and *My Son the Fanatic* (1997), and this is a period during which Kureishi himself has confessed that he became actively concerned with "the conflict between Islam and Western liberalism."[5] Admittedly this is something of a departure from his works of the late 1970s and 1980s, when it was easier to classify Kureishi as an "anti-Thatcherite" than as a critic of geopolitical alignments. Certainly, on one hand, Thatcherite capitalist ideology overwhelms his portrayal of young, disaffected British Asian men in works like *My Beautiful Laundrette* (1986). But on the other hand, we must recognize that Thatcherism is not the only ideological position under critique in those texts. In *The Buddha of Suburbia* (1990), for instance, Kureishi introduces another essentialist political identity to compete with Thatcherism.[6] When Karim realizes that the era's advocacy of meritocracy and entrepreneurship is nothing more than a rhetorical fig leaf, covering what remain to be essentially racist and exclusionary social structures, he is tempted by the prospect of joining a community organized around religious, cultural, and racial lines. In that novel Kureishi called this ideology "Islamic separatism"; by the time he wrote *The Black Album* and *My Son the Fanatic* it had evolved into "Islamic fundamentalism."

A comparison between the representation of these movements and the scrutiny applied to Thatcherism reveals the standard pattern of Kureishi's ideology critique. In each, the initial allure quickly dissolves as its chief exponents are shown to be little more than histrionic caricatures of political actors, whether it is the Thatcherite uncle of *My Beautiful Laundrette* or the separatist cousin of *The Buddha of Suburbia*. Herein lies the problem. Whereas for Kureishi fundamentalism must be dismissed for the same reasons we dismiss other essentialist ideologies, his critics worry that he is pandering to a mainstream audience by perpetuating the myth of an irreconcilable conflict between Western liberalism and Islam. Yet rather than interpret Kureishi's representations of Muslims as a conservative turn or a selling-out, I prefer to see them as consistent with his long-standing engagement with the ideals that belong to the liberal tradition. I say 'liberal' because as he posits dialogue against silence, engagement against withdrawal, revision against conservation, and change against stability, Kureishi very clearly articulates a liberal ideology. This observation has already been made by Ruvani Ranasinha in her recent monograph on Kureishi for the *Writers and their Work* series.[7] Yet Ranasinha also reminds her readers that liberalism can sometimes take hegemonic or oppressive forms. In particular, any form of liberalism that assumes that values such as secularity or rationality are *ab origine* and unalterable is essentially inconsistent with its own goals of inclusion precisely because it expresses "intolerance of any intolerance of itself."[8]

I do not intend to enter into the lively debates both within and between postcolonialism and liberalism. Instead, I want to demonstrate how, through *My Son the Fanatic*, Hanif Kureishi offers a glimpse of how liberalism might function positively in a postcolonial, multicultural, global society. The first section of this chapter outlines the postcolonial critique of liberalism only in order to evaluate Ranasinha's claim that Kureishi's later works complicitly "reinscrib[e] dominant liberalism as the norm."[9] Working with Jacques Derrida's understanding of hospitality, I will show how, through a series of accepted and rejected invitations, *My Son the Fanatic* expresses liberal ideals that are far from hegemonic, dominant, or canonical.[10] The second section takes up the challenges posed to this nascent postcolonial liberal philosophy by characters representing an encroaching globalization. Assuming as a point of departure that the goal of a liberal society is to make its members feel at home in their world, I ask what can be done when the institutions of one's world are being transformed by forces over which individuals – to say nothing of the nation – have very little power. The third and final section steps away from *My Son the Fanatic* in order to elaborate precisely what is at stake in the struggle to maintain and establish a postcolonial form of liberalism. By drawing into my analysis the official response to the 2005 London Transport Network attacks, I mean to emphasize the productive capacity contained within Kureishi's political vision, and its potential to counteract the violence that is caught in between the word and the bomb.

Culture and liberalism

In 1994, the short story "My Son the Fanatic" appeared in *The New Yorker* and in 1997 it was expanded into the screenplay for the film of the same name. Naturally, in the transition from short story to screenplay, much detail was added. Some aspects, such as introducing a mother caught between the father and son, or developing the father's extramarital love interest, will strike the reader as natural extensions of key narrative issues. Other developments, however, are less predictable: the introduction of the Schitz character, for example, or the attention paid to the post-industrial landscape. In what follows, I want to suggest that it is through these novel additions that *My Son the Fanatic* becomes increasingly politically relevant. As a short-story, "My Son the Fanatic" is a literary representation of the generational clash between a father "implicated in Western civilization" and his rebellious son.[11] As a screenplay, *My Son the Fanatic* attempts to grasp more fully the complexity of forces operating on subject-citizens seeking to find a home in the world.

The narrative in both texts turns on an ideological conflict between an assimilationist immigrant father and his separatist British-born son. Parvez, by his own admission, has had problems integrating into British society, but, like many immigrants who have felt excluded from their adopted homes, he has lived vicariously through his child:

> Ali excelled in cricket, swimming and football [...] [and was] getting straight "A"s in most subjects. Was it asking too much for Ali to get a good job now, marry the right girl and start a family? Once this happened, Parvez would be happy. His dreams of doing well in England would have come true.[12]

However, when Ali says that he is renouncing "Western civilization" in favor of a fundamentalist strain of Islam, the bewildered father can only ask, "Is there a particular event which has influenced you?" His son replies simply: "Living in this country."[13] This response encapsulates the same sense of domestic alienation present in Kureishi's earlier works, including *Borderline* (1981), *My Beautiful Laundrette* (1986), and *Buddha*.[14] Like the protagonists of these earlier texts, Ali belongs to a category I would call 'the partial citizen', a subject paradoxically constructed on the one hand by the principle of *de jure* equality and on the other by the anguish of *de facto* inequalities. The condition of partial citizenship arises because the basic unit of post-Westphalian political exchange is the nation-state rather than the pure state. The institutions of the nation-state can reflect the chauvinistic features of national discourse and therefore potentially exclude those citizens who inhabit spaces outside the traditional boundaries of the Herderian *volk*, the nation. Admittedly this problem is as old as the nation-state itself, but it becomes acute in the postcolonial space of recently dismantled empires, where large communities of formerly colonized subjects have immigrated to the imperial centre and have begun to demand the rights of citizens promised to them by the liberal rhetoric of imperial discourse.

In his introduction to *My Son the Fanatic*, Kureishi defines the feeling produced by this state of affairs as "unhomeliness":

> [A]ll along it was taken for granted that 'belonging,' which means, in a sense, not having to notice where you are, and, more importantly, not being seen as different, would eventually happen.
>
> *My Son the Fanatic*: xi[15]

Through the son (renamed Farid in the screenplay), Kureishi gives the lie to that mythology:

> Farid: Whatever we do here we will always be inferior. They will never accept us as like them. But I am not inferior! Don't they patronize and insult us?
>
> *My Son the Fanatic*: 66

Farid's frustrated defiance brings him into association with other disillusioned partial citizens and together they campaign against the "corruption of modern Britain" (p. 76). However, their political protest assumes a militant dimension as their demonstration against prostitution erupts into a mob, with men attacking women and Farid spitting in the face of his father's friend, Bettina. Elsewhere, Farid is rumored to have participated in the gang-beating of an underaged prostitute. In portraying Farid's transformation from a cricket-loving, guitar-playing, trainee accountant into a violent religious ideologue, Kureishi suggests that the rise of fundamentalism coincides with the political and social alienation of British Muslims in an avowedly (but not actually) secular liberal democracy. To the partial citizen of the nation-state, fundamentalism offers a positive form of identification, an alternative to those "being made to feel inferior in [their] own country"(p. xi).

Feelings of unhomeliness and alienation direct attention towards agency in the realm of political and social institutions, since the stability of a political society rests in its capacity to generate affiliative bonds between its members and its institutions. Agency, understood as the right to participate in the decision-making processes attached to the issues affecting one's life, is an essential component in even the most basic understanding of citizenship. Citizens are the architects and guardians of institutions, and a fully enfranchised citizen will feel at home in his or her world. Farid's frustration grows out of his sense that he cannot shape, but rather can only endure, the events and circumstances of his life. In an autobiographical essay, Kureishi elaborates this frustration when he insists that postcolonial realities necessitate "major adjustments to British society [...] [including] a new way of being British." He challenges Britain to show "what it really means when it describes itself as 'democratic.' "[16] The problem is rephrased slightly in another of his essays: "how could people feel themselves to be active participants in the life of a society when they were suffering all the wretchedness?"[17]

This demand for fully participative politics is underwritten by the realization that institutions on the one hand determine one's social identity, and on the other are mutable. The future of those institutions, the repositories of consensual values, is of particular importance in non-homogeneous communities and liberalism represents a political philosophy designed especially to resolve the problem. Given that states now consist of increasingly diverse individuals and groups, the disagreement within a community can be profound. From this, liberalism seeks to establish the terms and conditions under which such people can coexist both peacefully and in political association. Traditionally, liberal political engagement has been articulated through concepts of neutrality, impartiality, or secularity. According to Jürgen Habermas, modern liberal democracies express this neutrality best in the idea of a public sphere, a common space where members of a society meet either directly or through media to discuss matters of common interest.[18] Ideally, the public sphere is a secular space where members of a society congregate as equals; its legitimacy derives from the absence of private or prejudicial interest.

Of course, conditions of equality and neutrality are theoretical ideals and we quickly recognize: (a) that existing institutions nearly always overrepresent the interests of the hegemonic class; and (b) that certain individuals will always enter into the public sphere with greater influence than others. We must recognize that the claim to neutrality or detachment is always situated, and some of the strongest postcolonial critiques of liberalism make precisely this point.[19] One of the most serious challenges to liberalism is that the Habermasian split between public and private, the division that underpins classical liberal thinking, universalizes particular notions of secularity and rational political engagement. Charles Taylor expounds this view when he says that it is impossible for liberalism to "offer a neutral ground on which people of all cultures can meet and coexist" because it presumes, among other things, the divisibility of church and state.[20] Taylor calls this presumption problematic and potentially unjust because it is a product of an organic Christian intellectual tradition; for Taylor, the secularism of the public sphere cannot avoid its particular, private heritage. From a different perspective, Homi Bhabha has diagnosed this as the condition of unreconstructed liberalism, under whose aegis members of minority communities become "virtual citizens." The struggle to maintain aspects of their traditional culture in the putatively secular sphere of political activity places virtual citizens on the exterior of modernity, "relegated to a distanced sense of belonging elsewhere, to a there and then."[21] According to Bhabha, unless we begin to re-imagine secularity, the discourse of tolerance will only perpetuate existing social divisions and consolidate the interests of the dominant culture.

This rough sketch of liberalism and its postcolonial critique is relevant to Kureishi's politics because it stands behind Ranasinha's reading of *My Son the Fanatic*. In a chapter entitled "Muslimophobia," she argues that, whereas Kureishi's earlier works celebrate diversity as they satirize the foibles of both dominant and minority cultures, in *The Black Album* and *My Son the Fanatic*, Kureishi

> uncritically reflects and embodies [...] [the] predominant fears, prejudices, and perceptions of devout British Muslims as 'fundamentalists,' constructed as particularly threatening in the West. His caricatures further objectify this already objectified group, whilst reinscribing dominant liberalism as the norm.[22]

Ranasinha invokes Bhabha's critique to identify the shortcomings in Kureishi's work. She charges that, whereas Bhabha provides a "nuanced examination and approach" to the problem of Islamic fundamentalism, Kureishi creates a false dichotomy, with "unreconstructed secularism" as the only alternative.[23] Unfortunately, such claims cannot be made without strenuous contortions but, leaving to one side the inconsistencies in the argument, the hypothesis that "ultimately Kureishi's liberalism is close to its hegemonic forms" requires serious attention.[24] If true, it would leave Kureishi open to the charge that he exchanged the radical politics of his youth for the comforts associated with mainstream popularity. Certainly the coercive potential of pluralism to force minorities to engage with the dominant culture on unfair terms is, as discussed above, a strong critique of liberalism. It remains to be seen if much evidence can be found in Kureishi's works.

We can begin by identifying a critique of co-optive pluralism in Kureishi's *Sammy and Rosie Get Laid* (1988) through the character of Alice, a well-intentioned daughter of a colonial family and estranged wife of a Tory property developer, and her relationship with Rafi, a Pakistani politician and friend of hers from university. Alice expresses conventional middle-class liberalism when she learns of the violent reaction among members of the underclass to the shooting of a black woman by the police. She tells Rafi: "I hate their ignorant anger and lack of respect for this great land. Being British has to mean identification with other, similar people. If we're to survive, words like 'unity' and 'civilization' must be understood."[25] Later in the screenplay, when she discusses the eviction of the squatters with a community activist, she repeats her Thatcherite assumption that the liberal democratic state already constitutes a neutral entity:

Rani: The way Danny's lot were treated shows just how illiberal and heartless this country has become –

[...]

Alice: It doesn't affect the law. The law is to protect the weak from the strong, the arranged from the arbitrary –

Rani: But they are powerless just trying to find a place in this rotten society for themselves!

[...]

Alice: (*Out of shot*) Their place can only be found on society's terms, not on their own whim – [26]

Alice, who literally played the host to Rafi in his student days, is here shown to be incapable of genuine political hospitality. Her presumptions that British identity is either stable or somehow intelligible, that the institutions of the British state are already capable of delivering justice, and that minorities must accept all this, indicate an intolerance that Kureishi allegorizes by making her one of the few characters in any of his writings who is unable to establish an intimate relationship with the Other. As Rafi, for so long the object of her romantic attachment to the exotic, surveys the detritus of her unconsummated passion he can only muse, "How bitterness can dry up a woman!"[27] Before passing over this screenplay, we might also observe that, for Kureishi, inhospitality is not the provenance of either the Thatcherites or of well-intentioned middle-class liberals. "This is liberalism gone mad!" pronounces Rani, the lesbian journalist and political activist on the presence of Rafi, a corrupt Pakistani politician, in her country.[28]

In *Sammy and Rosie Get Laid*, Kureishi's liberalism is far from hegemonic but, then again, Ranasinha would claim that this screenplay is one of the earlier anti-Thatcherite rather than the more problematical anti-fundamentalist texts. By connecting illiberality with inhospitality, however, the screen-play shares much with *My Son the Fanatic*. Both texts insist that it is only through the gift of the welcome that individuals can begin to explore, together, the possibilities of a live culture. Given that individuals are always already constructed by a plurality of discursive engagements, cultural prac-tices, and institutions, requiring actual neutrality as a precondition of that welcome may be impossible. But this does not mean that neutrality is not worth pursuing *as an ideal,* and this is where I think Jacques Derrida's possible/impossible aporia of hospitality makes its contribution to the post-colonial liberalism that I would argue characterizes Kureishi's writings. Both Rafi and Danny, like the Pakistani *maulvi* and Herr Schitz of *My Son the Fanatic,* must be engaged despite their disagreeable backgrounds.

In *Of Hospitality* Derrida argues that what we might call hospitality's unconditional injunction – the altruistic willingness to surrender all for the guest – is compromised by the power dynamics that define the host-guest relationship. On the one hand, the very act of hospitality presumes an already established order, a state of affairs over which the host is master. On the other hand, welcoming the guest absolutely and without exception implies that everything the host owns, values, and controls could be under-mined, displaced, or destroyed. Therefore, for self-preservation, the host must always put conditions or limits on the offer. For Derrida that would be impossible, a traducement of the act. To resolve this problem, he suggests we emphasize the importance of emotion by admitting (alongside Emmanuel Levinas) that the terms and conditions of our ethico-political engagement with the Other are structured affectively. This amounts to a concession that the difference between the Self and the Other, the host and guest, cannot be resolved. Rather, the emotional reaction to difference must be confronted; fear and loathing replaced with, in the words of Kureishi's Parvez, "curiosity [and f]ascination" (p. 66).

Derrida's theory of hospitality illuminates the politics of *My Son the Fanatic*. We will recall that Kureishi's introductory essay to the screenplay links the rise of Islamic fundamentalism in Britain to a growing sense of unease among young British Muslims who are "being made to feel inferior in [their] own country" (p. x). Furthermore, the screenplay itself immedi-ately declares hospitality to be one of its central themes, since the opening scene takes place on an English country estate, in the home of the Police Commissioner (could there be a better site for the threat of state violence lurking underneath domestic gentility?). Parvez and his wife Minoo have been invited to the Fingerhut home because their son Farid is engaged to the Commissioner's daughter Madelaine. In an allegory of the struggle by British Muslims to integrate within their own country, the marriage between Farid and Madelaine will seemingly seal the social promotion of the family. Yet the printed directions of the screenplay tell us that the *"atmosphere is strained"* and everyone sits on *"hard chairs."* The host is *"bewildered"* and everyone else is *"uncomfortable"* except for Parvez who is *"both terrified and ecstatic"* (p. 3).

It should not be surprising that Parvez seems best equipped of the assembled characters to handle this cross-cultural encounter. After all, not only does he espouse integration but, as a taxicab driver, he has a professional interest in the cultivation of such skills: "The 'gentleman' is my code," he tells one client (p. 33).

However, despite his conscious attempts to assimilate by imitating the characteristic virtues of the hospitable and accommodating gentleman, Parvez has been frustrated. After 20 years in England, his only community consists of his work colleagues and his regular fares. He explains his frustration to Farid:

> *Parvez*: Anyhow, how else can we belong here except by mixing up all together? They accuse us of keeping with each other.
> *Farid*: Yes!
> *Parvez*: But I invite the English. Come – share my food! And all the years I've lived here, not one single Englishman has invited me to his house! – apart from Fingerhut who is a top-class gentleman! But still I make the effort.
>
> *My Son the Fanatic*: 65

If, on one hand, we understand Parvez as the character best trying to engage the Other in a liberal fashion, then we can understand his disappointment and admire his determination to continue offering hospitality. On the other hand, both the printed directions and Farid make it clear that Fingerhut, the putative top-class gentleman, finds Parvez "repellent" (p. 68). Placed at an ironic distance, then, it is hard for the audience to recognize Parvez as the model liberal.

Situated somewhere between the earnest liberal and its parodic counterpart, Parvez is a man whose mistakes have been encouraged by greed and self-interest. He is a professional rather than an altruistic host, one who seeks to extract profit out of every encounter. He plays concierge to the wealthy businessman, Herr Schitz, and endures public humiliation in search of money; likewise his gracious manner at the Fingerhut home is marred by his calculating mind. He hopes that the alliance with the Police Commissioner might produce a profitable career for Farid and, as his friends suggest, a reduction in parking tickets. With this in mind, Parvez advises his wife to "[l]eave matters of business opening to me" (p. 6).

When given the opportunity to behave altruistically, Parvez fails spectacularly. For example, when Farid asks whether they might host a visiting *maulvi*, Parvez meets the request initially with magnanimity: "Our house is open" (p. 67). Yet the limits of his tolerance are tested when Farid and Minoo disturb the pre-existing domestic order. *"The place,"* we are told, *"is not only preternaturally tidy, but the furniture has been moved around"* (p. 83). Likewise, the family tradition of eating meals together has been abandoned as Minoo

3

now eats in a separate room to accommodate the religious code of their guest. For Parvez, a man who derives much pleasure from the consumption of food, this is unbearable and he declares "I will not eat without you!" (p. 84). Rebuking the *maulvi*, Parvez, "*ostentatiously scrapes the food from his plate into the bin*" and retreats to his den where he deliberately sets out to annoy his guest by playing loud jazz music. Shortly afterwards, he turfs the *maulvi* out of his home altogether.

Such failings should not, however, be viewed as justification for dismissing Parvez's efforts in total. In this screenplay, the spaces cleared by the professional host are those that create the conditions for intimacy. Near the end, Parvez tries to explain his relationship with Bettina to his wife.

> Parvez: Friendship is... good, Minoo. I think it can be found... in the funniest places.
>
> *My Son the Fanatic*: 120, ellipses in original

These "funniest places," I submit, are shown by Kureishi to be professionally hosted spaces like Fizzy's restaurant, Parvez's taxicab, Schitz's hotel room, and the derelict factory (transformed by Parvez into a party venue). We can compare their productive capacity to the alienating effects of the traditional home. Fizzy's restaurant bears special consideration since the Indian curry house, with a menu that would not be recognized by consumers of traditional British or Indian food, is the pre-eminent example of the cultural renovation already occurring in Britain. It is at Fizzy's that Parvez and Farid have their crucial confrontation, and at Fizzy's that Parvez and Bettina have the intimate dinner that transforms their relationship. Like the restaurant, the taxi, the hotel room, and the disused factory are spaces that, while far from neutral, are symbols of old institutions put to new uses by the inclusion of political actors with new and varied ideas of the good.

Culture and globalism

Kureishi's fictional construction of spaces where postcolonial liberalism might emerge evokes what Edward Said would call a "worldly" text, something which cannot be examined in isolation from the world in which it circulates.[29] Accordingly, we cannot evaluate how Kureishi's liberalism might transform institutions, practices, or ideas without a clear idea about the forces shaping the spaces where it functions. As its partial citizens struggle for enfranchisement, *My Son the Fanatic* shows that the horizon of their world is no longer contained by the parameters of the (postcolonial) nation-state. But Kureishi is canny enough not simply to celebrate the ruptures caused by global forces within national discourse. In his descriptions of Bradford and Herr Schitz, Kureishi shows how globalization might just as easily entrap as liberate the partial citizen.

Saskia Sassen's work on global cities provides a useful point of departure. Sassen identifies hypermobility and ethereality as the predominant images found in mainstream accounts of globalization.[30] In contrast, in *The Global City*, Sassen argues that these images incorrectly turn globalization into a chimera.[31] Despite the frightening mobility of capital and instantaneous projection of power, Sassen shows how global forces are still perceptible in material geographies. As such when,

> global actors, whether firms or markets, overlap and interact with the national, they produce a frontier zone in the territory of the nation. Not merely a dividing line between the national and the global, this is a zone of politico-economic interaction where new institutional forms take shape and old forms are altered.[32]

To shift slightly into the language of Deleuze and Guattari, globalization achieves not an absolute, but a relative deterritorialization. The frontier zones have a palpable thickness where its deterritorializing processes can be perceived and scrutinized.

But Sassen's study limits the frontier zones to the upper echelon of global cities, the cosmopolitan sites of London, New York, and Tokyo, whereas *My Son the Fanatic* is set in Bradford, a town in the English Midlands. If, as I have been arguing, *My Son the Fanatic* considers the partial citizen in the age of the global, then why does Kureishi prefer Bradford to London? Engaging the phenomenon of Islamic fundamentalism could easily have been done in "Londonistan" – indeed, the short story *is* set in London. Likewise, if this is a screenplay about fluctuating identities and renovating institutions, then London's cosmopolitan energy, as John Ball and Sukhdev Sandhu have separately noted, seems better than a Midlands mill town.[33] With these questions in mind, I would propose that the move away from London has two critical effects. First, it supplements Salman Rushdie's aphoristic observation that "the Empire writes back to the Centre" by reminding us that other, non-metropolitan sites must be included in any complete analysis of that Center. Second, the dilapidated landscape of Bradford, with its empty Victorian mills, testifies that the benefits accrued during one period of colonial expansion do not necessarily carry over into its future stages.

Kureishi's choice of Bradford, making it and not London the site of fundamentalist violence, also challenges us to broaden our conceptualization of global forces. In Sassen's economic model, Bradford is not a thick frontier zone. Instead, the immiserated mill town belongs to a set of traditional manufacturing centers now in rapid decline. If the work once completed in Bradford's textile mills has since been outsourced to a foreign country, the people connected to that industry remain, for the most part, geographically fixed. Despite the best efforts of neo-liberal economists to convince us of the contrary, workers and their associated families and communities are not as mobile as the capital that shunts their jobs from country to country. Even in periods of economic depression, the economic immigrants remain in place. Parvez tells his wife near the end of the screenplay, "You can't go home, Minoo. It isn't like that now. This is our home" (p. 121).

However, just as chimeric economic globalization might be perceived in the thickness of a global city like London, Kureishi shows how social globalization is palpable in Bradford. Its first and most obvious influence is Islamic fundamentalism, a transnational ideology that draws on but does not directly emanate from the injustices of British colonial history. The postcolonial tension between Britain and Pakistan is but one small battle in the larger struggle between "the West" and the *umma islamiyya*. When Farid lambastes his father over dinner, he speaks not in terms of the British Empire but of "the capitalist dominated world we are suffering from! I am telling you, the Jews and the Christers will be routed! You have taken the wrong side!" (p. 69).

This fundamentalist ideology flourishes in Bradford because of its particular history. It is estimated that 90 percent of Bradford's South Asian population come from a single district in Pakistan: Mirpur. This concentration occurred because, as Kureishi says, "that was where the Bradford mill-owners happened to look for cheap labor twenty-five years ago."[34] Of course, this happenstance was affected by a pre-existing colonial relationship that formed economic and social bonds between Britain and Pakistan; this colonial coincidence lays the groundwork for transnational affiliations with other areas and communities suffering under the pressures of global capitalism. Parvez makes this depression clear when, on the site of a council estate, he drunkenly laments, "We have come from one Third World country to another" (p. 75).

However, the presence of Herr Schitz in the Bradford landscape indicates that Islamic fundamentalism is not the only global force operating on the local characters. Almost to the point of caricature, Herr Schitz is an allegory for transnational capital. An international financial speculator, Schitz descends from the sky into economically depressed Bradford as a messianic figure, fully aware that his investment can potentially revivify the dilapidated industrial town. Accordingly, he is *"received deferentially"* by the local businessmen (p. 28). As if to emphasize his connection with the new global order, the specific nature of his business remains elusive throughout the text. Early in their relationship, Parvez asks:

Parvez: What business are you doing here?
German: Out of town shopping, everything under one roof.
Parvez: The land and the labor is cheap, eh?
German: Like the women.

My Son the Fanatic: 31

Schitz is an undiscriminating figure who assesses the quality of British prostitutes with the same dispassionate objectivity he applies to the feasibility of a factory renovation. In a narrative organized around competing ideologies, his is baldly stated: "Respect," he advises Parvez, "is no substitute for pleasure" (p. 33).

Schitz's libertinism provides a foil for Parvez's moral liberal individualism. The men are approximately the same age, and when Schitz says that he has abandoned his family, Parvez offers a mixture of sympathy and moral upbraiding:

> *German*: What of life do you enjoy then?
> *Parvez looks puzzled.*
> Your family?
> *Parvez nods unconvincingly.*
> I left mine.
> *Parvez* That's not a very nice thing, sir.
> *German*: What do you know about it?
>
> *My Son the Fanatic*: 33

Nonetheless, Schitz seems to enjoy Parvez's company and invites him to continue their society outside the confines of the taxicab. With Schitz, Parvez participates in and consumes aspects of British culture – drugs, pubs, hotel rooms, and prostitutes – that were previously unknown to him.

For Parvez, a wage-slave whose marriage is characterized by mutual respect rather than passion, Schitz's unrepentant epicureanism reveals another way of being British. As Parvez begins to enjoy himself more, he embraces Schitz as a model. When Parvez takes Bettina on their first date, he chooses the restaurant where Schitz and Bettina had dined the night before, telling her that it is the "Best place. This is the bloody life, yaar!" (p. 87). Yet as Schitz and Parvez each fulfill their sexual desires through Bettina – "We boys are going to start to enjoy ourselves!" (p. 33) – it is Parvez who ultimately attains intimacy. For Schitz, as his name hints, there is only a continuous cycle of unfulfilling consumption. A waiter at Fizzy's restaurant describes him as "the German who ate the whole kitchen and then said it was too salty" (p. 87). Schitz's attitude places him within the small but influential class of global capitalists who now have little, if any, affinity with the particular geographies, localities, or communities implicated in their investment strategies. The disturbing fluidity of this disembodied capital is alluded to at least twice in the screenplay. First, Schitz discloses the arbitrary motivation for his choice of Bradford: "I wanted to try a strange and awful place where everything was new to me" (pp. 43–4). Later his whimsy reasserts itself when, tiring of the town and its prostitutes, he leaves Bradford without having invested anything.

At the same time, Schitz is not merely a rootless capitalist. He is German, a nationality Kureishi emphasizes in the published screenplay by identifying the character's speeches with "German" instead of "Schitz." There is nothing accidental in Schitz's nationality. Along with France, Germany has long been the continental Other against which Britain has fashioned its identity. The specters of the two world wars continue to haunt British collective consciousness, and the fear of a strong Germany is behind much British nationalist discourse. However, Schitz's power and influence in the English Midlands illustrates the hollowness of any appeal to the mythology of Britain as a self-contained, self-sufficient, and unified political entity.

Moreover, Germany is seen as the great champion of European integration, the bogey of Thatcher's crocodile handbags.[35] Schitz confirms these Euro-affiliations when he tells Parvez that he was formerly employed in "Munich, Lyon, [and] Bologna" (p. 43). These cities are obviously metonyms for the three largest European economies – Germany, France, and Italy – but we might ask why Kureishi did not use the more expected financial (Frankfurt-Paris-Milan) or political (Berlin-Paris-Rome) triumvirates. I would argue that, like his choice to change the setting from London in the short story to Bradford in the film, Kureishi's choice to orient Schitz within this network of second-tier cities suggests that his interest is not in measuring how the global order alters the spatiotemporal order of cosmopolitan centers, but rather in assessing how it affects other, secondary localities.

Another beneficiary of the deep-pocketed businessman is Bettina, who becomes the German's favored escort. Like Parvez, Bettina is a partial citizen caught in the hypocrisy of Thatcherite nationalism and European integration. Yet Bettina demonstrates that attachment to the national identity is secure only inasmuch as citizens consciously identify with national institutions. She willingly surrenders her English or British patriotism for the opportunity to escape the quotidian squalor of the Bradford sex trade and after one particularly profitable evening in the company of Herr Schitz, she tells Parvez:

> *Bettina*: We are doing very well in Europe. I am all for the Union. Next
>   stop Maastricht!
>   *My Son the Fanatic*: 51

Her allusion to the Maastricht treaty, the founding document of the European Union, signals her willingness to sublimate her English citizenship within the emerging European federation. But before we mistake such unionist federalism as emancipatory, Kureishi reminds us that the journey to Maastricht is not made without difficulty; Bettina's body, via Schitz's fists, allegorically stands in for the required national sacrifices.

Parvez too must surrender his personal dignity in the quest for Schitz's patronage. Schitz's power is confirmed in the pub scene where the entrance of a Pakistani, a German, and a prostitute into a seedy comedy club would seemingly provide a cornucopia of material for a *"fat vulgar Comedian [...] telling a stream of coarse jokes"* (p. 44). But nothing is said to Schitz – no mention of Jerrys, Colonel Klink, or Who Won the War. Instead, Parvez, the British citizen, is made to bear the brunt of the comedian's ire as the spotlight is, literally, turned on him, *"the only brown face there"* (p. 46). The injustice inspires solidarity between Bettina and Parvez, and she throws a pint glass at one of his tormentors. After the ensuing ruckus, even the otherwise selfish Schitz expresses empathy and consoles Parvez, "I feel sorry for you people, I really do" (p. 48).

The degree to which this momentary solidarity can lead to deeper understanding is mediated by economic and social factors. On the one hand, the pub brawl brings Parvez and Bettina into closer contact, and they end the night at her apartment philosophizing and sharing life stories. The stage directions unsubtly signal that the relationship is changing: *"For the first time we see her without wig or make-up"* (p. 52). As Parvez leaves, their intimacy is confirmed:

> *Parvez*: [...] See you tomorrow.
> *Bettina*: Call me Sandra – when we are alone.
> *Parvez*: That is the password, eh? To you.
> *She kisses him on the cheek and then on the mouth. He is surprised.*
> *Bettina*: I don't know when I last kissed a man. Sorry. Do you mind?
>
> *My Son the Fanatic*: 54

On the other hand, Schitz's empathy with Parvez is fleeting. The solidarity formed in the pub is forgotten as Schitz reverts to calling Parvez "little man" (pp. 70, 99, 113) and on one occasion kicks Parvez *"up the arse and laughs"* (pp. 99–100). These different outcomes testify that globalism *per se* neither improves nor erodes human relationships. In this, *My Son the Fanatic* suggests that we should interpret globalism's associated narratives as neither emancipatory nor oppressive. By providing new opportunities for previously disparate groups to come into contact with each other, they act only as the agent of change.

Culture and political action

In 2000, Hanif Kureishi discussed *My Son the Fanatic* with Geoff Gardner. As they broached the topic of the rise of Islamic fundamentalism, Gardner provided an Australian perspective:

GG: We haven't experienced the level of fanaticism that's shown in the
film.
HK: Well you've got something to look forward to.[36]

If Farid's renunciation of "Western civilization" seemed prophetic after the
September 2001 attacks, his warning that "there will be *jihad*. I, and millions of others, will gladly give our lives for the cause" ascended to the level
of the uncanny after the bombings of the London Transportation Network
on 7 July 2005.[37] At that point, Hanif Kureishi became a Cassandra of the
global "war on terror." While the press, expert commentators and the Prime
Minister's Office initially linked the attack to the nebulous foreign web of
al-Qaeda terrorists, Kureishi had already seemingly described the attackers in
detail. Like Farid, three of the four bombers were British-born and, to use the
euphemistic phrase preferred by The *Daily Telegraph*,of "Pakistani ethnic origin."[38] All three grew up in the British Midlands and, as the media scrambled
for their "back-stories," the portrayal of 22-year-old Bradford-born Shehzad
Tanweer seemed to be based on Farid. The header for Sandra Laville and Ian
Cobain's profile in The *Guardian* on 13 July 2005 neatly summarizes the *post
facto* image of Tanweer: "From Cricket-Lover Who Enjoyed a Laugh to Terror
Suspect."[39]
   The mainstream response to the "war on terror" illustrates precisely what
is at stake in the debate over the future of liberalism in a postcolonial, global
society. Pnina Werbner has already noted how quickly after the September
2001 attacks, "[Samuel] Huntington's clash of civilizations – or its denial –
had become the jargon of politicians and the media."[40] It is difficult to dis-
agree with Gilles Kepel's assessment that such arguments constitute "a kind
of Maginot Line of enlightened rationalism" which places Islam beyond the
pale.[41] Yet it remains true that these arguments' confrontational and inhos-
pitable tenor suffused responses to the London bombings. For example, the
speech delivered by Prime Minister Tony Blair, just hours after the attacks
of 7 July 2005, is characterized by the same rhetorical gambits made by
American President George W. Bush four years earlier:

> When they try to intimidate us, we will not be intimidated. When they
> seek to change our country or our way of life by these methods, we
> will not be changed. When they try to divide our people or weaken our
> resolve, we will not be divided and our resolve will hold firm. We will
> show, by our spirit and our dignity, and by our quiet but true strength
> that there is in the British people, that our values will outlast theirs. [. . .]
> This is a very sad day for the British people, but we will hold true to the
> British way of life.[42]

Like Bush, Blair very quickly placed the material violence of the attacks
within the frame of an ideological war between competing and irreconcilable

values. Even as evidence emerged in the following days that the bombers were British and not, as many had presumed, foreign operatives, Blair continued to pursue the "Us" versus "Them" line. Speaking before Parliament on 11 July 2005, Blair said:

> They will never destroy the way of life we share and which we value, and which we defend with the strength of belief and conviction so that it is to us and not to the terrorists that victory will belong.[43]

Such pronouncements of British values and positive assimilation, for Kureishi, are the hallmark of a tyrannical and co-optive pluralism which, in turn, is the harbinger of misunderstanding, miscommunication, and violence. In his essays, Kureishi has confronted the Blairite position directly: "If Blair's 'third way' implies consensus and the end of antagonism, our literature will sharpen and map out differences."[44]

As part of the effort to sharpen differences, Kureishi concludes *My Son the Fanatic* on an ambivalent note. Farid's disillusionment has helped Parvez realize that he will never be able to integrate into British society if the terms of engagement are dictated by men like Fingerhut. Just as the emblematic curry house simultaneously exemplifies the progressive transformation of British institutions and the provision of a professionally hosted space of engagement, a gastronomic metaphor organizes the debate over integration. Early in the screenplay, Farid asks his father, "Can you put keema with strawberries?" (p. 39). Near the end, Parvez has seen enough intermingling to declare, "And Farid says the cultures cannot mix. Jesus, they can't keep apart" (p. 87).

Yet, however inevitable contact between the majority and the minority may be, the terms of that engagement continue to fluctuate. Farid's participation in an intolerant, disrespectful, and violent mob represents one option; Schitz's retreat into a shell of protectionist privilege is another; Parvez's relationship with Bettina is a third. What began as an entirely platonic relationship has been thrown headlong towards intimacy by a son's turn towards fundamentalism and a foreign capitalist's callous disregard for human dignity. But at what cost has this desired connection been made? I would suggest that the answer is found in the space between filiation and affiliation. The transformative forces that make possible the union of a Pakistani taxi driver and a working-class white prostitute are the same that destroy traditional familial bonds. As the screenplay concludes, Minoo is on the next plane to Pakistan and Farid has left home to join the vanguard in the war of ideologies. Surveying the wreckage of his life, Parvez says, "I have managed to destroy everything. I have never felt worse...or better" (p. 123).

In this final moment, Parvez encapsulates the condition of the partial citizen. When Bettina asks him whether they should start afresh in India or Pakistan, Parvez demurs and instead declares his intention to stay in his empty house in case either Minoo or Farid decides to return. Before Minoo leaves, Parvez commits to this vigil: "Perhaps one day the boy will tire of his moral exertions and will need me. I will wait. You will come home?" (p. 121). Thus the final note of *My Son the Fanatic* sounds neither optimistic nor resigned. At a time when liberals are debating amongst themselves whether to pursue a politics of recognition, assimilation, tolerance, respect, or reconciliation, Kureishi through Parvez only commits to engagement.

What do we make of Parvez's final promise, made not to Bettina but to his wife and son? As Parvez returns to his empty home, his choices should not be mistaken for a selfish assertion of his individual desires. In making this Janus-faced commitment to Bettina, he refuses to abandon the past while simultaneously refusing to be overdetermined by it. In its stead, the only ethical statement Parvez can make is the commitment to engage and to welcome, whatever the consequences. In committing to engagement, Parvez stakes a liberal position appropriate to those who do not feel at home in their world; in a culture of inhospitality, of rigid ideologies, one gains a particular value by refusing fixed positions. It is noteworthy that *My Son the Fanatic* ends with an exterior shot of Parvez's terraced-home at night with *"windows blazing with light in the middle of a dark terrace"* (p. 125). His commitment to hospitality, I want to suggest, is summed in that final image. If earlier works like *My Beautiful Laundrette* and *Sammy and Rosie Get Laid* triumphantly reveal the hypocrisy beneath the institutions of the hegemonically liberal nation-state, *My Son the Fanatic* confirms that global narratives of emancipation are no panacea. They are not, however, without their consolations for the subject who approaches them in a liberal spirit. Through them, Kureishi demonstrates, partial citizens are offered the possibility of exploring new identities through the unexpected pleasures and pains produced by complicated but intimate contact with others.

Notes

1. In *How Novels Think: The Limits of Individualism from 1719–1900* (New York: Columbia University Press, 2005, 3), Armstrong claims that as an emerging genre, the British novel gave rise to a "class- and culture-specific subject" which "proved uniquely capable of reproducing itself not only in authors, but also in readers, in other novels, and across British culture in law, medicine, moral and political philosophy, biography, history, and other forms of writing that took the individual as their most basic unit."
2. Hanif Kureishi, "The Word and the Bomb," *The Word and the Bomb* (London: Faber, 2005), 10.
3. Vico suggests that the human condition is constituted by the history of our relationship with social institutions. His philosophical approach to truth and identity is summarized in his principle that *verum* is *factum* ("The truth is made"). The principle was first discussed in Vico's *De Antiquissima Italorum Sapientia* (1710) but received its fullest elaboration (although paradoxically without ever being named) in the *Scienza Nuova Seconda* (1730). See especially paragraphs 331 and

349 of *The New Science*, trans. Thomas Goddard Bergin and Max Harold Fisch (Ithaca, NY: Cornell University Press, 1948), 89, 93.

4. Hanif Kureishi, *The Black Album* (London: Faber, 1995).
5. Kureishi, "Word," 11.
6. Hanif Kureishi, *The Buddha of Suburbia* (London: Faber, 1990).
7. Ruvani Ranasinha, *Hanif Kureishi* (Tavistock: Northcote House, 2002).
8. Ranasinha, 83–4.
9. Ranasinha, 82.
10. Jacques Derrida, *Of Hospitality*, trans. Rachel Bowlby (Stanford, CA: Stanford University Press, 2000).
11. Hanif Kureishi, "My Son the Fanatic," *Love in a Blue Time* (1994; London: Faber, 1997), 125.
12. Kureishi, "Son," 120.
13. Kureishi, "Son," 125–6.
14. Hanif Kureishi, *Borderline* (London: Methuen; Royal Court Theatre, 1981).
15. All page numbers for the screenplay, *My Son the Fanatic*, will be indicated in the text.
16. Hanif Kureishi, *My Beautiful Laundrette and The Rainbow Sign* (London: Faber, 1986), 101–2.
17. Hanif Kureishi, "Bradford," *Granta*, 20 (1986): 149–70. Rpt. in *My Beautiful Laundrette and Other Writings* (London: Faber, 1996), 136.
18. Jürgen Habermas, *The Structural Transformation of the Public Sphere* (Cambridge, MA: MIT Press, 1991), 23.
19. The postcolonial critique of liberalism takes many forms. Some include liberalism in the litany of Eurocentric Enlightenment ideologies whose essentialist structures require radical critique. A good example of this anti-essentialist approach can be found in Gyan Prakash's "Writing Post-Orientalist Histories of the Third World: Perspectives from Indian Historiography," *Comparative Studies in Society and History* 32 (1990): 383–408. Others approach the problem practically, observing that the intellectual architects of liberalism, including John Locke and John Stuart Mill, have used it to justify the existence of the colonial state. Uday Mehta gives the clearest account in *Liberalism and Empire: A Study in Nineteenth-Century British Liberal Thought* (Chicago: University of Chicago Press, 1999). In a similar approach, which registers the continuing collusion between liberalism and colonialism, Dipesh Chakrabarty argues that liberalism enabled bourgeois visions of Indian nationalism to override subaltern interests and foreclose the possibility of other visions of community or nationality. See his *Provincializing Europe* (Princeton: Princeton University Press, 2000), especially pages 27–46.
20. Charles Taylor "The Politics of Recognition,", in Amy Gutmann (ed.), *Multiculturalism: Examining the Politics of Recognition* (Princeton: Princeton University Press, 2000), 62.
21. Homi K. Bhabha, "Cultural Choice and the Revision of Freedom," in Austin Sarat and Thomas R. Kearns (eds), *Human Rights: Concepts, Contests, Contingencies* (Ann Arbor: University of Michigan Press, 2001), 46.
22. Ranasinha, 82.
23. Ranasinha, 989. For what it's worth, and although it certainly cannot be taken as a refutation of Ranasinha's claim, Kureishi clarified his thoughts on Islamic fundamentalism in an interview with Amitava Kumar in 2000, calling it "modern," "a completely new phenomenon in the eighties," and "the invention of a brand

new tradition" (in "A Bang and a Whimper: A Conversation with Hanif Kureishi," *Transition: An International Review*, 88:1.4 [2000]: 127). He continues, "This wasn't some ancient tradition. I mean, there are all kinds of liberal ideas in the Muslim tradition, anyway. Pretending that this fundamentalism was the only Islam was definitely a modern thing. A kind of repossession of Islam" (p. 128).

24. Ranasinha, 83. Ranasinha accuses Kureishi of espousing "co-optive liberal pluralism" and claims that he "invents a polarity between Islamic fundamentalism and detached liberal individualism or secularism" (pp. 83, 88). Ranasinha says of *The Black Album*: "Apart from Shahid's family, who scorn religion [...], all the Muslims in this novel are extreme 'fundamentalists'. Kureishi's polarity ignores the range of different forms of Islam that are not extreme or aggressive" (88–9). Yet not three full pages later, Ranasinha readily furnishes evidence from the same novel of plurality within the community of practicing Muslims: according to her, Riaz and his father have fallen out over their differing opinions on religious practice (p. 92). She locates this religious dispute as the model for Farid and Parvez in *My Son the Fanatic*, and then goes on to identify the range of religious positions expressed by the Pakistani Muslims of Bradford. Says Ranasinha, "[the] articulation of dissension within the mosque undoes monolithic conceptions of the religious community" (p. 96). All this leaves one somewhat bewildered as one approaches the conclusion to her chapter and confronts the following statement: "[As] I have argued, Kureishi never explores any forms of Islam that are not 'fundamentalist" ' (p. 100).

25. Hanif Kureishi, *Sammy and Rosie Get Laid: The Script and the Diary* (London: Faber, 1988), 48.

26. Kureishi, *Sammy*, 56–7.

27. Kureishi, *Sammy*, 50.

28. Kureishi, *Sammy*, 30.

29. Edward Said, *Reflections on Exile and Other Literary and Cultural Essays* (London: Granta, 2000), 382.

30. One of the most prominent of such accounts is given by Michael Hardt and Antonio Negri. In *Empire* (Cambridge, MA: Harvard University Press, 2000, 295), Hardt and Negri declare that the emergence of global communication and transportation networks have annihilated the traditional understanding of place and distance.

31. Saskia Sassen, *The Global City: New York, London, Tokyo* (1991; Princeton: Princeton University Press, 2001), 96.

32. Saskia Sassen, "Spatialities and Temporalities of the Global: Elements for a Theorization," *Public Culture*, 12:1 (2000): 227.

33. See John Ball, "The Semi-Detached Metropolis: Hanif Kureishi's London," *Ariel* 27:4 (1996): 7–27, and Sukhdev Sandhu, "Pop Goes the Centre: Hanif Kureishi's London," in Laura Chrisman and Benita Parry (eds), *Postcolonial Theory and Criticism* (Cambridge: D. S. Brewer, 2000), 133–54.

34. Kureishi, "Bradford," 135.

35. Originally an economic community, the European Union has slowly expanded to claim pan-European sovereignty over diverse issues including human rights, trade, defense, and employment law. The palpable threat it poses to the sovereignty of its member nations was demonstrated in Britain recently when Tony Blair's terrorism bill, which would increase police powers to detain suspects without charge, was defeated in Parliament partly because MPs knew that the European Court would strike it down.

36. Geoff Gardner, "Nature of Keeping Awake: Hanif Kureishi and Collaborative Film-Making," *Senses of Cinema: An Online Film Journal Devoted to the Serious and Eclectic Discussion of Cinema*, 10 (2000), 15 Feb. 2006, 21 June 2007: http://www.sensesofcinema.com/contents/00/10/kureishi.html
37. Kureishi, "Son," 126.
38. John Steele, Paul Stokes and Ben Fenton, "They Were Suicide Bombers... and They Were British," *Daily Telegraph*, 13 July 2005, 12 Feb 2006: http://www.tele graph.co.uk/news/main.jhtml?xml=/news/2005/07/13/nbomb13.xml
39. Sandra Laville and Ian Cobain, "From Cricket-Lover Who Enjoyed a Laugh to Terror Suspect," *Guardian*, 13 July 2005, 15 Feb. 2006: http://www.guardian.co.uk/attackonlondon/story/0,16132,1527429,00.html
40. Pnina Werbner, "Divided Loyalties, Empowered Citizenship?" in Michael Waller and Andrew Linklater (eds) *Political Loyalty and the Nation-State* (London: Routledge, 2003) 109.
41. Gilles Kepel, *The Roots of Radical Islam* (London: Saqi, 2005), 19.
42. Tony Blair, "Downing Street Statement Following Terror Attacks in London – 7 July 2005," *10 Downing Street Website*, 7 July 2005, 15 Feb. 2006:http://www.number-10.gov.uk/output/Page7858.asp
43. Tony Blair, "Statement to Parliament on the London Bombings, 11 July 2005," *10 Downing Street Website*, 11 July 2005, 15 Feb. 2006: http://www.number-10.gov.uk/output/Page7903.asp
44. Kureishi, "Word," 9.

5

Decorated Death and the Double Whammy: Attempting to Erase the Excluded through Minstrelsy and Lynching

Barbara Lewis

> *Needless are Niggers.*
>
> Gertrude Stein, *Tender Buttons*[1]

> *Southern whites cannot walk, talk, sing, conceive of laws or justice, think of sex, love, the family, or freedom without responding to the presence of negroes."*
>
> Ralph Ellison, "The World and the Jug"[2]

This chapter begins and ends with consideration of a photograph linking minstrelsy and lynching. Taken in 1900, the photograph features a male corpse, a casualty of lynching. His face has been painted in the manner of a minstrel mask. His identity, whatever it once was, has been lost, redefined, and reconfigured in line with minstrel convention. Blackface minstrelsy was a nineteenth-century performance genre that, according to Robert Toll, "became an instant rage and a national institution virtually overnight."[3] In minstrelsy, males with access to power and possibility in American society, hereinafter called the included, took over and controlled the image of men and women who had been transported to the New World for the exclusive purpose of performing unpaid labor, and thus did not possess social, political, or economic capital, hereinafter called the excluded.[4] The excluded male in this turn-of-the-century photograph was violated and effaced. Lynching has taken away his mobility, breath, and future. Minstrelsy has given him another face. Daily domination imposed under the regime of segregation, which had become national law in 1896, has condemned him to a restricted existence; domination is in and of itself a form of living death.[5] The overemphatic nature of the photograph highlights the need for excessive erasure. Since denials and prohibitions in the routine course of segregated interaction, legally established throughout the country by 1900, precluded the black body from full, inter-active citizenship, what was the point of stressing that elimination unless the impress of the excluded body was so resistant to being expunged that it had to be repeatedly and ritually eradicated?

In 1900, W. E. B. DuBois curated an exhibit for the Paris Universal Expo-sition.[6] In that exhibit, which won accolades, medals, and the Grand Prix, DuBois, who had earned his Harvard doctorate five years earlier, highlighted the advances his people, the excluded, had made since Emancipation. What DuBois showed through photographs, maps, models, charts, graphs, books, and other printed and visual material was that the excluded were committed to learning and industry. They wanted to push forward. The gains in educa-tion had been rapid, such that many of the excluded, at the onset of the twentieth century, could read and write. This was no mean accomplishment for a group that not long before had been legally denied literacy, although individuals from the African Diaspora enslaved in what became the United States had been literate and publishing since the eighteenth century. There was also a significant increase in ownership of property and entrepreneurial activity among the excluded, who only recently had represented wealth and property for others.[7] Poverty and illiteracy still existed, but the excluded were on the move at the dawning of the twentieth century. Their bid for progressive change was not welcome by all.

In the South, the notion that the excluded were making strides cannot be extracted from the lynching mania that picked up and intensified the tendencies with which the Klan became identified. Every year from 1893 to 1904, an average of more than one hundred black bodies were lynched,[8] which means that for over a decade there were at least two to three lynch-ings every week. Not every lynching took place in the South, and not every casualty came from the ranks of the formerly enslaved, but most did. During the spectacle era of lynching, which lasted from the 1890s through 1940, the killing of the excluded and the subsequent partitioning and distribu-tion of the body parts and other associated paraphernalia, such as charred rope and photographs, were often carried out in front of huge audiences and treated as mass entertainment. Regularly in the nineteenth and early twenti-eth centuries, black autonomy and citizenship, meaning political inclusion and advantage, were visually and actually denied in the death grip of lynch-ing and under the mask of minstrelsy.[9] The minstrel legacy that is evident in the photograph points to protracted denigration, which was most virulent in its caricature from 1830 to 1900.[10] Also expressed in the 1900 photo-graph is the utter disposability (and its inverse, the total indispensability) of the black body in white perception, to which the above epigraphic quotes from Gertrude Stein, the literary doyenne of modernism, and Ralph Ellison, African American novelist and critic par excellence, testify.

Both lynching and minstrelsy were capacious in their appeal, draw-ing their clientele predominantly from the white working man of various stations who wanted to feel that he dominated his circumstance, that he was smarter than and lived better than and had more options than the excluded. Families patronized the minstrel show, especially at Christmas time and holidays, seeing it as a special excursion that was sanctioned fun, an occasion for bonding. They saw minstrel fare as clean entertainment that was not scandalous or offensive. The ribaldry and idiosyncrasy of black life was common property. Everyone could take a piece of it and feel instantly superior. Lynching and minstrelsy had their morally murky sides, but hardly anyone took note. Each assigned the black body to the category of profit and property; so did slavery.

Having broached some of the ways in which lynching and minstrelsy intersect, I will describe the 1900 photograph in greater detail, continue the discussion of the links between minstrelsy and lynching, and reveal how the photograph came to public notice, how the narrative it expresses carries significance in the twentieth and twenty-first centuries, how lynching, which is indeed a verified form of performance, persisted past the spectacle era and became a photographic spark in the Civil Rights movement, and how minstrel traces are still current in society. In the concluding part of the chapter, I will again emphasize the connection between lynching and minstrelsy by comparing the 1900 photograph with what we know of a lynching from roughly the same time that was highly stylized. I end with a narrative coda that attempts to give our avuncular forebear his due in history.

Taken over 100 years ago and posed in an alfresco setting, the photograph shows an old man seated in a rocking chair, seeming to nod and take his rest after a long day. His arms hang loosely at his sides, and his eyes are closed. His face, decorated and transformed into a mask, contrasts with the staid, traditional shape and neutral color of his vest and pants. His eyebrows and the bridge of his nose have been exaggerated and redefined with white paint. A moustache, emphasizing the fullness of his mouth, has been drawn on his upper lip. A slash of white demarcates his cheekbone, and a large white circle is visible just above his right jaw, which is turned toward the camera. Tufts of cotton have been affixed to his pate and to the sides of his face. He has been given a mask, which marks him with theatrical presence and significance. It is no ordinary mask, but a minstrel mask of exaggerated opposites, which, when applied to the face of a black man, captures him within a tradition that makes light of and ignores his humanity, featuring him as a clown or buffoon, fractional in stature, not fully human. Conversely, the men of action who have given him another face are additionally empowered. Through the intercession of the mask, they exercise extreme potency. Further, the scenario in which the reduced man is assigned a passive role is performed out in the open, absent any sheltering walls, and he is at risk of further erosion.

His body is slumped, and the old-fashioned spindle chair on which he is sitting is placed against the outside of a building, likely a barn or a shed, located on property that he may own or rent. The strips or planks of lumber out of which the building has been constructed are wide and rough-hewn. This building, serving as a backdrop, expresses function, not ornament. This is a working environment. It is autumn. At the man's feet, are leaves and twigs, brush and debris, evidence of a year's turning. The head of the decorated man, let's call him Uncle, is braced by a long stick, as though he were a puppet. A second man, let's call him Master, is standing just beyond the edge of the photograph. Only partially seen, he holds the stick that keeps Uncle in place. Master's left hand and a bit of shirt-cuff are visible within the frame, but the top half of his body, crowned by the brim of a hat, is seen in shadow against the back wall. Both men are dressed in white long-sleeved shirts, but Uncle's is speckled with blood along the forearm and chest. Unlike Master, Uncle has no hat, an obligatory costume for men of social standing. Perhaps his hat has been knocked off his head, one more indication of the disrespect to which he has been subject. Uncle wears boots, a buttoned vest and matching pants. The jacket to his three-piece suit lies across the chair back, where it cradles Uncle's neck. The reason for the stick or rod and jacket as supportive cushion is simple. Uncle has been lynched, and he cannot sit up on his own. Uncle's image, which establishes the corners and reach of the ensuing discussion, reflects a particular slice of political and cultural history characterized not only by racial stereotype but also by torture and humiliation performed as collective entertainment.

As a persistent American tradition, lynching is as old as the nation. In the revolutionary years, it was customary to punish traitors by lynching them, which did not necessarily mean death but meant making their punishment public. Later, in the aftermath of the Civil War, lynching took on special connotations in the hands of the Ku Klux Klan, which was founded by officers in the Confederate army; it focused its target on the emancipated, and death became more frequent. In the interruption of the quotidian that war initiates, what Agamben calls a state of exception is put in place, and its incidence is now more frequent in the aftermath of 9/11. Under such a regime, it is possible to eliminate "entire categories of citizens who for some reason cannot be integrated into the political system."[11] The campaign of lynching was an initiative of exception, which accepted as routine the ritual annihilation of the excluded. Although it bears the name the Civil War, that fratricidal struggle was by no means civil, at least not in the commonplace meaning of the term of extending courtesy to the many; but rather civil in the sense of describing the battle taking place between two geographical segments of the nation in the throes of determining what kind of civilization lay ahead and who exactly would be the preferred citizen. Uncle did not qualify for inclusion in the pantheon of the privileged.

Images, in addition to "other salient narratives," Wahneema Lubiano posits, "are the means by which sense is made in and of the world; they also provide the means by which those who hold power (or influence the maintenance of power) make or attempt to make sense of the world for others.

Such narratives are so naturalized, so pushed by the momentum of their ubiquity, that they seem to be reality. That dynamic is the work of ideology."[12] Images carry substantial weight, as is evident with Uncle. Along with words, images are the building blocks of language, meaning, narrative, and ideology. Narratives matter; so does ideology. Both enable an individual to sort through the jumble of sounds and impressions that come rushing down the pike moment by moment, at dizzying speed. Words, narrative, and ideology justify behavior, and words are themselves the constituents of "symbolic behavior."[13] Out of a welter of details, they structure meaning and communicate a moral, at the same time that they highlight the values of a given group. They serve as a verbal mask, an identity, for the people whose interests are being upheld, promoted. Words are ordering principles, but often they are discounted as throwaways, just like the maligned, excluded, apparently worthless and consumable people that Gertrude Stein references in the first epigraph to this chapter, which comes from a section of *Tender Buttons* called "Dinner". On a daily basis, we consume what is made familiar to us in narrative.

Sticks and stones, the well-known and often repeated childhood mantra warns, possess a staggering power to inflict pain and harm, the extent of which does not always permit recuperation. In 1900, when Uncle was killed, the excluded were suffering from a campaign of slander in the national media. This continuous mischaracterization was a key factor in crafting the Paris exhibit that DuBois assembled. In 1899, a colleague of DuBois, Thomas J. Calloway, wrote Booker T. Washington at Tuskegee in an effort to obtain funds for the exhibit. The Americans as well as the Europeans, he noted, "think of us as a mass of rapists [because of...] the constant references to us by the press in discouraging remarks."[14] Words have heft, and are capable of creating substance. They can inspire and create; they can also destroy. Lynching was associated with a narrative of criminality, and minstrelsy communicated through visual caricature. It also excelled in the linguistic. Into the mouths of the excluded, it put substandard language, and everyone understood the message. These were substandard people. They were to be avoided. And if they became troublesome, no one would mind if a few were eliminated. Language is potent. "If language can sustain the body," Judith Butler argues, "it can also threaten its existence."[15] Words can be lethal; they can poison and generate the conditions of fatality.

The combination of words and physicality packs a wallop. What is suggested in Uncle's photograph is operatic-scale violence that speaks volumes. In this image, we see a very different post-lynching involvement: not the usual scramble for charred remnants of the sacrifice, but investment in fashioning an icon of idealized, fetishized connection between black and white, expressed in the grammar of minstrelsy. Who could this masked man be? Did he step into or was he occupying a position of some authority? Well-dressed, this anonymous man who was brought low could have been an Othello of his particular locale, who might have attracted the eye of a Desdemona and was able to boast some degree of achievement, which gave him social advantages and made him the target of jealousy among those with the words and the weapons to hurt him. The narrative of lynching was responsive to a tale such as the one Shakespeare crafted about Iago and Othello. Even though there was no truth to the innuendo with which Iago filled Othello's ears, still the tale of infidelity and betrayal that Iago fabricated out of false words became reality for the once noble Moor, who was tricked into crime and then emotionally as well as mortally punished.

Throughout the South in the 1890s and beyond, fears of black men taking sexual liberties with white women were regularly advanced as the reason for lynching. One of the more famous came from Rebecca Latimer Felton, the wife of a southern politician, who claimed a following of her own. In 1897, Mrs. Felton gave a speech in North Carolina in which she said that if a thousand black brutes had to be lynched every week to protect white womanhood, so be it. Felton gave voice to the white resolve to shield the daughters and Desdemonas of the South. On that basis, it is not such a stretch to portray Othello as an example of the treatment awaiting those who cut through the racial fence. Although in Shakespeare Othello married Desdemona, his access to her was perceived as unlawful. He had transgressed, and so he had to die. Othello, turned into Shakespearean burlesque, was often performed in various versions on the minstrel stage, and was seen as text to be cut and tailored to current need. With the repetition and reiteration of a known and canonical story, the plausible and possible became certain.[16]

In Uncle's photographic tableau, which presents an image of order restored,[17] there are two male figures (Iago and Othello), separated by a stick. An emblem of authority and a symbol of distance and hierarchy as well as distaste (the ten-foot pole), the stick enforced the morality, distinctions, and prohibitions of the times. The person holding the stick, and thus the power, is represented only by his hand and the shadow of his hat and head. His image, like his power, is not fully seen; it is hidden, masked, but its traces and effect are evident. The other man, the one who is no longer a man in the living sense, is fully seen. He is on display. He is the trophy. The tension between the living and the dead, between the partially viewed man who is in control and the fully displayed man who is politically and socially flaccid lashes forth from the flat surface. A saga of imposed identity informs this photograph and countless others from the annals of lynching photography. The drama of subordination depicted herein, with always the same message of punished illicit ambition, was popularized in newspaper accounts, bruited about in casual conversation, and plotted out in back rooms and halls of power, thus assuming the force of a community's lingua franca, the trusted and spoken verbal currency of the nation.[18] As it moved from tongue to tongue, the lynching narrative seethed with rumors of guilt that criminalized an excluded and subordinated people and made them more and more vulnerable to attack and humiliation.

What is communicated in this photograph posed and taken over a century ago and others similar in intent and circumstance is a tenuous sense of triumph. For now, the intrusive and beastly Othello has been tamed, pushed back into passivity, and made to pay the wages of defeat and wear the face of the conquered. But who knows what tomorrow may bring, and so the quest for victory and absolute domination will begin again, conducted along a wide, twisting street littered with shattered bones and lives broken by the stick of violence. Any body in its path is prey to the stick that Master is holding, whether that body be young or old, male or female, sick or whole. The stick, which is also the staff of domain, has much work to do, and so it remains unsheathed and ever at the ready. It is the opposite of the carrot; it does not promise reward but punishment. To counter the defections of the excluded, the stick must be wielded often and in a variety of ways. It designates who goes to the right and is deemed acceptable and who goes to the left and is called undeserving. It is also used to start and feed a fire.

In addition, that stick draws a line, a boundary, becoming a medium of separation, somewhat akin, but only partially, to the rod that Moses used to part the Red Sea so that the Israelites might escape from Egypt. In the biblical story, Moses represented the aggrieved, subordinated group, and, through divine intercession, he assumed the rod of power that equipped him with the wherewithal to order the world for the benefit of his chosen people. In the saga that unfolds here, the stick is not controlled by the traditionally powerless. Rather, the stick of oppression is defined as the rod of justice. It cleaves the included, those with power and the prerogative of national belonging from the excluded, the disempowered, who are seen as excessive, expendable bodies. The drama that gets played out in Uncle's photograph, and in hundreds of other lynching photographs of the spectacle era, is situated at a point of transition, when the excluded, previously held in servitude like the tribes of Israel in the land of Pharaoh, sought a clear, unimpeded passage to a better tomorrow after centuries of slavery and trouble. But the crossover they attempted, from bondage to freedom, was policed by the potency of the threatening stick.

What kind of crossover did Uncle attempt? Why was he beaten back into minstrel submission? Was he a good farmer who, through careful husbandry, was able to pay his bills and buy a new suit? Did he educate his sons and daughters to professions? Did he forget to doff his hat to a white passerby? Did he set a high price on his labor? Did he brush up against a white woman in a public space?[19] We need to think about Uncle's identity. We don't really know who he is. He is Anonymous, a cipher, a man of unknown name. His null status provides latitude in naming him. He is both a specific person and a representative figure. He could be called Othello because the little we know about him suggests that he was ambitious and industrious and suffered an untimely death, like the Moorish general.

We could also call him Tom, after the slave beaten to death in Uncle Tom's Cabin, Stowe's famous novel, which also held the American stage in continuous performance from the middle of the nineteenth century through the birth of the twentieth. Whether warranted or not, the name of Tom has gotten confused with obsequiousness, and that doesn't quite fit what we know of Uncle. If he had been willing to bow down and efface himself, he would not have incurred the fate of an early death at the hands of persons unknown. We could also call him Jim Crow, the most famous figure on the nineteenth-century minstrel stage, but Jim Crow signifies segregation, and that might get confusing.[20] So I'm sticking with Uncle, which situates him in a minstrel tradition rife with a legion of uncles, older men who have been visually brought to heel. These various Uncles went by many different names. Uncle was killed over and over again, under different guises, for over 100 years, beginning not just in 1865, when the Klan started, but before that, during slavery, and after that, into the twenty-first century.

The lynching half of our focus revolves around denying autonomy and existence to a person who is represented in public as less than human. In that violent foreclosure, what has been denied redounds to the doer and definer of the violence. This arrogation of power is symbolized in the posthumous phase of lynching during which the sacrificial body is communally shared, pieces broken off and given to members of the crowd as keepsakes. Lynching is total theft of personhood, manifested in a show of power. In the way that it has evolved in America, lynching is exaggerated, enormous, and convoluted, unwieldy. The same is true of minstrelsy in which "white men caricatured blacks for sport and profit" on the minstrel stage.[21] They affected the look, sound, and style of black men and, less often, of black women. Minstrelsy proclaimed a hierarchical order, with blacks at the bottom of the social scale and whites on top. On stage, however, whites performing as black visually made blacks the top layer. Underneath the veneer of laughable blackness, the motive energy was the white body and mind. That relationship of white energy informing the imagery of minstrelsy is manifest in the photograph of Uncle as the plaything of the Master.

It took a while for Uncle to get noticed. One hundred years after his photograph was taken, an exhibit called *Witness: Photographs of Lynching from the Collection of James Allen* opened in a small gallery, the Roth Horowitz, on a side street on the Upper East Side in Manhattan. It was January 2000, for many people the beginning of the new millennium. To coincide with the exhibit opening, an anthology, *Without Sanctuary: Lynching Photography in America*, was published. Like the exhibit, the anthology, featuring Uncle's photograph on the half-title page, included lynching photographs covering an 80-year span from 1880 to 1960. James Allen and his partner, John Spencer Littlefield, had collected, over the course of 15 years, well over 100 such images. Ninety-eight were published in the anthology, and 60 were included in *Witness*. Placing Uncle's photograph as the frontispiece in the book indicated that, while it shared fealty with the other lynching images, it was also different, unique. "This is perhaps the most extreme photographic example extant capturing the costuming of a victim of extra-legal violence," Allen wrote in the anthology. "What white racists were unable to accomplish through intimidation, repressive laws, and social codes – namely, to mold the African American male into the myth of the emasculated 'good ole darkey' – they here accomplished by violence and costuming."[22] Allen does not use the term minstrelsy, but his identification of costuming and caricature as the distinctive characteristics in Uncle's photograph points to its minstrel heritage.

A self-styled picker or dealer in antiquities based in the South, Allen had not set out to collect photographs of lynching. Seeing Leo Frank, the Jewish northerner who was lynched in Georgia in 1915, depicted on a fragile postcard arrested Allen's attention and got him to thinking about the popular representation of a particular brand of American death in which the victim is hounded by the pack. At one point, Allen writes in the Afterword, "a trader pulled me aside and in conspiratorial tones offered to sell me a real photo postcard. It was Laura Nelson hanging from a bridge, caught so pitiful and tattered and beyond retrieving, like a paper kite snagged on a utility wire." In so many of these photographs and postcards, the violence depicted was pornographic, extreme, revealing. In addition to the images, the anthology contained commentary, a quartet of black and white voices, namely those of Allen; Hilton Als, an African American contributor to the *New Yorker*; Congressman John Lewis, a long-time advocate of racial equity; and Leon F. Litwack, the Pulitzer-prize winning historian. At the time, the exhibit was more of an event than the publication of the book. Given the controversy of the various exhibits since 2000, the book has drawn from that renown and become recognized, in its own right, as a trailblazer and the most important volume to date on lynching photography.

The Roth Horowitz Gallery, which could accommodate at most 30 people at a time, was too small to contain the explosion of down-the-street-and-around-the corner interest in lynching photography that an enthusiastic review in the *New York Times* as well as rapid-fire word of mouth encouraged, so the exhibit, renamed *Without Sanctuary*, like the book, moved to the New York Historical Society, where it stayed for eight months, throughout the summer and well into the fall of 2000. At the Horowitz, there had been a sense of intimacy that corresponded, in microcosm, to the type of thronged event that made people push and jostle to see and be seen and to get as near as possible to the body and its immolation. In the lynching space, as captured in photograph after photograph in the exhibit, a crush of bodies crowded together and took pleasure in the reduction of intervening distance. The faces of the crowd looking up at the camera were avid with excitement. They were now one people brought together with a specific goal in mind. Their senses could be fulfilled. They were conspirators in a political theater that privileged the eye, nose, ear. They could see the writhing body and smell it burning; they could hear the agony. The closeness, the oneness, was a large part of the reward, coming together and being united in collective power.

At the Horowitz, we could not escape the nearness of the photographs. We had to engage with them, look full in the faces of these men (and a few women) who had been made to suffer, their bodies tortured, scourged, damaged. One photograph broke with pattern. It was of a white woman named Ella Watson. Unlike the others whose bodies were on display, Watson was photographically spared the indignity of exposure during her final moments. Watson, who was lynched in Wyoming in 1892 during the scramble and fight for land known as the Johnson County Wars, was not shown in the posture of death. A ballad that vowed revenge on her killers was placed next to her photograph. The disparity in representation between black and white, between male and female was stark. True, the corpse of Laura Nelson, a black woman, hanging in midair from a bridge in Oklahoma, was clothed. Neither she nor Ella Watson was stripped like so many of the black men. However, the onlookers, including women and children, collected end to end at the top of the bridge, came not to identify with Nelson and her dead son but to stare at the oddity of their plight, mother and son dangling at the end of the line. Unlike Watson, Nelson was not seen as a beloved member of the community, someone worthy of being memorialized in verse, someone whose death was to be avenged.

This lack of empathy and identification between black and white, which to some extent, on a lower frequency (to quote Ellison) still obtains today, calls into question the possibility of witness, which implies that the person who is viewing the outrage that happens is willing to risk the repercussions of stepping out from the corpus of the crowd and speaking up on behalf of the victim. Perhaps the greater fascination for the viewing public, even at the Horowitz, with the white faces in the crowd rather than with the black suffering bodies indicated the difficulty of sympathy across the distance of race. All we know is that, when it became clear that the show of lynching photography was a hit and needed new quarters, the name changed. Instead of *Witness*, the exhibit became known as *Without Sanctuary*, the title of the anthology, which expresses the fugitive status of African Americans.

A suit for justice that was brought shortly after the Civil War illustrates the fragility of witness for the excluded. In Kentucky in 1868, two Klansmen, Bylew and Kennard, went into a cabin late one August evening and using a broad ax, killed three people who belonged to the same black family. A fourth member of the family, Richard Foster, who was 17 years old, was badly wounded. He survived for two days, but before he died he identified his killers. Laura, a younger sister, hid in a trundle bed, escaping detection.

When the two Klan members came to trial, Laura corroborated what her brother had said. Under the Black Code in many southern states, including Kentucky, blacks could not testify against whites. So the lawyers for the family took the case to federal jurisdiction, and argued that Lucy Armstrong, the 90-year-old blind grandmother who had been killed, was not covered by the Black Code and so qualified as a citizen. Thus, on her behalf, suit could be brought.

Bylew and Kennard won on a technicality and were set free. On appeal, the case went all the way to the Supreme Court, where the earlier ruling was upheld. The justices determined that Lucy Armstrong may have been a citizen, but she lost that status with her death. Even though there were witnesses, they were black and could not be credited. For the excluded, the status of citizen was limited and the status of witness had no real weight; in the end, the law offered no sanctuary, no place of belonging or protection. What is made clear in this judgment is the separate treatment of the included versus the excluded. Law shelters the included, giving them sanctuary. They are untouchable, exempted, situated beyond the sphere of liability. The excluded, on the other hand, can find no shelter under the law. They can be murdered in their beds. Like Uncle, they can be turned into a cautionary spectacle. Wherever they are, they can be injured by any and all, without defense, subject to any interference, even to the point of fatality.[23]

The person who has no sanctuary or haven is isolated and friendless, an outsider to community, dropped off a tree or a bridge and left to hang in the air. Like the bandit that Agamben writes about in *Homo Sacer*,[24] blacks in the aftermath of the Civil War had no legal roof or umbrella under which to stand. To be without sanctuary is to be geographically and legally vulnerable, to be consigned to the ranks of displaced and scattered peoples, without access to a zone of acceptance; it is also to be positioned as prey, without support, animalized and shorn of possibilities, capable of being axed in the middle of the night with no penalty for the hand that brought death. "Witness" suggests sympathy and connection; "Without Sanctuary" connotes dispassion and distance. Witness, a legal term, had little force in the lives of the excluded. What was gained in the move from the Horowitz to the New York Historical Society and in the change from *Witness* to *Without Sanctuary* was increased spectator capacity and wider coverage of the subject. The question remains, in the reordered perspective that emphasized restricted geography rather than the human bond across the boundary of race, was the emotional impact of the show still strong?

In spotlighting the subject of lynching, which was often dismissed as an unimportant area in the American past, *Without Sanctuary* stirred up a wide range of responses, including a refusal to attend. Many of those who did visit were aghast that some of these images were postcards. For years, until 1906, these postcards were sent through the mail. They were social currency, connecting friends and family at a distance, as several written notes on the back of the postcards reproduced in the anthology attest. Even when they could no longer be posted, these portrait postcards were bought and sold, traded and exchanged. They served as vivid reminders of communal festivals that drew hundreds and thousands together in conviviality. That the federal government once gave a nod to or at the very least for years ignored the barbarity of the practice was astonishing to many exhibit goers. It was hard to believe that activities like these had been implicitly condoned in the United States. Anger was a frequent response. So was consternation. How could this have happened, people wanted to know, week after week. Some people walked through the halls of the exhibit with tears in their eyes, others stared stony-faced, disbelieving, but somehow enraptured by the mystery of it all. The spectacle of the denigrated black body, scourged and purged, still riveted the eye.

In its day, minstrelsy elicited strong reactions from its audiences as well. It involved them, made them feel, made them laugh and cry and throw things when they were disappointed and angry. Like minstrelsy, lynching was not a cool medium; it generated the heat of emotion. In the book, Hilton Als, a regular commentator for the *New Yorker*, admitted that his reaction to the exhibit was visceral. He was a black man, and all he could do to protect himself from being overcome by rage and shame was to conjure up a barrier of distance. He needed a stick of his own, something to ward off the demons. He was far from alone. Docents were hired to hold regular sessions to help viewers of the exhibit express and come to grips with the intensity of their reactions.

That lynching, mass festivals of public togetherness and group definition, can indeed be categorized as performance has been established and explored by Kirk Fuoss, who writes: "extralegal public executions conform to contemporary construals of performance" and were understood and observed as such by "contemporaries of the phenomena." H. L. Mencken, an early twentieth-century master of the satiric, assessed lynching as a purveyor of excitement that was akin to the stage show, the circus, and fairground, pleasing young and old. Not only participants in the lynching but also journalists, who on occasion overlapped, subscribed to the notion that lynching was entertainment. Two of the many theatrically tinged quotes that Fuoss has found come from the *New York Times*, the first in 1875 and the second ten years later: "All of the actors in the midnight tragedy of one act had started for their rural homes, leaving their victim hanging to the court-house tree." Referencing a lynching that was staged in the Midwest in 1885, the reporter for the *Times* commented on the change in public approval that brought lynching into the open: "A lynching in broad daylight is a feat of Kansas enterprise. Missouri in her palmy old days never undertook a matinee performance of the lynching act." For me, the most chilling quote connecting performance and lynching comes from a man who was thrilled by what he saw and heard during the lynching of a woman. "It was, he said, 'the best show I ever did see, Mister – you oughter seen that nigger wench fight and heard her howl when we strung her up.' "[25]

What occurs in both minstrelsy and lynching is the performance of the power and pleasure, the privilege of whiteness achieved through the auspices of an imposed and denigrated, despised, and disposable blackness. In *Scenes of Subjection*, Saidiya Hartmann re-evaluates "performance in terms of the claims made against power."[26] Here, in this chapter, I am concerned with how blackness is made the ground of whiteness, how whiteness is constituted from the elements of blackness that are taken in and absorbed ritually in the ceremonies of lynching and minstrelsy. In the 1900 photograph, Uncle occupies the foreground. Part within the frame and part outside it, Master defines Uncle to buttress his significance, to give him not only integrity and wholeness but also magnitude. Without Uncle, Master is not master. In both minstrelsy and lynching, the included draw their sense of self-worth from the excluded. The included can do whatever they choose. The excluded can do only what they are allowed. The included are virtual gods. The excluded are underlings. In competition with black, white must always win and prevail; that is the minstrel contract.

From Judith Butler, I adapt the notion of a minstrel contract. Butler writes about a bondsman's contract in which the bondsman becomes the substitute, or mimetic double, of the lord or master for the purposes of productive labor. The slave becomes an extension of the master's body. Master absorbs Uncle's energy. In the minstrel contract, then, the master becomes the mimetic extension of the slave, and the slave becomes the mimetic double of the master, consigned to perform what the master dictates. In taking on a second body, the master's body is made greater; he is aggrandized, becoming not only himself but other, thereby erasing the need for the independence of the other, and able to achieve more than is possible for one human. In slavery, the terrain that minstrelsy expresses, the slave loses control over his body. In minstrelsy, the imitated, fixed and circumscribed loses control to the imitator, who is active. Minstrelsy expresses the relationship that obtains between the imitative controlling body and the imitated controlled body, and that is the eclipsing of and eradication of the controlled body by the sovereign or master body.[27] The minstrel contract is fundamental to the American compact. Minstrelsy is the good guard enforcing its promise, using comedy and mockery to advantage. Lynching is the mean and marauding guard, bringing out the full force of the huge and punishing weapon that collective command can wield. To demonstrate its aggregated weight, one or more lives might be sacrificed at the same time or within rapid daily sequence of one another, to show the dire cost of deviance and dissent.

Minstrelsy designates the social scapegoat who must pay the price of suffering, thus allowing the included to distance and deny fallibility, culpability. Failings are projected onto the excluded, who are inferior; the included are superior. The excluded are pariahs, sitting, like Uncle, outside the civilized world. As such, they are entitled to be the occasion of the joke. The pointed, mocking insult of minstrelsy belittles not only males but also females, who were featured in wench performances. The excluded female enjoyed no protection. During slavery, she was beaten, sometimes to the point of extinction, and made to suffer other violent onslaughts, including the repeated sexual occupation of her body. After slavery, she was not exempt from the predations of lynching. As a group figure, the excluded female assumes increased significance, since the male is often neutered and denied masculine expression. The persona of the female thus takes on greater connotation, often encompassing the status and range of the entire group. In lynching, it sometimes happened that the woman was taken as a replacement figure for a man. This substitution of female for male underlines their perceived equivalence in the eyes of the included, their utter fungibility; one black body will do as well as another; what matters is that the body is black and excluded, and thus guilty.

"After a barn burning near Columbus, Mississippi," Litwack writes in "Hellhounds," "suspicion fell on the son of Cordelia Stevenson. Unable to locate him, a mob of whites settled on his mother, seized and tortured her, and left her naked body hanging from the limb of a tree for public viewing."[28] Nothing shows the divide between the included and the excluded as does the treatment of women. Included women are privileged; their death or defilement is cause for revenge. Excluded women are reviled; there is no limit to the rough treatment they can incur; their gender often earns them no dispensation. On occasion, the excluded woman was lynched along with her child, evidence that the animus underlying lynching did not stop with the present but also targeted the future. Whether the culprit who is ritually punished in public happens to be male or female, the body of the offender is excised from the community. Ouster, symbolic or actual, is the object of the public performance of lynching as well as of minstrelsy. Neither gender nor age, as we see in the case of Uncle, protects anyone in the camp of the excluded from the menace of the stick, whether it is the slapstick of minstrel comedy or the spirit-breaking stick in the hand of the lynchmaster.

By the 1890s, minstrelsy, slavery's handmaiden, was on the downswing, although minstrel performances persist as late as 1920, and, in amateur versions, fun in blackface was had, primarily in town venues, through the 1960s and beyond. Clearly, minstrelsy did not die with the nineteenth century.[29] Rather than submit to the threat of extinction, it simply reconfigured itself, first in vaudeville, then in film and television.[30] On amateur stages, including those associated with lodges, fraternities, roasts, and community parades, and on college and high school campuses, blackface performances were performed throughout the twentieth century and into the twenty-first. Some of these blackface performances included lynching scenarios as well, an indication that minstrelsy and lynching are yoked in the public imagination.

One of the most famous references to lynching in the latter part of the twentieth century came from Clarence Thomas, who was then being confirmed as a Supreme Court Justice. The impact of his charge that he was being subjected to a "high-tech lynching" pulled the covers off a dirty little secret, that lynching had not yet been totally put to bed in the American mind.[31] Neither had minstrelsy. As recently as 2004, fraternity brothers at Georgia State University went to a hip-hop party with their faces slathered in blackface. Just before the last century ended, students at Auburn in Alabama staged a lynching scene with the alleged culprit wearing blackface; photographs of their "black fun" were put on the web. A few years before that, some New York policemen and firemen in cork re-enacted the 1998 lynching death of James Byrd in Texas as they cavorted in blackface on a float piled high with fried chicken and watermelon. Clearly, for these paid upholders of social order the sign of blackface linked violence and victimization, a lesson that some in the younger generation have also absorbed,[32] most likely through the media, which has inherited the nineteenth-century currency of the minstrel show.

On popular stages blackface minstrelsy makes an occasional appearance, most prominently at the Wooster Group in New York, which reprised its controversial, late twentieth-century cross-gender blackface version of *The Emperor Jones* by Eugene O'Neill in March 2006. In the 1980s, the Wooster Group, a leading avant-garde New York theatrical troupe, landed in funding hot water when it used blackface in *Route 1&9*, which paired Thornton Wilder's *Our Town* with routines from Pigmeat Markam, a black comic with a vaudevillian persona who performed on Ed Sullivan and was popular in the 1950s and 1960s. Kate Valk, the group's main actress found the blackface mask not only releasing but enormous fun, an experience not unlike the license that the minstrel mask lent white performers in the nineteenth century. Elizabeth LeCompte, administrative director, admitted there was something pornographic about it, something immediately communicative and provocative. When the Wooster Group, out of which Willem Dafoe emerged and with which Spalding Gray was associated, produced Eugene O'Neill's *The Hairy Ape*, which they took to Broadway in 1997, they brought the blackface along, but they called it sootface,[33] believing, perhaps, that another name would soften the connotation. Sootface, though, also connects minstrelsy to earlier European roots and the reviling of the dirty chimney sweep. In 2000, Spike Lee, the nation's leading African American director, released *Bamboozled*, which questioned current media fascination with blackface. What the minstrel form, and its more active cognate, lynching, represent, the subordination and excising of the other, still enjoys circulation in the world of the contemporary.

The modern iteration of lynching, its marketing and promotion and its huge crowds, was helped immensely by the camera. The narrative of lynching, essentially a moral tale in which the punishment of hell was meted out on earth, demanded authentic illustration. That was photography's contribution. Further, the lynching photograph, often made into a postcard, allowed those at a distance to have a virtual hand in the killing. Some of the photographers who captured the lynching scene had reputations and worked on consignments. Gildersleeve was the photographer who got the call to record the lynching of an 18-year-old Jesse Washington, which took place in the business and administrative district of Waco, Texas in 1916. Right outside the mayor's office window and in front of 15,000 people, Washington was incinerated into a charred scarecrow, a literal stump of a man. Gildersleeve, who cranked out the desired images on the spot, sold the postcards he made for "ten cents apiece to those unfortunate enough to have missed acquiring their own portion of Washington's body," Grace Elizabeth Hale writes. "Ten cents, after all, was significantly cheaper than the five dollars that by day's end Washington's teeth were reportedly fetching and less even than the links of the chain that were trading for a quarter."[34] The lynching photograph served as a reminder for those who had been there, but it also insured that the lynching was viewed by some who were not in attendance; it was passed from hand to hand and eye to eye and extended the impact of the event beyond the immediate moment.

The year before Jesse Washington was lynched, *Birth of a Nation* (1915), directed by D. W. Griffith, was released and became the first American blockbuster. The film, which was based on *The Clansman* (1905), a novel by Thomas Dixon that started its creative life as a stage play, retold the narrative of lynching, complete with castration, and culminated in the triumphant ride of the Ku Klux Klan. The graphic excitement of photography was magnified in film. No less than Woodrow Wilson, the president, and Edward White, Chief Justice of the Supreme Court, were impressed with how admirably the story of the Klan was told. White even admitted his earlier membership in the Klan, that "uprising of outraged manhood," as he termed it.[35] The American saga of lynching appealed to all classes of Americans, from highest to lowest, and was relayed in the press, in photography, on stage, in the novel, and on the wide screen.

After World War II, the story of lynching lay quiescent. Then, in 1955, Emmett Till, a teenager, left his home in Chicago to visit family in Mississippi for the summer. Out with some friends, he made a purchase in a small country store. Later, it was alleged that he flirted with the white female clerk. For that unsubstantiated breach of southern decorum, he was pummeled into disfigurement, saddled with a cotton gin and sent to the bottom of the river. Uncle's face was impacted by addition, and so was Till's; Uncle was given a second face, and Till's face was so swollen that it seemed as if his head had been doubled. When Till's body was discovered and then shipped back to Chicago for burial, the coffin was nailed shut, with implicit instructions to keep it closed. Instead, his mother, who was determined to have justice, ordered the coffin opened, and said that she would not be complicit in hiding any southern sins. The horrific face of her beloved and only son would be seen, and the example of his suffering would serve notice that the southern brand of racial retaliation was savage in the extreme. The media, particularly the black press, printed the images and carried the story of the outrage that the excluded expressed with their feet. In the *Washington Afro-American*, a banner headline proclaimed "10,000 View Dixie Lynch Victim's Body." The *New York Amsterdam News* reported "50,000 Line Chicago Streets to look at Lynch Victim." The *Pittsburgh Courier* suggested an even more impressive figure as a four-inch headline exclaimed, "100,000 View Battered Body of Lynch Victim."[36] The excluded were drawn together and united, made puissant, through a spectacle of criminality that was external to the group. The photograph of a bludgeoned Till, and the sight of the inherited debt that he had assumed galvanized the Civil Rights Movement. His was the face that launched a thousand protests.

Many African American parents did not want to tell their children of the horrors they had seen and endured under the lynching regime. Emotionally and physically, they wanted distance from a painful memory. Lynching was not a topic of regular or easy conversation in the black home. In the white community, however, lynching was easily discussed within the family. Children were visible in many of the crowds. Some were there strictly as observers. Others took on roles in the lynching event, for example, helping to gather fuel for the bonfire. Often, their parents encouraged their presence so that they could learn the ways of the included vis-à-vis the excluded. In his study of lynching, originally published in 1929, Walter White, an executive at the National Association for the Advancement of Colored People (NAACP), who looked white – given his blond hair and blue eyes – recounts going to a town in Florida where several white children thronged around him, thinking he was one of them. A little girl of ten, somewhat older than the others, was the ringleader. Let's call her Rosie. Led by Rosie, the eager, fresh-faced children offered to take him to the place where the niggers, five of them, had been lynched recently. "Animatedly," White writes, "almost as joyously as though the memory were of Christmas morning or the circus, she told me, her slightly younger companions interjecting a word here and there or nodding vigorous assent, of 'the fun we had burning the niggers.'"[37] The excitement they had enjoyed bubbled up irrepressibly into their words and memories. Lynching had left its impact, and its trace was pleasurable and strong, vital enough to last a lifetime; vital enough to motivate the keeping of a photographic reminder of a lynching in the family album for years and years where it could be periodically fingered in an evocation of nostalgic memory. And years later, it could be taken out and circulated, marketed, and made part of a collection that would again achieve public presence and international stature.

For whites, more often than not, the lynching scene evoked feelings of camaraderie, perhaps even reverence and awe. Together, they were doing the work of nation, assembling around a core set of values that required allegiance and served to separate insiders from outsiders. Together, they were policing their society, insuring its safety from threat. Together, they were banishing the enemy. Together, they were defending their borders, engaging in cohesion building by dismantling an alien body. Together, they tasted the power to undo and redo, to create and destroy. Perhaps there are still a few people alive who, as children, witnessed a spectacle lynching in the service of nation, but even if so, those numbers are quickly dwindling in the twenty-first century. How do we feel about this ritual of our past? Are we revolted? Are we intrigued? What can we learn? In restoring lynching to wide public attention, was there a paradigm shift? Or was it merely that enough time had passed so that we could look back with sufficient dispassion to begin to unravel the tangled skein of emotion attendant on lynchings like that of Jesse Washington in Waco, which transpired in the heart of a southern city and brought so many of the included together, as though in rapt observance of a service of significance.

In the photograph of Uncle, which was taken as an extension of the brutality directed against him, someone has played fast and loose with him, turned him into a cruel joke. Even in death, he is consigned to the role of providing amusement. Fingers and hands that did not honor him have touched him, transformed him, messed with his face and hair, rearranging them and making him a disfigured and seated buffoon, stymied in time. Uncle seems to be present, but only the hull remains. He is a shell of a man, a husk to be filled by an external agency that literally has put designs on him.

Beyond the cruelty and the evacuation of life, there is another reason why the photograph of Uncle matters. This image is exceptional. In its expression of the tie between minstrelsy and lynching, it is unlike the other lynching photographs in *Without Sanctuary*. Uncle's body has not been stripped. Whatever torture he may have suffered is not blaringly legible in the photograph. His body is intact, not butchered and roasted. No crowds throng the victim; he is virtually alone as though in a space of tranquility. He is exhibited in an attitude of repose, as though he merely slipped away from the world in slumber. Uncle's exceptionality provides a keyhole into a view of lynching that is usually concealed: that it is not just destructive but creative as well, and what is created is an effigy or trace of the power of whiteness. The artifact of Uncle's image, which sits at the crossroads between centuries, ties theatre to tableau to photography.[38] Situated in the space of meeting between life and death, Uncle's photograph recalls the timeless and also expresses the modern and the postmodern; it is simultaneously old and new.

To contextualize Uncle's death in the tradition of its time, Uncle's lynching is contrasted with that of Henry Smith, who was lynched seven years prior in Paris, Texas. Smith's execution, according to Grace Elizabeth Hale, was the "first blatantly public, actively promoted lynching of a southern black by a large crowd of southern whites," and it "modernized and made more powerful the loosely organized, more spontaneous practice of lynching that had previously prevailed."[39] Uncle was killed in an era of evolved practice. Like Uncle, Smith was costumed and mocked as part of the lynching ceremony. To accentuate his ignominy, Smith was dressed in regal robes and led to death in a stately horse-drawn carriage. The minstrel significance of that mockery and of the mask superimposed on Uncle's face post-mortem is inescapable. Through disguise, Uncle and Smith are both made to assume another identity, consistent with minstrelsy.

Questions persist. What happened to Uncle after the photographer left? Was his body pumped full of bullets, like a target in a side show? Was he burned? Was he dragged on the back of some vehicle through the black neighborhood as an example of the treatment that misbehaving could incur? Did they call his relatives and tell them to collect the remains before sundown?[40] Did some local farmer need a scarecrow? All we know from this image is that, momentarily, at least, Uncle is alone, or almost alone. Master is there with him, partially visible but still propping him up and casting him into a nostalgic finale, which caricatured or masked the person that he was or wanted to be. For whites, the mask, which they called upon regularly in minstrelsy and sometimes invoked for their lynching activities, meant freedom and latitude; for blacks, the mask meant buffoonery and limitation. In the modern era of lynching, however, the mask was no longer materially necessary for whites, although it continued to mark blacks; being marked and made inferior was their assigned status, which, as in the case of Uncle, followed them through the doors of death.

According to theorists such as Peirce, Sontag, and Barthes, the photograph is not simply a mask; it bears an uncanny, almost indistinguishable connection to the real; it is an implicated but mute witness. Still, in some instances, as is true with Uncle, it provides the only documentation we have of what happened long ago, in the decades when spectacle lynching reigned. If the photograph is a mask at all, it is akin to the death mask.[41] In fact, the idea of a death mask is appropriate to Uncle's situation. Except that, instead of taking the impress of his actual features in the climactic moment of his life, the mask that has been applied to Uncle's face denies his personal story. It falsifies his history, wrenches it from one tradition of masking, the African, which connected him to a heritage that nourished him, and places it within another tradition that disregarded him, his ancestors, and descendants. It erases his lineage, significance, and autonomy as a person, his right to his bodily and spiritual self. Violence has taken away his legitimacy and authenticity and all he is left with is a name that is not his own. Photography, the avant-garde medium of the time, took the image of black humiliation into the pages of the nation's press, into the leaves of the family album, through the mails, into the home, and into the stores, in all of which places it could be consumed and become part of daily life, accepted as given.

What would never be accepted, at least not as long as lynching kept its iron grip on the American mind and emotions, was the transformation of the black body from property to person, its release of the mask that denied it standing within the familial group of its own blood. Lynching was intended to insure that the black body remained static in status, and that it did not make the leap "from chattel to citizenry."[42] That, like a photograph, it remain framed within and by a specific time. That Othello, however much service he had rendered to the state, was precluded from the flesh and favors of Desdemona. That Uncle, and all his relatives, would remain visually captive inside the corners of a faded, closeted photograph, extracted from the flow of history, virtually hanging from the limb of time, incomplete, subject to a grand narrative that misinterpreted, belittled, dispossessed, and tricked the body of the other. Expressed theatrically, any discrepancy between a higher and lower station would be met with the sure and violent offices of the stick. The excluded body would be beaten into submission and be ever out of step, hobbled so as to impede the pace of success in the contemporary world. Defectors would be taken out to the woods and shot, posed in the guise of a nostalgic ideal of subordination and dependence.

For Uncle, and the other black and excluded men and women, intimidated by the knowledge that ambition marked them as fodder for lynching without provocation, just on the basis of skin and hearsay, the future became territory into which the few rather than the many chose to venture. But for those few taking the challenge, and they did, in time, become many, violence and the campaign of terror to wrest away control of every aspect of their lives, waking and sleeping, living and dead, left them with no alternative but to make a way out of the desert of no way. And they did. They banded together and fought back with whatever they could manage, and some of their skills and talents became honed. As the exhibit DuBois presented at the Paris Exposition made clear, by 1900, when Uncle was killed, the excluded had long been mounting efforts to be included, on the basis of accomplishment. With more than 200 books to their credit, a number of them had mastered the word. And they knew story. They had their feet and their families and their belief in deliverance. And they found new locations for their presence. Some, but not all were caught in the land of pain, back on the terrain of minstrelsy, where they were expected to adhere to the minstrel contract of no black representation except what the master group chose to allot.

Both lynching and minstrelsy attempted to control and legislate the bordered space that blacks could fill in the public sphere; they were visible only on certain occasions that portrayed them at a disadvantage. That was the history out of which they came, and it was the history that kept beckoning them back. But the past is not past. It is still with us, even in the twenty-first century. The patterns of interaction established within the enduring minstrel contract ground our behavior today, even if we don't always recognize that its force is present. We are stuck on the same narrative track. As Susan Mizruchi writes in *The Science of Sacrifice*: "This is because postmodern America – the problems that it faces, the solutions that it imagines – remains, in some fundamental sense, a product of the late nineteenth century. It is also because national thinking on race, in particular, has been narrow and circular."[43] By no means have we come as far as we would like to believe.

Laura Nelson and her son, Henry Smith, and Uncle form our representative triptych from the host of men and women, known and unknown, who have been sacrificed on the scaffold of the American nation. They and their counterparts in the twentieth century, for example, Emmett Till and Jasper Byrd, were not only apprehended and lynched, mutilated, and murdered by the giant stick of bias, but also maligned by words, incendiary and rumor-filled. In death, they acceded to the immortal. In death, they escaped the ignominy that had dogged their heels. As Ellison writes in the essay from which the second epigraph is taken, the artificially divided world, the borders of which lynching and minstrelsy maintained, was never absolute. This inability, this impotence, to fully police the gate, accounts, to a great extent, for the exaggeration of minstrelsy and the grotesque overkill in lynching. There were spaces that the excluded crafted in which they could exercise autonomy and pursue their own forward movement, particularly in the realm of culture. The signature that the excluded have written across the American landscape has persisted despite lynching, despite minstrelsy, despite all efforts to deny political presence and entitlements to the excluded. Uncle can be viewed as the patriarch, the elder of the tribe. He may have been killed, but he was not obliterated. The image of his endurance is etched on our eye, whether or not we choose to bear witness. In homage to the craftiness that took him to the brink, let me end with a refrain that, in deference, I would etch on a memorial to Uncle (and all the uncles and aunts, grandmothers, sisters and brothers, mothers and fathers, whose death was precipitated and mocked and unavenged, lost in the leaves of history):

> Sticks and stones
> Splintered my bones
> And story
> Framed and harmed me.
>
> Nevertheless
> I must confess
> Not even
> History or
> Memory
> Can erase me.

Notes

1. Gertrude Stein, *Tender Buttons: Objects, Food, Rooms* (1914; New York: Haskell House, 1970).
2. Ralph Ellison, "The World and the Jug," *New Leader*, 46 (1963): 22–6.
3. Robert Toll, *Blacking Up: The Minstrel Show in Nineteenth-Century America* (London: Oxford University Press, 1979), 21.
4. "The fundamental categorical pair of Western politics is [...] that of [...] exclusion/inclusion." Giorgio Agamben, *Homo Sacer: Sovereign Power and Bare Life* (Stanford, CA: Stanford University Press, 1998), 8.
5. Judith Butler, *The Psychic Life of Power: Theories in Subjection* (Stanford, CA: Stanford University Press, 1997), 41.
6. Rebecka Rutledge Fisher, "Cultural Artifacts and the Narrative of History: W.E.B. DuBois and the Exhibiting of Culture at the 1900 Paris Exposition Universelle," *Modern Fiction Studies*, 51:4 (2005): 741.
7. Two examples demonstrate how entrepreneurship and prosperity within the ranks of the excluded were sometimes met. In 1892, Thomas Moss, Calvin McDowell, and Henry Stewart were co-owners of a grocery store in Memphis, Tennessee, that competed for business with a nearby white-owned business. When plainclothes policemen came to their store unannounced, the proprietors defended their property. For their trouble, the three were taken to the outskirts of town and shot. Ida B. Wells, a journalist who was godmother to Moss's daughter, investigated the murder of these three black businessmen, and was herself threatened with lynching if she dared return to Memphis. See James Allen (ed.), *Without Sanctuary: Lynching Photography in America* (Santa Fe: Twin Palms, 2000), 28.
8. Thomas Gossett, *Race: The History of an Idea in America* (New York: Oxford University Press, 1997), 269.
9. My interpretation of minstrelsy shares with Michael Rogin a belief that blackface banished blacks from self-representation, a banishment that approximated social and political death. I stand with Susan Gubar in thinking that minstrelsy and lynching are twins, with one, lynching, given to much greater and more overt extremes than the other. Scholars such as Dale Cockrell, W. T. Lhamon, and Eric Lott do not see minstrelsy in such dire terms. Led by Eric Lott, who resuscitated academic interest in the field, they promote the idea that white minstrel performers were not necessarily malign in their intentions and impact. The motivation of whites to imitate blacks, they argue, included a sense of celebration and appreciation of black difference as well as a desire to possess and replace black presence. Lhamon shows the reach of minstrelsy into today, seeing its repercussions and persistence in the world of hip hop. See Michael Rogin, *Blackface, White Noise: Jewish Immigrants in the Hollywood Melting Pot* (Berkeley: University of California Press, 1996); Susan Gubar, *Racechanges: White Skin, Black Face in American Culture* (New York: Oxford University Press, 1997); Dale Cockrell, *Demons of Disorder: Early Blackface Minstrels and their World* (Cambridge: Cambridge University Press, 1997); W. T. Lhamon, Jr., *Raising Cain: Blackface Performance from Jim Crow to Hip Hop* (Cambridge, MA: Harvard University Press, 1998); and Eric Lott, *Love and Theft: Blackface Minstrelsy and the American Working Class* (New York: Oxford University Press, 1993).
10. David R. Roediger, *The Wages of Whiteness: Race and the Making of the American Working Class* (London: Verso, 1991), 123.

11. Giorgio Agamben, *State of Exception* (Chicago: University of Chicago Press, 2003), 2.
12. Wahneema Lubiano, "Black Ladies, Welfare Queens and State Minstrels: Ideological War by Narrative Means," in Toni Morrison (ed.), *Race-ing Justice, En-Gendering Power: Essays on Anita Hill, Clarence Thomas, and the Construction of Social Reality* (New York: Pantheon, 1992) 324.
13. Hortense J. Spillers, *Black, White, and in Color: Essays on American Literature and Culture* (Chicago: Chicago University Press, 2003), 91.
14. Fisher, 753.
15. Judith Butler, *Excitable Speech: A Politics of the Performative* (New York: Routledge, 1997), 5.
16. "To say the same thing over and over again not only induces [...] conviction, but creates a fellowship of belief. The ritual narrative in this way bears an element of infection; everyone is compelled toward the same story" (Spillers, 264).
17. The tableau, which became popular in the nineteenth-century theatre, borrowed the older and more traditional legacy and cachet of the world of established fine art and brought it into the theatre. In this tableau, which is not *tableau vivant* but rather *tableau mort*, the borrowing comes from the minstrel stage and is absorbed by the newer and more modern medium of photography. Bruce McConachie writes about the significance of the tableau, how it visually communicated a moral message, in *Melodramatic Formations: American Theatre and Society, 1820–1870* (Iowa City: University of Iowa Press, 1992).
18. "[T]he media, along with other public and private entities (including institutions, churches, schools, families, and civic organizations, among others), constantly make available particular narratives and not others. In turn, such consistently reinforced presences reproduce the world in particular ways: what we see becomes what we "get," what we believe" (Lubiano, 329–30).
19. The questions that I ask here reflect some of the excuses given for lynching the excluded. Litwack in "Hellhounds," *Without Sanctuary: Lynching Photography in America*, ed. Allen (Santa Fe: Twin Palms, 2000), 24–5, provides a partial catalogue of some of the provocations for lynching: "Many of the transgressions by blacks would have been regarded as relatively trivial if committed by whites and were not grounds anywhere else for capital punishment: using disrespectful, insulting, slanderous, boastful, threatening, or '*incendiary*' language; insubordination, impertinence, or improper demeanor (a sarcastic grin, laughing at the wrong time, a prolonged silence); refusing to take off one's hat to a white person or to give the right-of-way (to step aside) when encountering a white on the sidewalk; resisting assault by whites; [...] testifying or bringing suit against a white person; being related to a person accused of a crime and already lynched; political activities; union organizing, conjuring; discussing a lynching; gambling; [...] refusing to accept an employment offer, [...] refusing to give up one's farm; conspicuously displaying one's wealth or property; and (in the eyes of whites) trying to act like a white man."
20. "That the legal doctrine of separate but equal took its name from minstrelsy may or may not intimate public awareness about segregation as a farcical form of equality. However, from today's perspective, that tenet enacts a kind of unconscious repetition of the minstrel's insulting portrait of the less-than-equal darky" (Gubar, 78).
21. Lott, 3.
22. Allen, 165.

23. Russ Castronovo, *Necro Citizenship: Death, Eroticism, and the Public Sphere in the Nineteenth-Century United States* (Durham, NC: Duke University Press, 2001), 247–8.

24. Agamben, 183.

25. Kirk Fuoss, "Lynching Performances, Theatres of Violence" *Text and Performance Quarterly*, 19:1 (1999): 7–9.

26. Saidiya Hartmann, *Scenes of Subjection: Terror, Slavery, and Self-Making in Nineteenth Century America* (New York: Oxford University Press, 1997), 56.

27. Judith Butler follows Hegel in discussing the appropriation of the bondsman's body by the lord, how the body of the bondsman is merely an extension or tool of the lord's body, not an autonomous entity. This simultaneity or overlapping of bodies, one with much greater power than the other, is present in minstrelsy, where the lord's body is hidden beneath but the agency behind the body of the bondsman. An audience made up primarily of men who would be lords, who wish status and standing in their world, accrue an augmented identity from minstrely. They take pleasure and enjoyment in the extension of their personal grasp and in the covert quality of that extension, which is committed secretly and in the dark, sanctioned as fun. See Butler, *Power*, 34–5.

28. Litwack, 16.

29. "Despite the common assumption that minstrelsy declined at the turn of the century, it profoundly influenced America's first movies and musicals, raising questions about the degree to which these apparently new and innovative genres contained and sustained conventional structures of oppression" (Gubar, 55).

30. The film that launches American film-making as an international phenomenon, *The Birth of a Nation* (1915) fuses "the two basic rituals of the postbellum South – the minstrel show and the lynching – [exposing] the essential kinship between them." Clyde R. Taylor, *The Mask of Art: Breaking the Aesthetic Contract – Film and Literature* (Bloomington: Indiana University Press, 1998), 115.

31. Lynching has been a fairly frequent subject in literature, written by black and white authors. "The legacy of lynching persists in literary texts such as James Baldwin's 'Going to Meet the Man' and William Faulkner's *Light in August* that continue to be read and in literary texts such as David Hill's *Sacred Dust* and Mark Childress's *Crazy in Alabama* that continue to be written. It persists in the celluloid images of *Birth of a Nation, Mississippi Burning*, and *Rosewood*. It persists in political ads, such as the infamous Willie Horton spot, that perpetuate the related lynching practices of not only staging the black criminal but also of staging the criminal as black. It persists in the skepticism and fear with which many non-whites approach this country's judicial system. It persists in endlessly replayed video footage showing a contingent of LA police officers brutally beating Rodney King, and it persists in the verdict of jurors who found the police, just as their historical counterparts so often found those accused of participating in a lynching, 'not guilty'. It persists in contemporary hate crimes intended to terrorize not only the immediate victim but also the entire group of which the immediate victim is but an unlucky representative – groups such as blacks, Arabs, gays and lesbians, persons with AIDS, women" (Fuoss, 28–9).

32. "In October 1993, two teenagers dressed in Ku Klux Klan costumes attended a Halloween party at Yosemite High School (In Oakhurst, California) with a friend in blackface – whom they pretended to lynch. They won prizes for their costumes, though the school's principal soon regretted the decision to allow them to attend" (Gubar, 85).

33. James V. Hatch and Eroll G. Hill, *A History of African American Theatre* (Cambridge: Cambridge University Press, 2003), 445–6.

34. Grace Elizabeth Hale, *Making Whiteness: The Culture of Segregation in the South, 1890–1940* (New York: Vintage, 1998), 218.

35. David M. Chalmers, *Hooded Americanism: The History of the Ku Klux Klan* (Durham, NC: Duke University Press, 1987), 27.

36. Christopher Metress, *The Lynching of Emmett Till: A Documentary Narrative* (Charlottesville: University of Virginia Press, 2002), 31.

37. Walter White, *Rope and Faggot: A Biography of Judge Lynch* (Notre Dame: University of Notre Dame Press, 2002), 3.

38. "[T]he first actors separated themselves from the community by playing the role of the Dead [...] this same relationship [...] I find in the Photograph [...] Photography is a kind of primitive theater, a kind of *Tableau Vivant*, a figuration of the motionless and made up face beneath which we see the dead." Roland Barthes, *Camera Lucida: Reflections on Photography*, trans. Richard Howard (New York: Hill & Wang, 1981), 31-2.

39. Hale, 207–8.

40. When Rufus Moncrief was lynched and his body exhibited on the branch of a tree, along with the body of his dog, after failing to "pull off his hat" and generally be deferential to a group of white men, his wife, who was 80 years old, was informed of the whereabouts of her husband and told that she had to immediately remove both corpses from the tree where they were hanging or "the farm would be burned to the ground" (Litwack, 26).

    For each of the questions raised in this paragraph, I could cite references, but doing so would become cumbersome. I chose this example because the thought of a woman of advanced years being forced, in the sundown of her life, to suffer such an insult to her family and home struck me as particularly poignant. The questions I raise are by no means idle or without precedent.

41. Ana Douglass and Paul A. Vogler, *Witness and Memory: The Discourse of Trauma* (New York: Routledge, 2003), 190.

42. Robyn Wiegman, *American Anatomies: Theorizing Race and Gender* (Durham, NC: Duke University Press, 1995), 82.

43. Susan L. Mizruchi, *The Science of Sacrifice: American Literature and Modern Social Theory* (Princeton, NJ: Princeton University Press, 1998), 275.

6
Sacrificial Practices: Creating the Legacy of Stephen Lawrence

Mary Karen Dahl

> *But whatever the boy's story becomes, it stopped being about the child himself the moment the first camera shutter fell.*
>
> Libby Brooks, *Guardian*[1]

I Preamble

Ali Ismail Abbas lost his arms and entire immediate family to a bomb during the initial phase of the US-led Coalition's assault on Baghdad in the spring of 2003.[2] When pictures of the wounded boy surfaced on television and in the press, coverage in the United Kingdom rapidly construed him as representing all children that Coalition attacks displaced and deformed. Reporters and commentators emphasized the beauty of his face and extensive damage to his body. In the case of one leading tabloid, the *Mirror*, pictures published over a series of days focused on that beautiful face with its large brown eyes and Ali's badly burned torso with stumps where the arms had been amputated. On the interpretive horizon is the outline of the crucifixion. A metal structure poised to keep bedclothes from touching his burned body haloes his torso.[3] Writing in the Easter Sunday edition of London's *Guardian* newspaper, Mary Riddell made explicit the embedded discourses of martyrdom and sacrifice: Using the idea of resurrection, she addresses not only the use of Ali's image to raise funds for Iraqi children, but the need to restore the culture of the country devastated by the violence of the invasion and the chaos that ensued.[4] Indeed, Ali's image and narrative – an unprovoked attack on a family of innocents – had triggered similar impulses in ordinary people throughout the United Kingdom. Readers of the *Mirror*, for example, immediately began sending donations to a fund established by the paper to support UNICEF's efforts to aid children in the war zone. Contributors explained that they identified Ali with their own children and grandchildren.[5] Some hoped to redeem the actions of their government through their own generosity: Without us he will not survive. If we help, our collective responsibility for the bombings that devastated his family, our guilt, will be diminished. Others struggled to balance the losses sustained by one child, Ali, against the larger good of ridding his homeland of a tyrant.[6] Still others saw Ali's suffering as endowing him with a unique power. A Kuwaiti doctor involved in Ali's treatment articulated a vision of redemption: "Ali will be a messenger of peace with a mission. He will grow up to the challenge to shed the message all over the world from the Iraqi people through his own experience."[7] Putting such hypotheses aside, in fact large sums of money were raised for Ali's rehabilitation and for groups that came to be associated with him. *The Age* reported in September 2003, less than six months after the first images of Ali appeared, that 725,000 British pounds sterling had been raised through appeals sponsored by three newspapers: the *Evening Standard*, *The Sun* and the *Daily Mirror*.[8] But the question remained, could any sum raised, any good performed, any country liberated redeem the boy's violated body and spilled blood?

I encountered the story of Ali Ismail Abbas as I was looking into the way that theatre and film-makers had represented Stephen Lawrence, another victim of unprovoked violence who had achieved wide and enduring public notice in the United Kingdom. Reading coverage of the tenth anniversary of the Lawrence murder at the same time that the Abbas narrative played out in print in the spring of 2003, I was struck by similarities in the discursive treatment of the two young men, the ways that ordinary kinds of writing and speaking circulated through tropes of sacrifice and redemption. While the case of Stephen Lawrence will remain my primary focus in this chapter, I have begun with the story of Ali Ismail Abbas in order to reclaim the sense of public shock that sometimes greets images of violated humanity and to bring into view the common recourse to an ill-defined notion of sacrifice in order to frame such an event and potentially generate good out of evil. My larger purpose is to open up some of the questions and ethical dilemmas suggested by these everyday ways of thinking, speaking, and acting – what Chantal Mouffe and Ernesto Laclau might refer to as "articulatory practices" – and explore their implications for those who make art to rectify injustice.[9] Of these, one of the most troubling is the possibility that in the creative processes of making meaning we violate the victim anew. Libby Brooks put the problem succinctly, writing in a British daily about the iconic status thrust upon Ali: "But whatever the boy's story becomes, it stopped being about the child himself the moment the first camera shutter fell."[10] That is, not only is it impossible to undo the damage done to Ali and his family, but also the very fact of taking his image and narrating the circumstance of his wounding in our own words circumscribes, fixes, his choices and his future. Or runs that risk. This is the scandal that lurks in the act of transforming violence into art – the scandal that underpins works addressing the case of Stephen Lawrence in theatre and film.

This chapter looks at the articulation of the original violent event and its aftermath with notions of sacrifice and related social dynamics in order to

clarify some of the expectations (and consequent limitations) of acts of creativity undertaken in the name of the victim to redress the violence that has
been done. It takes for granted that efforts to create meaning, including acts
of scholarly analysis, distort and fix the subject, and in that way always are a
kind of little violence done in the name of the victim/subject, driven by the
impulse to do good, to create in the face of destruction. And in that way, the
critical act itself is like the making of art and is bound by similar concerns
and constraints. I suggest that we who make art – whether as writers, directors, actors, spectators, and so on – like those of us who perform criticism, are
always suspended between creation and destruction, balancing on a knife's
edge in the awareness that we destroy in order to create. We objectify, rend
asunder, in order to construct meanings of the fragments. In this way, acts
of criticism, like acts of art, are sacrificial. I begin, then, modestly.

II   The violent event

By now the basic narrative is well established: Stephen Lawrence was a young
black Briton stabbed to death by white racist youths in southeast London on
the evening of 22 April 1993. The probable killers were identified immediately (by 23 April), but still, more than 15 years later, no one has been
convicted of the crime, despite criminal proceedings, civil litigation initiated by his parents, a coroner's inquest, and sporadic official efforts to renew
the search for evidence. As a result of these repeated, unsuccessful attempts
to obtain justice for Stephen, during the years after the murder the criminal justice system itself became the target of investigation. Was more than
run-of-the-mill incompetence at work?

No satisfactory answer was forthcoming until 1997 when Labour, led by
Tony Blair as Prime Minister, replaced the Conservative government of John
Major. Almost immediately Jack Straw, the new Home Secretary, charged
Sir William Macpherson with presiding over a public inquiry into actions
taken by police at the scene of the murder and during the ensuing investigation. Macpherson issued an official report on the findings in Spring1999:
Institutional racism in the Metropolitan Police force was a root cause that
justice had failed Stephen Lawrence. Sweeping changes in policy and law
would be required to eliminate racist attitudes and practices in the force.
The report ended with a list of 70 measures that should be undertaken to
remedy the problems that had become evident in the hearings.[11] One far-
reaching result was that the provisions of the original Race Relations Act
1976 were amended and expanded by The Race Relations (Amendment)
Act 2000.[12] The killing itself, now coupled with the Macpherson Report on
the Stephen Lawrence Inquiry, has taken its place alongside Enoch Powell's
famous "rivers of blood" speech (21 April, 1968), the Brixton Riots (10–12
April, 1981) and the consequent Scarman Report (1981) as a watershed event
in the history of violence and race relations in the United Kingdom.

And so Stephen's place in popular culture is secure. In the United Kingdom his name and image are as recognizable as those of pop stars.[13] The tenth anniversary of his murder became a national event. On 22 April 2003, 700 people gathered at St Martin-in-the-Fields. Attendees included government figures such as Trevor Phillips, then the newly appointed head of the Commission for Racial Equality (CRE), and the Minister for Social Exclusion and Equality, Barbara Roche MP, who read a message from PM Tony Blair that paid tribute to the "courage and dignity of [Stephen's] parents and family" and reaffirmed his government's commitment to creating an "equal, tolerant multi-cultural society."[14] Reporters and columnists marked the date with assessments of race relations in Britain: They reviewed the state of the institutional reforms the Lawrence case had generated,[15] and voiced alarm at the increase in black-on-black violence[16] and the rise of the British National Party (BNP).[17] One of the Lawrence family's lawyers, Imran Khan, used the lens provided by the case to accuse the same Labour Government that had promulgated the Race Relations (Amendment) Act 2000 of having wiped out its effects by "the most racist asylum and immigration legislation this country has ever seen."[18] Encoding discussions of all forms of public policy and official conduct in terms of Stephen when racism was at issue had become normal in public discourse.[19] Today, the practice continues with no diminution in its descriptive efficacy. In 2005, a *Guardian* headline efficiently summed up the findings of a study of institutional racism sponsored by the Home Office: "Racism still blights police despite post-Lawrence improvements."[20]

III Re-articulating Stephen: sacrificial tropes and social practices

Numerous Britons – citizens, reporters, scholars, and policy-makers – have joined Stephen's family in their effort to make meaning of his death. This chapter discusses two of many contributions artists have made. First, London's Tricycle Theatre presented *Guardian* journalist Richard Norton-Taylor's edited versions of the transcripts of the Stephen Lawrence Inquiry as *The Colour of Justice* in 1999. This tribunal drama won the Time Out Live "outstanding achievement award for theatre" and has become an exemplar of the genre.[21] Second, Paul Greengrass wrote and directed a television drama for ITV that was produced by Mark Redhead and broadcast on 8 February 1999. Titled *The Murder of Stephen Lawrence*, it won the 2000 British Academy of Film and Television Award (BAFTA) for Best Single TV Drama and a Special Jury Prize at the 2000 Banff World Television Festival. It aired to positive reviews in the United States on 21 January 2002.[22] As these dates make clear, both play and film came to fruition years after the killing. They are embedded within the much larger complex of actions performed by diverse social agents who raised Stephen to his current iconic status. My discussion assumes that this larger complex shaped the projects themselves and their public reception and considers them in relation to this wider array of activities with particular attention to those that participate in dynamics that – as in the case of Ali Abbas – explicitly or implicitly rely on tropes of sacrifice.

Thinking through sacrifice 1

Libby Brooks points to the violent act of appropriation on which representation depends – the act that distances the living boy from "his" story. The representation takes on a life of its own and shapes the choices that follow, not only for the boy, but also for those who make or consume the story. Ali Abbas survived his ordeal and soon began to "talk back" to his interpreters, actively shaping his own message.[23] For Stephen Lawrence that was not possible. His death has produced widespread public effects as an ever-widening circle of interested citizens created what is now widely termed "Stephen's Legacy." But while actions performed in his name honor him, they cannot bring him back to life. Indeed, they risk obscuring our view of the actual teenager whose murder triggered these events. The social productivity of his legacy depends on the original and continuing violation of the actual boy.

This contradictory movement enacts what Marxist literary analyst Terry Eagleton describes as "the paradox of the truth that destruction is also creation."[24] Recognition of this paradox, that violence creates as it destroys, is not unique to Eagleton among Western scholars.[25] In the modern era, for example, the perceived duality of violence is key to the view of sacrifice promulgated by two associates of Emile Durkheim, Henri Hubert and Marcel Mauss, who aim at describing "the social function of sacrifice."[26] To do so, they treat religious beliefs as "social facts"(p. 101). From this perspective, they hypothesize that the function of sacrificial rites is to effect communication between the sacred and profane in order that human society can prosper. Some rites expiate the guilt of a person or community while others bring strength and health to them through communion with another realm. That which is sacrificed – the victim – mediates between the two realms and has – or has conferred upon them – special properties to ensure their ability to serve that purpose. As for the act of violence itself, it consecrates even as it destroys the offering. It is both sacrilege and sacred act (pp. 32–3), and the ritual involves procedures to contain and direct this volatile force and protect those who invoke its terrible power.

Although contemporary scholars may view their intent to uncover a common, culturally non-specific sacrificial mechanism as a fundamental flaw in their project, the "grammar" of sacrifice (as British anthropologist E. E. Evans-Pritchard called it[27]) that Hubert and Mauss developed helps me name some of the taken-for-granted notions that inform the articulatory practices shaping the Lawrence legacy. As I extrapolate from and transpose their analysis to the predominantly secular discourse of the present day, these include:

1. Identification of social conditions that require re-mediation.
2. Expectation that construing a death as a sacrifice will benefit society.
3. Attribution of special characteristics and status to the victim, perhaps including links to the community the sacrifice serves.
4. Concern that actions taken on behalf of the victim are consistent with its function as the center of sacrifice.
5. Anxiety as to the efficacy of the sacrifice, coupled with acute sensitivity to omnipresent social violence.
6. Desire that society achieve redemption.
7. Ambivalence towards violence as the perceived agent of death and social redemption.

Many of these notions, highlighted with the help of Hubert and Mauss, seem to circulate under and through actions taken on Stephen's behalf, art made about the case, and analysis of its effects by critics and scholars.

Thinking through sacrifice 2

While the language of sacrifice is not always spoken aloud, the tropes that mark the surface in the reporting of the Ali Abbas story form a substrate for the actions performed in Stephen's name. The underpinnings of sacrifice and related social rituals are sufficiently evident that two media scholars have framed the case in these terms. They choose different explanatory paradigms and differ sharply in their conclusions. Their differences hint at the anxieties that circulate through actions taken on behalf of the violated subject.

In *The Search for Justice in a Media Age*, Siobhan Holohan joins an established conversation in media and cultural studies about "law, order and deviance."[28] As her subtitle *Reading Stephen Lawrence and Louise Woodward* indicates, she examines media and legal discourses in two criminal cases that were widely reported in the United Kingdom. Her exploration of the Stephen Lawrence case is interesting especially because of her focus on issues of multiculturalism in a Western democracy. Holohan poses a question with considerable resonance for theatre scholars and practitioners: Do media discourse and symbolization empower minority voices and activism at the micro-political level? She reminds readers that some cultural theorists view rapidly proliferating new communications technologies and more media savvy consumers as combining to enable micro-political debates that can effectively undermine or shift dominant ideology, but Holohan herself comes down on the side of Baudrillard: Currently, "media representations of crime and justice fall back on stereotypical constructions of otherness which routinely fail to fully engage with the political motivations of the story" (p. 8). Her argument suggests that engaging with a full range of political motivations might lead to real debate or even allow for the possibility of privileging minority over majority interests. In the Lawrence example, however, she decides that media worked to maintain the existing arrangement of power relations.

To reach this conclusion, Holohan builds on scholarship that has shown how "the media fits into a larger ordering system that structures meaning around agreed norms and values." By normalizing and demonizing members of marginalized communities, representation effectively excludes them from power. Among the studies she cites is *Policing the Crisis*,[29] which she points out "makes links between public sphere debates on law and order and the construction of racial stereotypes that are suggestive of scapegoat theory." She develops that line of analysis by joining René Girard's model of scapegoating to Michel Foucault's ideas about deviance and social control (pp. 5–6). As I read her, these theorists help to describe hidden processes that impel social groups to single out and kill or expel an enemy other or discipline them into conformity. This framework emphasizes dynamics that operate independent of human decision-making. The effect is to underline notions of hegemonic stasis, where sacrifice reinforces the status quo and leaves an inevitably racist liberal democracy intact (p. 142).

Whether or not one completely agrees with her, Holohan raises valuable questions for *The Colour of Justice* and *The Murder of Stephen Lawrence*. Does either engage a full range of political motivations rather than flattening issues and individuals into two-dimensional stereotypes? Do they privilege minority over majority interests? Empower minority voices at the micro-political level and encourage substantive debate? As we shall see, the performances and broadcast fuelled discussion at not only micro-, but macro-political levels.

In contrast, *The Racist Murder of Stephen Lawrence: Media Performance and Public Transformation* deploys an analytic paradigm that allows for the possibility that public moral pressure and media activism can instigate social change.[30] Here media and communications scholar Simon Cottle draws on ritual theorists Emile Durkheim and Victor Turner, putting to work his own concept of "mediatized rituals" to show how British print media helped to single Stephen out as a worthy representative of a larger pattern of injustice and fix his iconic status in the public imagination (p. 31). He argues that the Lawrence case defied "ideas about elite-dominated news production and representations" (p. 24). Instead, "Dominant state institutions and elites increasingly had to struggle for public legitimacy within the mediatised field populated not only by elite actors but also by a profusion of signs and symbols, at the heart of which was the symbol of Stephen Lawrence himself." Cottle endows the symbol itself with agency: "As symbol, Stephen Lawrence breathed life into the struggles conducted in his name" (p. 25). Cottle analyzes those struggles using Victor Turner's concept of "social drama," arguing that Stephen's murder constituted the "breach of some social relationship regarded as crucial in the relevant social group" that initiates such processes.[31] More precisely, Cottle says, it was as the media began to focus on the racist nature of the killing that it came to serve this function (pp. 71–3 ff). He interprets Macpherson's recommendations for sweeping policy changes as evidence the drama ended with "reconciliation," in Turner's terms (p. 153 ff.).

Unlike Holohan, Cottle emphasizes sources of agency that propel the social drama: activist individuals, diverse voices from both majority and minority media outlets, and the symbolic representation of Stephen himself. He argues that individuals choose to initiate or engage in the social drama. He rejects hegemonic stasis, argues for the potency of the symbol, and implicitly joins the many critics who claimed that the play and television drama could change British society.

As the gulf between Holohan's and Cottle's assessments indicates, it is difficult to reach a single definitive assessment of actions taken on the victim's behalf. Efficacy is a matter of interpretation: In a rapidly moving field of social actors, what might success look like? Who or what holds the key to success? Is it possible to weigh the benefits that accrue against the loss of the victim's life? Not unexpectedly, doubts about actions taken in Stephen's name repeatedly surface in the Lawrence case. Writing about it in the *New Statesman*, commentator and playwright Tariq Ali warned, "The politics of empathy, so dominant in our culture today, are by their very nature emotional and volatile. They have no lasting impact. The people who have been temporarily moved by a particular tragedy soon sink back to a trusting acquiescence."[32] Cottle disagrees. He claims that here the "politics of pity" theorized by Luc Boltanski "merged with feelings about 'race' " and "widespread liberal sentiments about the murder of innocents" to produce a "more socially engaged politics" that he calls "a politics of shame" (p. 192).[33] But even though he asserts the beneficial effects of the Lawrence case, he reminds us of the dual aspect of the victim. That is, he explicitly dedicates his book not to the icon, but to the boy himself. The potential he attributes to the symbol depends on the original act of violence that converted the living boy into inert object. The icon draws strength from (is created as powerful by) the social agents who insist on endowing that object and that original violence with sacrificial potency. The boy, however, is lost. Recognition of this central fact destabilizes any settlement or reconciliation.

Articulatory practices 1: the expectation of benefits

Stephen Lawrence was neither the first nor the last to die at the hands of racists in Britain. In the previous two years, three black or Asian men had been killed in the same neighborhood. Between 1991 and August 2006, the Institute for Race Relations reports nearly 90 murders in mainland Britain that are confirmed or suspected of being racially motivated.[34] Over time, however, actions taken by his family and by anti-racist activists who appropriated the case, along with prolonged media attention, created Stephen as one victim representing many others and interpreted his death as a signal that virulent racism continued to be practiced actively in Britain.

The process of elevating Stephen to the status of exemplary victim has led to observable social effects. It is a reciprocal process. Actions, whether positive or negative, taken in his name demonstrate and enhance his iconic potency. The range of activity is considerable: In addition to official inquiries, reports, institutional assessments, policy recommendations and investigations, the case has given rise to informational, educational, and analytical products including numerous articles and interviews in the popular press and media, internet dialogue, books for the casual adult reader and for children, and scholarly contributions to the field of criminology and media studies. The Stephen Lawrence Charitable Trust rewards good students who, like him, are members of racial or ethnic minority populations and aspire to professional training as architects. The Prince of Wales has established a scholarship in his honor and the Goldschmied Trust gives a prize in his name.[35]

At the individual level, too, the killing has produced significant effects. Characters in the real life drama have become national figures. For some, including members of the Metropolitan Police, the outcome has been public humiliation. For the presumed perpetrators, the consequence has been notoriety. They and their mothers complain of harassment.[36] On the other side of the equation, Stephen's parents have received widespread praise and, in 2002, were awarded OBEs for their efforts to obtain justice on behalf of their son and other victims of racial violence.[37]

Those of us who took no part in the killing, the pursuit of justice, or its policy outcomes also can take part in producing Stephen's legacy – and creating its value – both as consumers of press and media coverage and more actively. Followers of the case in the popular press might have seen, for example, that the *Daily Mirror* gave special recognition to Doreen and Neville Lawrence at their May 1999 Pride of Britain Awards. The *Mirror* facilitated reader involvement (in this as in other national events) by making available for online purchase images of the figures they covered. The possibilities included the Lawrences posing with Cherie Blair at the Awards and photographs of Doreen visiting her son's grave in Jamaica. We could compose our own electronic archive or hardcopy scrapbook, perhaps developing a narrative about the heroic mother who successfully pursued justice straight into the halls of power or, alternatively, selecting, cropping, and editing images in a way that countered dominant media coverage and privately, at least, denied her further public acclaim. Indeed, we could compile a villains' or heroes' gallery of the putative murderers. The *Mirror* even gave us the option of sending our selections as e-cards "to friends and family." In this way, it provided its consumers with the means of participating in and directing the cultural activity that grew from Stephen's loss.[38] We still can visit the place where he fell, mortally wounded, on the street, either as part of a demonstration against racism or as individuals. London now literally is imprinted by his death. Maps locating the site are easily found on the web (along with chronological accounts of major events the murder precipitated). A memorial stone marks the spot and the repeated attacks on it bear witness to the existence of those who wish to deny his death meaning, or to literally write over and deface his (physical) legacy, thus re-enacting the violation of his body – actions that perversely enhance the marker's effectiveness as a sign of martyred innocence.

Articulatory practices 2: art and expectations

Stephen's story, his killing, and the unsatisfactory search for justice have produced real effects in art, law, and institutional reform, to name the most obvious. In the realm of art, critics emphasized the social productivity of presenting the story on stage and screen. They endowed each with the power to transform attitudes and institutions. About *The Murder of Stephen Lawrence* Andrew Billen asserted, "The film's peak-time showing on our most watched channel could prove to be crucial in turning the murder into one of those occasional real-life parables that actually changes how a people thinks about itself."[39] Stan Hey wrote, "we were told, quietly and convincingly, the story that is likely to change the shape of policing in Britain."[40]

Similarly, reviewers of *The Colour of Justice* noted the coolness of the presentation as a whole, while working hard to articulate its emotional impact on them and other audience members.[41] So moved were they that several launched a public campaign for the stage production to be taped for television and broadcast on BBC2. Some called for the play to move to the West End, tour the country, and run at the Royal National Theatre. All of which came to pass.[42] The shared assumption was that having large numbers of British citizens see the production would be beneficial for addressing racism not only in the police force, but in the culture at large. Consistent with that premise, many performances were followed by post-show discussions with panels including the actual players in the historical events – perhaps Doreen or Neville Lawrence and one of their legal representatives, Imran Kahn or Michael Mansfield. Expecting that future spectators would want to support the quest for justice, press articles explained that, although tickets would be more expensive when the production transferred from the Tricycle to the Victoria Palace, a portion of each ticket would go to a charitable fund.[43] Underpinning these actions was the determination to make good of ill and the expectation that other members of the community, strangers, would join in that effort. Thus many individuals treated the play and television drama as exceptional events, greeting them with high expectations and sweeping them into the diffuse, widespread movements that reframed the killing using tropes of sacrifice and redemption. Whether intentionally or not, Greengrass and Norton-Taylor participate in that activity: Each walks a careful line between normalizing the victim and making him extraordinary, constituting him as the One who can signify both the racism in society and the solution to it.

Articulatory practices 3: creating/consecrating a victim

The paradoxical qualities that Terry Eagleton attributed to the violence at the heart of sacrifice are critical to the theatre and video treatments of Stephen's story. Silenced by violence, Stephen is described again and again by the living in terms that preserve his status as a victim worthy of elevation to martyr status. Young, gifted, and black, in death he is endowed with a nearly magical power to heal a racially divided nation. As his mother testified to the Lawrence Inquiry, "maybe he would have been the one to bridge the gap between black and white because he didn't distinguish between black and white. He saw people as people."[44] Although his mother phrases this as a possibility that might have been achieved "had he been given the chance to survive," he achieves redemptive status by being killed.

So, as Hubert and Mauss suggest, violence consecrates. But as noted earlier, not all violence is construed as sacrificial; not all victims of violence achieve sacrificial status. Articulatory practices – ways of thinking, speaking, and acting – recuperate the victim and rewrite his death in tropes of sacrifice. The stories that have come to be told about Stephen frequently rehearse the attributes that constitute him as a worthy object of sacrifice. He was a good student, had professional career goals, respected his parents, and did nothing to invite the attack that killed him. Not only in media accounts, but also in *The Colour of Justice* and *The Murder of Stephen Lawrence*, qualities of innocence, "likeness," and even "likeability" come to the fore. Stephen becomes one black victim who represents other black victims of racial violence, and achieves this by being seen to be "like" ordinary youngsters from the middle-class white families that form the majority population in the United Kingdom. His reputed blindness towards racial difference ironically allows him to be seen as representing British youth in general at the very same time that he functions as the victim of a larger, ongoing process of racialization. He stands for British society at the same time that he suffers at its hands.

Stephen's representational efficacy is readily apparent. When activists or grieving parents invoke Stephen's image to characterize other victims, they use a complex of terms like those associated with him: Killed by white racists in an unprovoked attack in July 2005, Anthony Walker, we learn, regularly attended church, had a steady girlfriend, and was a good student and gifted athlete (basketball) who had decided to pursue law as a career instead of sports.[45] Another whose name often is coupled with Stephen's, Damilola Taylor, like him was denied complete justice because of incompetent policing. At a ceremony attended by Prime Minister and Mrs. Blair as well as

Doreen Lawrence on the anniversary of his murder, the father of this ten-year-old boy chose to emphasize his love of British football (soccer), popular films, and drawing. Described as a fine student, he planned to become a doctor and look for a cure for the epilepsy that afflicted his big sister.[46] As noted earlier, likeness to Stephen may earn young black Britons scholarships. Here, linking these victims to Stephen underlines the fact that these deaths were undeserved and a loss to society. More than that, looking at the cases one after another, one sees a pattern in the speakers that suggests their deep desire to ensure that the victims are seen as exemplary of British youth. Each one is – apart from his color, of course – the son that every good middle-class British family hopes to have. The descriptors bring them into a familiar framework: they are not the enemy other. As Gee Walker asserted, "I have to tell you about Anthony, he was everybody's son."[47]

Articulatory practices 4: representation versus racialization

Anthony Walker's mother is speaking against, both resisting and attempting to reverse, a complex of social and historical processes captured by the concept of racialization. Like Gee Walker, advocates for Stephen present him as "everybody's son." By insisting on his status as "one of us," they claim for him the rights accorded to members of the majority white culture. Similarly, both film and play, *The Murder of Stephen Lawrence* and *The Colour of Justice*, take racialization for granted and devise strategies to resist its mechanisms. The film deliberately sets out to make Stephen "one of us"; the play exposes racialization in action. Both implicitly rely on tropes of sacrifice to achieve their ends.

A brief look at the notion of racialization itself will clarify how the film and play go to work. Frantz Fanon employed the term in the 1950s to set the stage for his analysis of ways that African and Arab intellectuals could best contribute to the creation of national cultures in states then emerging from colonial occupation. Fanon does not fully explicate the term, but says a "racialization of thought" has occurred whereby European colonial powers collapsed the diversity of their African subjects (for example) into a single description of an inferior Other without culture – the Negro. Now, Fanon argues, that notion, incorporated by artists and intellectuals, "leads them up a blind alley": They undertake equally reductive gestures (the celebration of negritude or making a cult of the past) instead of creating art in the midst of current resistance movements against European ideological and actual oppression.[48] Already in Fanon's analysis it is apparent that racialization is a two-way street that constrains choices made by both oppressor and oppressed. Current usage, according to Robert Miles and Malcolm Brown, assumes similar dynamics. Deciding that the concept fundamentally "denote[s] those instances where social relations between people have been structured by the signification of human biological characteristics in such a way as to define and construct differentiated social collectivities," they repeatedly describe it as "process." It is a "process of categorization, a representational process of defining an Other, usually, but not exclusively, somatically." As Fanon suggests, it "is a dialectical process of signification." In defining the Other we define the Self, so when Europeans defined others in terms of skin and coloration, they defined themselves according to those same attributes. When they conferred a negative valuation on those they saw as Other, they effectively valued themselves in positive terms. The dialectic is longstanding and ongoing, dating from before the European discourse on 'race' emerged in the eighteenth century. Most significantly for the art works under discussion, Miles and Brown make the point that "the racialization of human beings entails the racialization of processes in which they participate and resultant structures and institutions." It both entails ways of seeing and the actual lived policies, rules and procedures that instantiate and reinforce those ways of seeing Self and Other.[49]

Paul Greengrass's television film *The Murder of Stephen Lawrence* traces Stephen's story from the evening of the attack through the family's quest for justice from the Metropolitan Police to the coroner's inquest that conclusively found Stephen's murder to be the act of white racists. The formal declaration by the inquest jury that racism was the issue not only confirmed the nature of the attack (and the victim's innocence), but also repudiated official police claims that color had not played a part in their initial investigation, their (failed) pursuit of the perpetrators, or their treatment of the survivors. In effect, the film depicts a society contaminated by racism, purified by the death of an innocent, and redeemed by individuals working to maintain the sanctity of the victim.

The sacrificial discourse that threads through the film and its reception is readily apparent in the marketing of the docudrama to US consumers. The PBS website associated with the ExxonMobil Masterpiece Theatre airing by Boston's Public Broadcasting Station (WGBH) emphasizes the significance of the story for race relations in Britain, but the descriptive blurb on the video packaged by WGBH justifies its relevance to US viewers by construing us as parents: The brief narrative first uses the fear of violence against children to suture together all parents, black and white, in the United States or United Kingdom, then interprets the killing and its aftermath as signaling a level of "institutionalized racism and violence" that must horrify parents everywhere. The packaging celebrates Doreen and Neville Lawrence as heroic representatives of the common man who took on corrupt state institutions and redeemed both themselves and (by extension) Britain as a whole by exposing racist practices to view. The cover promises a real life story that is about us, too, and suggests that the outcome will uplift viewers just as the Lawrences' campaign benefited British subjects.[50]

Looking back at ideas I extrapolated from Hubert and Mauss, these discursive moves have a familiar ring. The two theorists drew attention to links between ritual offering and beneficiary. If the victim is to mediate between the sacred and profane, it must have, or must acquire, a special status; it must either be, or become, connected to those who will gain from the rite. Any good that comes of violence against the victim depends on this connection (and careful adherence to procedures that respect/maintain its status). Here, in the video drama, Greengrass puts analogous strategies to work. The film-maker takes pains to present Stephen and his family as a/kin to the hard-working white majority population. The effort is not hidden – quite the contrary. The film-makers attribute the style of the piece to their desire to position spectators with the black family. *The Murder of Stephen Lawrence* uses a hand-held camera and "very fast 500 ASA film stock" to record the actors re-enacting the events, sometimes improvising dialogue, without the benefit of "tracks, dollies, even tripods" or artificial lighting. Paraphrasing Greengrass, Redhead explains: "It'll be just as if we are observing it all. Like the Lawrences, we will never quite have the perfect view, we'll be fighting to see everything."[51]

In actuality, however, the film constructs a more complex set of identifications. The drama begins with events leading up to the attack. Initially it follows two simultaneous lines of action: We see Neville picking up and driving Doreen home then settling in for the evening; We see the teenager playing video games at his uncle's, then rushing to get home before his curfew. He and his friend Duwayne Brooks board one bus, then get off to wait for another that will take them across town more quickly. Shots of the boys looking for their connection, the attack, and its aftermath alternate with shots of Stephen's parents worrying about his absence. Thematically, the choice efficiently establishes the tight knit family unit, the parental dynamic based in close supervision of the children, the normal teenager caught up in a typical pastime who is eager to stay out of trouble with his folks. It is an ordinary day and these are ordinary Britons working to secure their place in the middle class.

In addition, the choice to alternate scenes depicting both perspectives splits our identification and generates a complex interplay of tensions that draws us into the situation of each pair. That response accelerates as we watch the attack. Spectators see what the parents do not. The camera positions viewers with Duwayne as he looks back and sees the swarm of attackers surround his friend and run off. It shifts to place us behind Stephen as he tries to run after Duwayne and falls. We move with Duwayne as he desperately calls for an ambulance from a phone box and attempts to stop cars and passersby. Then the lens places us with other citizens who happen on the scene and attempt to make sense of the situation.

The filming technique, the documentary-style re-enactment that Greengrass selected to give a sense of our being present at the scene, struggling to see events in real time, conditions our viewing. On the street, for example, the camera restricts the perspective so that we cannot discern the boy's condition in the darkness of the night. It shows us his still profile, glistening black liquid on pavement, his unconscious form on a stretcher. From the dimly lit, confused scene on the street we shift to a brightly lit emergency room at the hospital where for the first time the extent of Stephen's injuries is indicated. In an angled shot from below the waist of his supine form, we see a blood-soaked layer of clothing pulled off to reveal his torso, then views of Stephen's face and a cluster of medical workers administering electric shocks to restart his heart. The hospital scenes, like the street scenes, alternate with shots of the Lawrences opening their door to an unidentified neighbor telling them Stephen is hurt, their arrival at the hospital, anxious waiting, and response to news of their son's death. During this action, views of the boy – the red-stained chest, the blood encrusted face – mark Stephen's movement from ordinary active teenager to still life. He is present. We see him pass from life to death. From our privileged perspective, we know the boy is dead before his parents do. Now our sympathy shifts to them. The bloodied corpse, the inert object, displaces the living boy in the mind's eye.[52] The parents' struggle to define the meaning of the boy and his life gradually occupies the screen and our attention.

As this suggests, the film-makers went to great lengths to bring viewers into the action. Their efforts reflect the effects of racialization as a determining societal condition that shaped the killing, its aftermath, and its depiction in art. Describing how the project developed, for example, producer Mark Redhead recalls that one of the first questions that he and writer/director Paul Greengrass faced was whether white artists could make a story about the experience of black Britons. Although already commissioned to make the film by ITV's head of drama, Nick Elliott, he reports that he and Greengrass sought the consent of Neville and Doreen Lawrence and engaged them in "extensive interviews." He takes care to frame this rather routine procedure in terms of race, noting that Greengrass explicitly sought guidance for "what it is like to be black in Britain" from the actors, Hugh Quarshie and Marianne Jean-Baptiste. Certainly the artists' attention to the ethics of representation is not misplaced; the issue is rather their attention to racial difference as a potentially distancing factor that must be overcome through measures that help members of the white majority to more closely identify with an/Other. They create an ordinary family and a normal boy so that the effect of his death and the parents' frustration will engage us in the narrative and, potentially, in the quest for social justice.

The film-makers choose tactics that connect the victim and family to "us" and encourage us to see Stephen as an ordinary, yet exemplary youth. If we see Stephen as "everybody's son," it becomes possible to reframe his death as benefiting the community. Here the film-makers encounter the dilemma at the heart of Eagleton's paradox: the victim dies; we live. More than that, we make something of the broken body. The choice to focus on the parents' actions after his death underlines the cost of sacrifice: the destruction of the living being that produces an actual absence at the center of the story. Rather than attempting to retell the story of Stephen the boy, here the artists concentrate on the actions made possible by his killing. On the one hand, that choice is understandable – that is where the narrative of social redemption lies – but the fact of the lost life of one specific child retreats into the background. They attempt to compensate for this by including Neville's anguished monologue at Stephen's graveside, but the boy is gone and the father's grief takes center stage. From a certain perspective, in fact, the narrative shows how the family defines and defends what will become the iconography associated with Stephen. Neville Lawrence enumerates the fundamental elements of that iconography in the press conference sponsored by the Metropolitan Police soon after the killing. I do not mean to suggest that their memories of the actual boy are false, or that their grief is tainted, but rather that in order to claim justice "in his name," they must insist on attributes that qualify him for the exemplary status he achieves as an ordinary Briton who is at the same time a representative victim of racial violence. The film-makers depict this process without irony. Their technique naturalizes the process without opening up questions about its workings in the context of a racialized society. The film shows that Stephen is everybody's son; his death has served the general good. His parents have done what any good parent would do. We can become angry at the system that has failed them, but we are not asked to interrogate the very nature of racialization as an enduring, historically based dynamic.

The Colour of Justice performs just such an interrogation. This tribunal drama uses actors to re-enact portions of the inquiry chaired by Sir William Macpherson. To make the script, journalist Richard Norton-Taylor selected "less than one per cent" from in excess of "eleven thousand pages" of testimony, whittling down "69 days of public hearings" to an evening of theatre.[53] The style of the production, directed by Nicholas Kent with Surian Fetcher-Jones, was bare bones. The set was "an anonymous committee room of filing cabinets, desks, computers, witnesses and lawyers and the constant clicking of the stenographer."[54] Members of the cast (numbering nearly 30) were selected to resemble the person they were playing. Those who could, met with the individual whose words they were presenting.[55] The performance mode was understated. One reviewer described it as "acting-that-isn't-acting"; another noted the skill involved in attaining that effect.[56]

The hearings focused neither on the guilt nor on the innocence of the probable perpetrators, nor on the exact cause of Stephen's death, but rather on uncovering the actions and attitudes that led to the failure by police to pursue the attackers, promptly arrest them, and collect sufficient evidence to undertake successful prosecutions. Consistent with that focus, the play looks at actions taken by individuals at the murder scene and in the investigations that followed. We watch as lawyers representing the inquiry, the Lawrence family, Duwayne Brooks, and police officials elicit testimony from witnesses including citizens who stumbled upon the scene by accident, one of the alleged attackers, numerous members of the police force, and finally from the Crown Prosecutor, who explains why in July 1993 the Crown Prosecution Service dropped the initial attempt to prosecute the suspects then in custody. The evidence against them was adjudged insufficient.

Over the course of the 2-hour-and-40-minute performance, a series of questions are raised and left unanswered. Some first night reviewers enumerated a paragraph-long list of political and cultural issues that struck home for them as individuals. But as the evidence accrues, it suggests not just incompetence or corruption, but deep-seated, systematic bias against black citizens simply because of their color. Witness after witness denies their prejudice even as they describe behaviors that demonstrate its presence whether unconscious or firmly repressed. Several actually display it on the stand, perhaps by failing to see the difference between labeling someone "coloured" rather than "black" (pp. 74–5) or by asserting that, although white, they were raised in Africa and therefore "understand black people" (p. 93).

The re-enactment of testimony, individuated and materialized through the witnesses, reveals how racism manifests in daily decisions and actions. Certainly the Met was under the microscope, but the behaviors were so "normal," their rationales so "ordinary," so unremarkable, that it became possible to extrapolate from the police to citizens in any institution, any walk of life. British policing has been criticized for its actions towards black Britons at least since the 1981 Scarman Report. But here, despite the outrage audience members may have felt, and reportedly expressed at witness statements, it also became possible to recognize in the testimony ideas and rationalizations that are widely held and expressed without concern or correction. Taken as a whole, the behaviors and attitudes could be seen to be actively engaging in the process of racialization. So, although the preponderance of the testimony describes the official disregard for one young black victim and his family, newspaper critics single out the white passerby who, with his wife, stopped to render aid to Stephen. Some reviewers label this character the "Good Samaritan"; many hope that they would be as honest as he in admitting to racial bias when he first saw black youths running towards them. They must also hope they would have acted as this couple did, ingrained cultural bias or not.[57]

But where is Stephen in this? All of this outpouring of praise, all of this creative energy is exerted in the name of the one whose annihilation makes it possible. How does the play address the absence at the center of its meaning? No actor represents Stephen onstage. Instead the characters evoke him through statements and testimony. The first of these is Edmund Lawson, counsel to the inquiry, who at the outset of the proceedings pauses to "remind the inquiry" of the victim. For this purpose he chooses to quote a policeman who knew the boy socially because his own son and Stephen were members of the same club: "I never saw him display any form of aggression and would describe his temperament as the same as his father – quiet and unassuming, exemplary character." Lawson takes care to say that Stephen had never been involved in any trouble, never had "contact with" police acting in any official capacity. He underlines Stephen's innocence and good character by framing them within Stephen's performance as an "active and successful member" of the Cambridge Harriers Athletic Club (pp. 19–20). So the boy, like his father, was the sort of citizen we want for Britain, a fact reinforced by his sportsmanship. Again, near the end of the first and at the beginning of the second act, statements by Neville Lawrence and Duwayne Brooks summon images of Stephen as he was in life. These reinforce and embellish Lawson's description. His father speaks of his artistic gifts and thoughtfulness as a son while making it plain that Stephen belonged to a church-going, law-abiding family (pp. 63–4). Duwayne adds to the picture of peaceable innocence: "I never knew Steve to fight no one. Steve wasn't used to the outside world. [...] Steve didn't understand that the group of white boys was dangerous" (p. 97). His very innocence cost him his life.[58]

Truthful though they may be, the iconography of the worthy and innocent victim shines through these accounts. As witnesses strive to concretely describe Stephen's qualities, his precise and unique presence in their lives, the template of sacrificial victim threatens to overlay and obscure the portrait. It begins to dissolve the individual as a material and historical human being even as it elevates him into the pantheon of culturally potent figures. Contributing to this dynamic is the repeated evocation of Stephen as he lay dying. The testimony is chronological and begins with those initially on the scene. The first and third witnesses, Constables Bethel and Gleason, typify what will emerge as the pattern of the official response: Each tries to explain why they failed to identify the exact nature of Stephen's injury or render first aid; Gleason is caught in outright lies as to his behavior towards the grieving father at the hospital. As they are questioned, we might imagine Stephen bleeding to death on the pavement while official Britain stands by and his friend looks frantically for the ambulance. Evidence given by the second witness, the white Catholic Conor Taaffe, adds specifics.

During his testimony Taaffe taps deep into Western Christian iconography. The details are additive, they guide the spectator in imagining the scene. Prompted by Lawson, he confirms that he had gone "straight over towards where Stephen had fallen" and that he had "bent down" and then "carried on holding Stephen's head" and "praying over him [...]" (pp. 34–5). Under questioning from the Lawrence family's legal team, Taaffe acknowledges that his wife "cradled Stephen's head" and supplements the pieta reference with the information that "she spoke in his ear." Her words complete the image: "she said: 'You are loved. You are loved.'" Unprompted, Taaffe then reports his own actions: After he "washed the blood off my hands," he had "poured the water with Stephen's blood in it into the bottom" of what he has just described as a "very, very old, huge rose bush – rose tree," in fact, that stands in his back garden. "So in a way I suppose he is kind of living on a bit" (pp. 36–7). "Lo, how a rose e'er blooming," as the song commemorating the birth of Jesus goes. The simple recital of facts brings Stephen's dying moments vividly to mind, and Taaffe's testimony stayed with onlookers. Its potency derives both from the actions described and from the cultural encoding. But while Taaffe draws on his own Christian idiom, he articulates a desire that crosses sectarian boundaries: From destruction, creation. Life exceeds death. Good men and women can overcome the evil that contaminates the polity. The contrast between official and private responses to the wounded youth reverberates through all the testimony that follows.

As the hearings unfold, the emphasis shifts from Stephen's death to the search for justice. By the final scene, when Macpherson requests a moment of silence to close the public inquiry, instead of asking us only to "remember Stephen Lawrence," he expands the focus to include the "courage of his parents" (p. 143). The victim has become an agent of change through actions undertaken on his behalf by the survivors. Additionally, while the proceedings literally invoke Stephen, the testimony frames him as an innocent violated first by racist thugs and then by racism woven through the polity. Although officials and witnesses strive to breathe life into his memory, the framing redefines the living boy as an object (no matter how hallowed). The original violation reverberates in operations performed through representation – as we move on.

This double violation of the subject victim goes to the heart of the collective good that performance potentially engenders. Erika Fischer-Lichte reminds us that theatre relies on "actors and spectators" coming together in the same time and space. Through this "bodily co-presence," they create the event and its meanings.[59] In *The Colour of Justice* the boy who is not present is the subject of, the reason for, all other efforts to be present. When the stage Macpherson invites attendees at the inquiry to observe a moment of silence, audience members joined the actors in standing – an action that many reviewers treated as uniquely powerful in their experience of the theatre. Their emphasis on the communal experience suggests a profound desire for the performance to have momentarily succeeded in re-membering Stephen, summoning him into the presence of the community formed in his name and forged by the telling of what happened on the night of his death.[60]

But despite its emotional power, the act is limited in its effect. It may inspire to the performance of good deeds (or good art), but Stephen is gone. More than that, unless coupled with analysis and strategic planning, the act, like the play itself, risks falling into Tariq Ali's "politics of empathy." Like those who assume that ritual sacrifice is inevitably conservative, Ali worries that art enables viewers to release emotional energy, thus preserving the status quo, rather than propelling them into action. Here, despite the obvious, that the community cannot actually rematerialize Stephen, it is critical that the performances were tied to activities in support of the institutional changes underway as a consequence of the Lawrence's extended campaign for justice. Opening just as the Macpherson Report on the Inquiry was to be released, the play concentrates the evidence and, through its communal invocation of the victim, intensifies the impact of the official findings, potentially building support for its judgments and commitment to the more radical recommendations. The many modes of delivery – television broadcast and tour – enhanced its potential to facilitate social change.

IV Balancing on the knife's edge

Taken as a whole, the killing of Stephen Lawrence has generated wide-ranging effects from political decisions to scholarships for disadvantaged youth to highly regarded artistic products and theoretically informed critical analysis in diverse disciplines. One can argue that the benefits are significant, even if justice has not been done and society has not been cleansed. Racism continues. Racist violence continues.[61] More important, if we think of the boy himself, the paradox reappears: Social benefits accrue as a consequence of a life destroyed. That fact produces anxiety as a companion to activism. Scholars and cultural workers alike face that paradox.

From the perspective I have derived from Hubert and Mauss, it would seem that, in the main, the actions that have served to configure Stephen Lawrence as a representative victim have worked to transform the violence committed against him into benefits for society at large, with material consequences for specific individuals. That they have worked in this way is due at least in part to what I would term "sacrificial expectations." That is, we construe a violent death as sacrifice in the expectation that benefits will accrue. These benefits may take the form of purging society of corruption or of bringing blessings or redemption into the social whole. But do such benefits balance the destruction of the offering, the loss of a single life? Of course not. As Doreen and Neville Lawrence have said, nothing, neither honor nor financial award, can compensate for the loss of their child. No work of art, no social goal can justify changing the live boy to the dead offering. To begin to weigh possible benefits against the living one lost – put thus crassly – is to write an obscene equation. In fact, stated so baldly, the question seems simplistic, but I merely echo the questions British columnists raised regarding Ali Abbas: Could the liberation of any nation "redeem" the boy's broken body and spilled blood? And while this is the sort of equation national leaders routinely make in their means-ends arguments regarding the formation and implementation of state policy, it also has relevance for cultural activists and commentators. If representation is also a kind of violence, it calls for great care in the everyday practice of making meanings – or rendering death meaningful.

At this point it is critical to affirm that while artistic representation – and even critical analysis – may violate the original subject, this is a violation of a different order than the act that kills, cripples, or maims. As the practitioners of sacrifices that Hubert and Mauss theorized might caution us, it is wise to approach either violence or victim with care lest we ourselves be consumed by violence or spoil the sacrifice, but conflating the original violence with representational violations is foolish. Here, both the creators of *The Murder of Stephen Lawrence* and of *The Colour of Justice* select modes of representation that resist tipping over into sentimentalizing the struggle for justice or indulging in pornographic depictions of the violated body, while finding moments that hold Stephen up to view, either by prompting us to construct a personal image of him from word pictures or by careful scripting of scenes depicting him. They take pains to demonstrate their desire to productively shape the public conversation about the need for radical institutional and attitudinal change. Indeed, critics compared *The Colour of Justice* in its power and civic function to drama at the moment of its birth in the West: Like the *Oresteia*, it provided the polity with a public forum to discuss its most significant concerns.[62] In both works, however, the move is away from the victim. In that sense, the violence remains Stephen's alone. The pain is his; the death is his. Gestures towards it are appropriations, no matter how well meaning. His violation has produced endless streams of interpretations by and for those who did not experience it. His image floats above the nation, a disembodied symbol of the concrete effects of race hatred and the state's differential treatment of its citizens.

So, in empathy with Neville and Doreen Lawrence, we may long for the impossible, the return of the dead. Although we "know better," we undertake the risk of violating the subject anew, giving way to the desire to make the victim whole again and to experience his presence through the lie of art. Through the actual experience of the video or live performance we may come to know the victim whole/wholly transformed, which is the closest we can come to actually re-membering him. That urge, matched with the desire to honor his life through meaningful action, forces us to engage the paradox Eagleton, like Hubert and Mauss, identified, that "destruction is also creation." In making art as writers, film-makers, performers, spectators, and critics, we live *within* that contradiction.

While no work of art – or political action – can compensate for the loss of one life, by sustaining the contradiction we maintain the potency of the victim as the instigator of change. Tariq Ali critiques the "politics of empathy," but the notion that empathy is needlessly indulgent and produces only momentary responses is, I think, not always accurate. I suggest instead that in the aftermath of the Lawrence killing, a wide array of social actors drew on sacrificial tropes to define and value the victim; they engaged in the politics of empathy in order to unsettle public trust and forestall acquiescence. In effect they engaged in complex enactments of the politics of empathy that incorporated targeted social and political activism.[63] At the same time, however, despite the radical creativity they assert in response to social and institutional violence, their actions always fall short of the goal, they do not raise the dead; they do not prevent every such death. The outcome of the Lawrence case is inevitably mixed: his killing produced both endless pollution because the boy is dead and boundless good as a result of acts taken in his name. Those of us who create out of the breakage of life (take that as you will) sustain that tension, that state of both/and, without resolution. Resolution is death.[64]

Notes

1. "Guilt and innocents: We turn child victims of wars into icons but ignore the deeper realities," *Guardian*, 26 Apr. 2003, 20 June 2007: http://www.guardian. co.uk/comment/story/0,3604,943934,00.html

2. Thanks to Aaron Thomas for his thoughtful comments on the chapter, which is an expanded version of a paper that used notions of sacrifice to explore the ethics of representation with specific reference to Ali Ismail Abbas and Stephen Lawrence for a seminar convened by Jisha Menon and Janelle Reinelt for the American Society for Theatre Research 2003 Annual Meeting.

3. Andy Rudd and Ryan Parry, "£25,000 for our Ali Appeal," *Mirror*, 10 Apr. 2003, 29 May 2003: http://www.mirror.co.uk/search
   Because electronic access to *Mirror* archives is limited, see Mark Jones, "A thousand words," *Guardian* 4 Dec. 2003, 20 June 2007, for a similar image with commentary on using Ali's image to raise funds: http://www.guardian.co.uk/Iraq/Story/0,,1099902,00.html

4. Mary Riddell, "Blinded by the myths of victory: The fight to save Ali Ismail Abbas offers us illusion of hope to soothe our consciences," *Observer*, 20 Apr. 2003, 20 June 2007: http://www.observer.co.uk/comment/story/0,6903,9401 21,00.html

5. Jill Palmer, Ryan Parry and Andy Rudd, "GULF WAR 2: HORRIFIC IMAGE THAT MADE THE WORLD WEEP: I saw the picture of Ali lying there so helplessly and burst into tears-MUM-OF-TWO FAITH GOODALL, 37, WINCHESTER," *Mirror*, 9 Apr. 2003, 29 May 2003: http://www.mirror.co.uk/search

6. "Now Prove it was a War Worth Winning," *Mirror*, 14 Apr. 2003, 29 May 2003: http://www.mirror.co.uk/search
   See Deborah Orr, "It is still possible to hate the war, but to give thanks for regime change," *Independent*, 8 Apr. 2003, 21 June 2007: http://comment.independent. co.uk/columnists_m_z/deborah_orr/article 114205.ece

7. Peter Fray, "Symbol of peace or devastated boy," *The Age*, 18 Sept. 2003, 21 June 2007: http://www.theage.com.au/articles/2003/09/17/1063625093108.html

8. *Metro* (London) raised an additional £280,000 (see Fray). For the *Sun* appeal, see Nick Parker, "Our man meets brave Ali," 17 Apr. 2003, 21 June 2007: http://www.thesun.co.uk/article/0,,2-2003171928,00.html

9. Ernesto Laclau and Chantal Mouffe, *Hegemony & Socialist Strategy: Towards a Radical Democratic Politics* (1985; London: Verso, 1994), 96.

10. Brooks.

11. Sir William Macpherson, *The Stephen Lawrence Inquiry – Report of an Inquiry by Sir William Macpherson of Cluny*, Advised by Tom Cook, the Right Reverend

Dr John Sentamu, Dr Richard Stone, 1999, The Stationary Office, London, 21 June 2007: http://www.archive.official-documents.co.uk/document/cm42/4262/4262.htm To see the recommendations, go to: http://www.archive.official-documents.co.uk/document/cm42/4262/sli-47.htm

12. "Summary: The Race Relations (Amendment) Act 2000," *Guardian*, 22 Feb. 2001, 21 June 2007: http://society.guardian.co.uk/print/0,3858,4140596-106186,00.html

    For the relationship between the Stephen Lawrence Inquiry and the 2000 Act, go to: http://homeoffice.gov.uk/. Search on Act or Inquiry by name to find policy changes and progress reports. For example: http://police.homeoffice.gov.uk/community-policing/race-diversity/stephen-lawrence-inquiry?version=2

13. Andrew Anthony, "The crime that Britain won't wake up to," *Guardian*, 24 Apr. 2003, 21 June 2007: http://www.guardian.co.uk/g2/story/0,3604,942163,00.html

14. Tom Geoghegan, "Service marks Lawrence anniversary," *BBC News* (UK), 23 Apr. 2003, 21 June 2007: http://news.bbc.co.uk/1/hi/england/london/2967985.stm

15. Tania Branigan, "Stephen's Legacy," *Guardian*, 23 Apr. 2004, 21 June 2007: http://www.guardian.co.uk/lawrence/Story/0,2763,941710,00.html

16. Anthony.

17. Branigan.

18. Imran Kahn, "Labour's hypocrisy on race: Asylum laws neutralise the gains made since the Lawrence inquiry," *Guardian*, 22 Apr. 2003, 21 June 2007: http://www.guardian.co.uk/lawrence/Story/0,2763,940797,00.html

19. Vikram Dodd, "Stephen Lawrence mourners pledge to carry on the fight: Hundreds mark 10th anniversary of racist attack," *Guardian*, 23 Apr. 2003, 21 June 2007: http://www.guardian.co.uk/lawrence/Story/0,2763,941588,00.html

20. Alan Travis, "Racism still blights police despite post-Lawrence improvements," *Guardian*, 28 Oct. 2005, 21 June 2007: http://www.guardian.co.uk/lawrence/Story/0,2763,16026tml11,00.html

21. Richard Norton-Taylor, *The Colour of Justice: Based on the Transcripts of the Stephen Lawrence Inquiry* [2nd printing with corrections] (London: Oberon, 1999). "Prize for Lawrence case drama," *Guardian*, 3 Dec. 1990, 21 June 2007: http://www.guardian.co.uk/lawrence/Story/0,2763,194772,00.html

    Norton-Taylor's theatre credits include *Nuremberg: The War Crimes Trial* (1996), *Justifying War: Scenes from the Hutton Inquiry* (2003), *Bloody Sunday: Scenes from the Saville Inquiry* (2005), and *Called to Account: The Indictment of Anthony Charles Lynton Blair for the Crime of Aggression against Iraq – A Hearing* (2007). All premiered at the Tricycle Theatre.

22. *The Murder of Stephen Lawrence*, dir. Paul Greengrass, perf. Marianne Jean-Baptiste, Hugh Quarshi, Granada Television and Vanson Productions, 1999. Greengrass's subsequent credits include *Bloody Sunday* (2002), *The Bourne Supremacy* (2004), *United 93* (2006), *Imperial Life in the Emerald City: Inside Iraq's Green Zone* (retitled Green Zone, in post-production).

23. Sam Jones, "'Does he understand why war took place?': The Iraqi boy cast as the human face of war," *Guardian*, 19 Apr. 2003, 21 June 2007: http://www.guardian.co.uk/editor/story/0,12900,939509,00.html

24. Terry Eagleton, *Sweet Violence: The Idea of the Tragic* (Malden, MA: Blackwell, 2003), 27.

25. Carolyn Nordstrom rejects the tradition in Western philosophy that views violence as functional, that is, productive, and linked to creativity in *A Different Kind of War Story* (Philadelphia: University of Pennsylvania Press, 1997), 15.

26. Henri Hubert and Marcel Mauss, *Sacrifice: Its Nature and Function*, trans. W. D. Halls (1898; Chicago: University of Chicago Press, 1964), 1.
27. E. E. Evans-Pritchard, Foreword, *Sacrifice: Its Nature and Function*, by Hubert and Mauss, trans. Halls (1898; Chicago: University of Chicago Press, 1964), viii.
28. Siobhan Holohan, *The Search for Justice in a Media Age: Reading Stephen Lawrence and Louise Woodward* (Aldershot: Ashgate, 2005), 1.
29. Stuart Hall, Charles Critcher, Tony Jefferson, John Clarke and Brian Roberts, *Policing the Crisis: Mugging, the State, and Law and Order* (Basingstoke: Macmillan, 1978).
30. Simon Cottle, *The Racist Murder of Stephen Lawrence: Media Performance and Public Transformation* (Westport, CT: Praeger, 2004), 195.
31. Victor Turner, *Dramas, Fields and Metaphors: Symbolic Action in Human Society* (Ithaca, NY: Cornell University Press, 1974), 78–9. Quoted by Cottle, who structures his argument in terms of the social drama (p. 39). For a related, insightful analysis that focuses on the interplay of reality, theatrical tropes, and live theatre, see Janelle Reinelt, "Towards a Poetics of Theatre and Public Events: In the Case of Stephen Lawrence," *TDR/The Drama Review*, 50:3 (2006), 69–87.
32. Tariq Ali, "Lost boy," rev. of *The Case of Stephen Lawrence*, by Brian Cathcart, *New Statesman*, 21 June 1999: 46.
33. See Luc Boltanski, *Distant Suffering: Morality, Media and Politics* (Cambridge: Cambridge University Press, 1999).
34. See Institute of Race Relations (IRR) websites: http://www.irr.org.uk/2002/november/ak000002.html;http://www.irr.org.uk/2002/november/ak000008.html
35. For the Trust mission "to promote diversity in architecture and associated professions" in full, see: http://www.stephenlawrence.org.uk/flash.html
36. Simon Baker, "Lawrence suspects jailed for racist attack on detective," *Independent*, 6 Sept. 2002, 21 June 2007: http://news.independent.co.uk/uk/crime/article176145.ece
37. "Lawrences awarded OBEs," *Evening Standard*, 31 Dec. 2002, 21 June 2007: http://www.thisislondon.co.uk/news/article-2698316-details/Lawrences+awarded+OBEs/article.do
38. Although the site has now changed, a full array of images was available at: http://www.mirrorpix.com/ (accessed May 29, 2003 and November 20, 2005).
39. Andrew Billen, "The Courage and the Menace," rev. of *The Murder of Stephen Lawrence* (ITV television), *Evening Standard* (London), 19 Feb. 1999. Online. LexisNexis Academic. 21 June 2007.
40. Stan Hey, "All because he was black," rev. of *The Murder of Stephen Lawrence* (ITV television), *Mail on Sunday*, 21 Feb. 1999. Online. LexisNexis Academic. 21 June 2007.
41. Alastair Macaulay, "No happy endings in real life," rev. of *The Colour of Justice* by Tricycle Theatre, *Financial Times* (London), 14 Jan. 1999. Online. LexisNexis Academic. 21 June 2007.
42. For example, Macaulay; see also, Robert Butler, "Trial by Tricycle Theatre," rev. of *The Colour of Justice* by Tricycle Theatre, *Independent* (London), 17 Jan. 1999. Online. LexisNexis Academic. 21 June 2007; and Michael Billington, "Arts: Guilty as Charged," rev. of *The Colour of Justice* by Tricycle Theatre, *Guardian*, 14 Jan. 1999. Online. LexisNexis Academic. 21 June 2007.
43. "Lawrence play to go on-air during run," *The Stage*, 12 Feb. 1999: 3.

44. Quoted by Dr Richard Stone in "Do 'People Like Us' get the best out of grants?," The 2000 Allen Lane Foundation Lecture, Ismaili Centre, London, 8 Feb. 2000. Stone was an advisor to the Lawrence Inquiry. See: http://www.allenlane.org.uk/2000.htm

45. For example, "Student dies in racist axe attack," *BBC News* (World), 31 July 2005, 21 June 2007: http://news.bbc.co.uk/2/hi/uk_news/england/merseyside/4730559.stm

46. "Damilola remembered," *Daily Mail*, 27 Nov. 2001, 2 June 2007: http://www.daily mail.co.uk/pages/live/articles/news/ news.html?in_article_id=86401&in_page_id =1770

47. "Mother's axe killing vigil appeal," *BBC News*, 3 Aug. 2005, 21 June 2007: http://news.bbc.co.uk/1/hi/england/merseyside/4740095.stm

48. Frantz Fanon, *The Wretched of the Earth*, trans. Constance Farrington, preface by Jean-Paul Sartre (1961; New York: Grove, 1968), 212, 214 and 206–48 *passim*. In *Black Skin, White Masks*, trans. Charles Lam Markmann (1952; New York: Grove, 1967), Fanon analyzes mechanisms by which members of such Othered subject populations incorporate – and can reject – the valuations constructed by their oppressors.

49. Robert Miles and Malcolm Brown, *Racism*, 2nd edn (London: Routledge, 2003), 101–2. Chapter 1, "Representations of the Other," traces the early history of the dialectic in the West (pp. 19–53). See also Gurchand Singh, "The Concept and Context of Institutional Racism," in Alan Marlow and Barry Loveday (eds), *After Macpherson: Policing after the Stephen Lawrence Inquiry* (Lyme Regis: Russell House, 2000), 29–40.

50. For PBS website, see: http://www.pbs.org/wgbh/masterpiece/lawrence/ei_case. html. For cover notes, WGBH Boston Video G36193, packaging copyright 2002, WGBH Educational Foundation.

51. Mark Redhead, "Media: The justice game," *Guardian* (London), 15 Feb. 1999. Online. LexisNexis Academic. 21 June 2007.

52. See Simone Weil, *The Illiad, or The Poem of Force*, trans. Mary McCarthy (1945; Wallingford, PA: Pendle Hill, 1956), for the effects of violence.

53. Norton-Taylor, *Colour* 5. Additional citations from the play will be noted in the text.

54. Michael Coveney, *"The Colour of Justice,"* rev. of *The Colour of Justice* by Tricycle Theatre, *Daily Mail*, 15 Jan. 1999: 50.

55. Dominic Cavendish, "Theatre: And nothing but the truth," *Independent*, 6 Jan. 1999. Online. LexisNexis Academic. 21 June 2007.

56. Benedict Nightingale, "Questions that need to be asked," review of *The Colour of Justice*, Tricycle Theatre, *Times* (London), 14 Jan. 1999. Online. LexisNexis Academic. 21 June 2007; see also Macaulay.

57. Paul Taylor, "Theatre: Guilt in all its subtle shades," review of *The Color of Justice*, Tricycle Theatre, *Independent* (London), 14 Jan. 1999. Online. LexisNexis Academic. 21 June 2007; see also Nightingale.

58. Spectators also could draw upon knowledge of the victim accumulated from media reports over the six years since his killing and the numerous media events scheduled to coincide with the anticipated release of the Macpherson Report. See Billen.

59. Erika Fischer-Lichte, *Theatre, Sacrifice, Ritual: Exploring Forms of Political Theatre* (London: Routledge, 2005), 23.

60. Nicholas De Jongh, "Compelling trial by theatre," rev. of *The Colour of Justice*, Tricycle Theatre, *Evening Standard*, 13 Jan. 1999:60; see also Nightingale, Taylor.

61. Arifa Akbar and Owen Walker, "Family of suspected race killing victim mourn 'bright, loving son,'" *Independent*, 25 July 2006, 21 June 2007: http://news.independent.co.uk/uk/crime/article1195248.ece

62. John Peter, "The week's theatre," rev. of *The Colour of Justice*, Tricycle Theatre, *Sunday Times* (London), 17 Jan. 1999. Online. LexisNexis Academic. 21 June 2007. See also Billington; director Nicholas Kent also references the Greek example, in Cavendish.

63. Ali's views complement this form of activism. Although he distrusts the efficacy of empathy as a political motivator, he praises the human agents – specifically Stephen's parents – who work to effect change.

64. Articulated in a telephone conversation with playwright/scholar Irma Mayorga, 5 May 2007. Thanks also to Carrie Sandahl and Laura Edmondson for their words of encouragement and support.

7
Violence Makes the Body Politic(al): Technologies of Corporeal Literacy in Indian Democracy

Maya Dodd

The political community [. . .] is always a community of remembrance.[1]

Despite amendments to the Indian Constitution since 1975, the performance of political community in India has increasingly been underpinned by material acts of violence. Given the non-violent aspects of the freedom struggle, the recurrence of violence in staging modes of political belonging marks a curious departure. Enacting violence in staking communitarian claims is becoming a troubling aspect of Indian democracy and this inquiry follows the processes that bind individual bodies to the body-politic.

By scrutinizing seemingly discrete events that occurred in India across the decades since 1975, the reasons for an amplification of individual publicity being premised on bodily violence become evident. As democratic performance, through a taxonomy of infamous, famous, and anonymous bodies, I map how bodies serve as sites for a contemporary creation of communitarian identity. Individual acts of violence are politicized through the collective logic of communities. The centrality of the physical body to a national project reveals that *literacies of corporeality* are as much a response to the daily violence of the state in fixing social meaning as they are to conflict resolution by other means.

Since 1975, sacrificial violence not only seeks to constitute community but also embodies responses to the Indian Constitution's avowal of secularism and socialism. Consequently, to comprehend where the precedents of such performances lie, the discourses around what historically constituted the private sphere grants an understanding into recurrent violence in the post-Independence era.

The body of history

The dawn of Indian independence arrived in 1947, marked by the deaths of over a million people conscripted into citizenship. The violent ruptures

152

dramatizing the imagining of nation into state were performed on bodies of the famous, the infamous, and the anonymous. After 12 million people crossed a newly drawn border, the assassination of Mahatma Gandhi in 1948 signaled the end of one struggle and the beginning of another. This new beginning was inscribed in the Indian Constitution which came into effect in 1950. As postcolonial nation-statism separated peoples and land, the resilience of earlier forms of community revealed another narrative.

In raising the question of how "individual biography becomes social text," Veena Das identifies the use of "detailed local knowledge through which people are recognized, named and their individual misery transformed into the misfortunes of the community."[2] This essay proceeds from the premise that notions of community do not merely persist as transhistorical legacies, but are actualized in the present through particular means. This argument demonstrates the idea that frequently, and particularly in post-Independence India, the performance of political community is underpinned by material acts of violence. Given the amendments to the Indian Constitution since 1975, the recurrence of violent modes of political belonging prompt this interrogation: Notwithstanding a long legacy of nonviolent struggle against the British, why did violent death (by assassination, murder and suicide) become the preferred mode to stage victimhood and politicize political community after the formation of the postcolonial state?

On a deeper probing, one sees that highly publicized deaths have increased since 1975, often *manifesting* the very gaze of the postcolonial state even as such individual publicity stakes group claims against the state. By terming the instances I will describe below as *literacies of corporeality*, my argument focuses on the fact that these violent events acquire primary signification through their performance of community. The deaths become performances of a rhetorical process explicitly seeking political ends. In *The Body In Pain*, Elaine Scarry writes, "At particular moments when there is within a society a crisis of belief [...] the sheer material factualness of the human body will be borrowed to lend that cultural construct the aura of 'realness' and 'certainty.' "[3] In the instances of publicized death I will go on to describe, there is a concomitant obfuscation of the "materiality" of the violence, as the individual body's pain is extended to function within a broader discourse: the democratic provisions of the postcolonial state.

By scrutinizing seemingly discrete events that occurred in India across the decades since 1975, it becomes increasingly apparent that in postcolonial history there is an amplification of a public aspect of democracy that is premised on bodily violence. Even as practices of making the state visible have served to activate democracy, democratic maneuvers have often taken on violent performances to activate community. To query this, it is important to understand the peculiar nature of violent episodes in the time-frame *following* the establishment of legal rights to protect citizens. Although the guarantees of a democratic Constitution sanction legal conciliation, the recurrent resort to violence in staging communitarian claims prompts the following inquiry into the relations between community, privacy, and possession that forge the bond between individual bodies and the body-politic.

The following events demonstrate how in these violent enactments "incidents" that seem exceptional become national events because they index the anxieties of negotiation between communities and the state. The following episodes, which occurred in recent Indian history, will delineate the ways in which a continued activation of such notions of community has occurred through specifically violent events in post-Independence India. I employ a tentative taxonomy of infamous, famous, and anonymous bodies to emphasize the aspects of publicity in these sites, but with the purpose of seeking convergences between their enactments of community, privacy and possession.

Legal reinscriptions

The relations between the ideologies of possession, privacy, and community can be seen as constitutive to the logic of politics in the post-Independence period, though the origins of these dynamics are not coterminous with *only* this phase of Indian history. If the story of Indian democracy is specifically that of a *postcolonial* democracy, it needs to be related not only through persistent legacies of anti-colonialism, which after the event of Independence are directed at the state, but also to acts of democratic publicity and individual performance that draw from local repertoires.

Since 1975, post-Emergency India is distinctly characterized by crises of governability, declined state legitimacy, and a distinct souring of the dream of postcolonial freedom through undemocratic practices and the failed translation of procedural democracy to substantive democracy. In this regard, the changes to the Constitution effected by legislation during the Emergency also render the year 1975 as especially noteworthy. In reminding us of the distinctive cornerstones of Indian democracy, the authors of *Reinventing India: Liberalization, Hindu Nationalism and Popular Democracy* write:

> In the event, the Constitution which they finally promulgated on 26 November 1949 did say that India was to be a "sovereign democratic republic," but the words "socialist" and "secular" were only introduced into the preamble of the Constitution much later, as a result of cynical amendment passed during the period of 'Emergency Rule' of Prime Minister Indira Gandhi between 1975 and 1977.[4]

Even after the period of the Emergency, that these particular changes to the Constitution were not revoked is of vital consequence to the tale of democracy's mutations. In outlining how the state's addition of "socialist" and "secular" to the Constitution translates into the cultural logic of the post-Emergency period, these Constitutional changes foreshadowed a transformation of electoral politics with substantial implications for narrations of democracy.

Corporeal communiqués

Perhaps it is fitting that what begins as a story of democracy, through a center-staging of the postcolonial state and its promise of freedom for postcolonial subjects, now transforms into a miniaturized story of practices of individual freedom. I am prompted to turn the lens over because the chronicles of democracy and the state both exceed the formal aspects of constitutional relations, and also adhere to state formations in unanticipated ways. The elusive category of "infamous bodies" inaugurates the embodiments of democracy's excesses when met with its exceptions.

1 Infamous Bodies

Is fear still fundamentally an emotion, a personal experience, or is it part of what constitutes the collective ground of possible experience? [...] There is, however, such a general consensus that we cannot separate ourselves from fear, that it is necessary to reinvent resistance. Fear, under conditions of complicity, can be neither analyzed or opposed without at the same time being enacted.[5]

In 1988, the then Prime Minister of India, V. P Singh, or the *Raja of Manda,*was popularly immortalized as the "Raja of Mandal" following his efforts to implement policies of affirmative action as recommended by the Mandal Commission Report.[6] Appointed under Moraji Desai's Janata government in 1979, the Mandal Commission released its inquiry into identifying socially and educationally backward castes in 1980, and advised redress for OBCs (Other Backward Castes) through an increase of quotas to 49.5 percent. The controversial report gathered dust for nearly a decade, till Singh announced its implementation through a proposal to increase occupational and educational reservations up from the standard 27 percent figure put in place by Constitutional provisions in 1950 . This form of affirmative action sought to cover an unprecedented number of government jobs, and academic admissions (to technical and other institutions), thus reducing the number of "open" category positions available to the rest of the population (who were not identifiable as belonging to a "backward caste").

The fact that Singh, an upper-caste *Rajput,* would seek to restrict the access of "his own kind," albeit through democratic provisions, was first viewed with suspicion (as being a blatant attempt to decisively capture new votebanks for his party), and then met with widespread outrage as a betrayal of the upper castes. Not since the 1970s had the country's "youth" been so publicly politicized, as demonstrations by high-school and college students took North India by storm (to the extent that academic sessions had to be canceled for several months). Effigies of V. P Singh were publicly burnt; the streets and walls were covered in political graffiti (with bitter slogans like *"V P Kutta"* damning the Prime Minister as a dog), and everywhere one looked, the signs of young discontent were all too apparent.

The one event that crystallized the extent of discontent among upper-caste youth was a strangely barbaric one, whose main actor has now all but disappeared from public memory. Rajiv Goswami, a student at Delhi University, in joining the demand of many upper-caste students staged what would become the most potent symbol of the agitation. He convened a meeting at Deshbandhu college and, in front of thousands of teenage students, declared that if student demands to reduce reservation quotas were not met, he would set himself on fire. A prior announcement was made and the media arrived. At first there was skepticism of Goswami's intentions, but as he began to douse himself with kerosene and set his body alight, the media quickly realized this was no effigy being burnt and were right on hand to capture images of a real, live, young body on fire.

The image of a burning Rajiv Goswami was caught on tape, seizing headlines around the country and in turn sparking off a series of self-immolations as imitations of Anti-Mandal protest. Though Goswami survived his attempt (despite sustaining severe damage, he was rehabilitated after extended hospitalization), his burning body became the emblematic sign of and site of resistance to the state.[7] Rajiv Goswami's act of violence performed on his burning body became the ultimate stage on which to perform political claims against caste reservation.

It is crucial to note that Goswami's act was performed without any active "assistance" and this can be viewed through Emile Durkheim's proposed category of the "altruistic suicide."[8] The category defines an act marked by an excess of integration, wherein the individual dissolves his/her sense of self into the broader claims of group identity. In Durkheim's formulation, members of the military were especially prone to this tendency. Goswami's act of threatened suicide may be viewed as a civilian act aspiring to a military mode. Using his own body as a weapon of commentary, he was answering the call of duty of his "community," where a perceived threat of job losses was personified in Goswami's very being. Thus, in sharp contrast to V. P. Singh's distance from community interests, Goswami's attempted self-immolation gestured to his one-ness with his community. Also, given the state's legal proscriptions against suicide, the act itself embodied a defiance of V. P. Singh's government. In attempting to invoke annihilation, Goswami was claiming for himself the risks faced by his community and the need for a dramatic solution was routed through his performance of a visible violence.

This was a period in India's history where even in urban settings that usually went unmarked by caste, who you were increasingly began to include a reference of your last name: a transparent naming of your caste community. The rhetoric of the threat of caste could ultimately be seen as an economic one powerfully vocalized as: "They will take our jobs away; Who wants to be operated on by inept doctors who only gained admission to medical college because of their caste," and the like. Such utterances combined the worst fears of an insecure middle class who had frequent cause to wonder if feudalism had ever yielded to modernity.

The conjunctural coincidence of the Mandal Report's release in 1980, at the instance of a government opposed to the authoritarian revisions of the Emergency, bears reflection. That more than 30 years after Nehru's capitulation to dissent against the inscription of the word "socialist" in the Constitution, Indira Gandhi succeeded in amending the Constitution in 1976 to retroactively include the words "socialist and secular" is significantly proximate to the Mandal Report.[9] At least the formal imperative to acknowledging the need for redistributive justice ran across party lines. Unfortunately, the burning body of Rajiv Goswami offered the lesson that even in urban settings in modern India, caste was more durable than flesh.[10]

2 Famous bodies

It is one singularity among others which, however, stands for each of them and serves for all.[11]

Dead bodies that adamantly refuse to "be absorbed into the background," may be deemed, after Giorgio Agamben, as "exemplary" bodies. He writes, "One concept that escapes the antinomy of the universal and the particular has long been familiar to us: the example. In any context where it exerts its force, the example is characterized by the fact that it holds for all cases of the same type, and at the same time it is included among them."[12] The body of a Prime Minister aptly serves as such an exemplary body and so it follows that twice in post-1975 India, the outpourings of national dissent have taken place on the actual bodies of these exemplary figures, through the tragic staging of political assassinations. After Indira Gandhi's assassination in 1984, her son Rajiv Gandhi was swept to power in a landslide victory.[13] Unfortunately, he was also to be the second Indian Prime Minister to suffer a similar fate.

It was in the year 1991 when history took some decisive turns in India: the national markets were opened up to the fast-approaching flows of global capital, and it was finally time to "say with pride that one was Hindu" (as evidenced in the popularity of the BJP slogan *"Garv se kahon hum Hindu hain"*). But before India could fully break loose of its nominal secular socialism, the living legacy of Nehru's dynasty, embodied in the figure of the Prime Minister, Rajiv Gandhi, was assassinated at an election rally.[14]

Rajiv Gandhi paid with his life for sending Indian Peace Keeping Forces to Sri Lanka during a civil war, and the LTTE (Lankan Tamil Tigers Elam) responded through the barrel of a different gun: the act of a suicide bomber. In gaining prominence through the "exemplary" dead body of a prime minister, as had been done by the Khalistani movement, the LTTE sought to publicize their cause of a separate Tamil state through a definitive performance. The ex-Prime Minister's handling of the Tamil-Sinhala conflict in Sri Lanka lay at the essence of this act of violence. However, the brutal annihilation of the Primer Minister's body exemplified the state's failure to perform its primary duty: the preservation of the life of its citizens.

To view this incident from the opposite side of the lens, one could say that the suicide-bomber had challenged the state's claim to even secure "bare life"[15] for its subjects. In choosing to die for her cause, she fixed the interpretation of her death: as a defiance of the Indian state's policies in Sri Lanka. This final sacrifice did as much to dissolve her corporeal limits within the larger community of the LTTE as it did in extracting from the Prime Ministerial body the ultimate requital.

As the nation, then, tried to avert its gaze from the *Frontline*[16] photographs, which tantalizingly displayed the graphic photographs of pieces of Rajiv Gandhi's scattered body, today, one can't help but notice that the "event" has exceeded his corporeal destruction. In retrospect, more than four decades of "secular" rhetoric lay mangled in the blown up body of the one-time Prime Minister. Just as Mahatma Gandhi's assassination in 1948 signaled the end of a certain possibility, these key episodes rendered the fragility of even the politically elevated corporeal form. It is striking to view the figure of Rajiv Gandhi as the embodiment of the "secular" nature of his party's professed credo, since his death was dramatically followed by the death of a certain type of secularism. The 1992 attack on the Babri Masjid accentuated the aged body of Prime Minister Narasimha Rao, as he seemed to represent the tired and exhausted secular credo of his party, the Congress-I. The rise of Hindutva forces from then on seemed an event propitious in its divine "embodiments" after such corporeal frailties.

3 Anonymous bodies

Those excluded from the capitalist relation incarnate its form directly in their bodies: they fall, they were [...] They directly embody the ungraspability of the capitalist present: disaster.[17]

The "communalization" (i.e., an increased presence of religion in politics) of India persistently defies the Nehruvian ideal which projected the image of a "secular" India, organized on principles of social equity and enlightenment rationality. It is not that the aspiration towards a horizon of modernity was abandoned, as much as the fact that regularly intermittent episodes of communal riots never quite allowed a deeply divided populace to forget the possibility of war in their midst. Again, just as with the conjunctural case of Mandal's occurrence soon after Indira Gandhi's inclusion of "Socialist" in the Preamble to the Constitution, her inclusion of "Secular" was an almost ironic foreshadowing of what was to come. It is interesting to note how subsequent to these Constitutional amendments the postcolonial nation-state was most brutally challenged by violent separatist movements in Kashmir, Punjab, and Assam, to name a few. Haunted by the ghosts of the two-nation theory that divided India from Pakistan, the "secular" state sought to rid itself of communal associations. In Punjab, almost every other family harbors tales of violence and death in their family biographies. Some of these narratives have been chronicled in *The Other Side of Silence* by Urvashi Butalia.[18] In Bengal, a deafening silence surrounds the survival of refugees from East Pakistan. After the partitions of Pakistan and Bangladesh, the 1984 Sikh riots in Delhi, which ensued in the wake of Indira Gandhi's opprobrious Blue Star operation in the Golden Temple, blasted away the state's secular credentials and returned those ghosts to the Indian state's front door. The riots of 1984 injected a strategic urgency for the need to create separate country for a "martial Sikh community." Underlying these manifestations of religious community, are anonymous un-enumerated subjects whose solidarities are premised on future-resistances to repetitions of the past. Every single time, they said, "Never again."

In the un-naming of specific individuals through a categorical invocation of "community," one fact becomes apparent: the intimate knowledge of "embodied risk" underpins community formation. The especial failure of the state to guarantee security to individuals in communities (prompting secession or retaliation) can be productively related to the suggestion that in the context of ecological risk, there is a breakdown of "the classical conflict model of public versus private interest."[19]

When presented with imminent dangers, the dissolution between subjecthood and the context of its performance attracts the subject into the refuge of community. Communal riots render victims anonymous (unlike the identification of agents of violence such as the LTTE), but anonymous claims resound even more loudly when faced with a deafening silence from the state. The failure of the state in these instances could be related to its deferral to a communitarian arbitration of retaliatory justice.

In reviewing the case of the trial of Rajiv Gandhi's assassins, the suicide-bombers' defiance of the Indian state's sovereignty poses a different picture: The state is predicated on a toleration for this formation of violence so long as it is enacted by subjects within its legally enforced territorial limits. The Tamil Tiger separatists were subject to indictment through a criminal trial, as the case concerned a political assassination instigated by a community of *non-Indian citizens*. It was imperative that the state pursue prosecution as a display of territorial sovereignty. Given that this was not a group that the Indian state had formal jurisdiction over, the LTTE differed from other "communities" in that it had enacted an extra-national violence countering the hegemony of national civil boundaries.

Unlike the "justiceability" of the LTTE, in the case of the Sikh massacres of '84, the demolition of the Babari Masjid in 1992, the nation-wide riots that occurred in 1993, and particularly the riots in Godhra in 2002, the negligence of the state to deliver justice to the aggrieved parties raises the question of complicity. The resurgences of violent conflict in the above cases point to two things: the original claims of community appeals addressed to and through the state, as well as the failure of the state to meet these claims. Was turning a blind eye to the violence a method of granting sanction to community demands?

In Bombay, even those legally proven of being guilty of atrocities committed during the 1993 riots in Bombay (namely Shiv Sena chief, Bal Thakeray) remain unpunished by the institutions of the state. In the memorandum of the Action Taken Report (ATR) in which the ruling Shiv Sena-BJP government responded to the report of Justice Srikrishna, the following comments were made:

> The Government broadly agrees with the observations of the Commission about the background of the riots. The Government also feels that the Special Civil Code for the minorities, reversal of decisions in the Shah Bano Case, opposition to the singing of Vande Mataram, use of loudspeakers for Namaaz and the inconvenience caused to the public because of the obstructions on streets created by Namaaz offering mobs, the honorarium granted to Maulvis, the concession granted for Haj pilgrimages also led to further bitterness between Hindus and Muslims. This alienation and mutual distrust is responsible for the occasional occurring riots and the riots started on 6th December 1992 and thereafter and 6th January 1993 and thereafter.[20]

Given that the comments were issued from the party in power, which itself was one of the key accused, questions of complicity are hard to ignore. What is telling about these comments though is their emphasis on aspects of civil life and the stress on the incommensurability of civic practices. Upon noting the perceptions of ineffectiveness, justice delayed, and suspicion with which the state has been viewed by different communities in these cases, I turn away from the stage of the state to a public parallel justice system where conflicts between communities are often settled in an equally deadly way.

Love and the liberal state

The need to interrogate the relations between possession, privacy, and community in post-Independence India is epitomized by the reportage of an event which occurred in August 2001: the circumstances surrounding the murder of two lovers of different castes who were publicly hanged to death by their parents. On 7 August 2001, the police reported that at Alinagar ka majra village (in Muzaffarnagar district in Uttar Pradesh), Vishal a 20-year-old Brahmin youth, and Sonu, his 18-year-old beloved, of the Jat community, were hanged to death due to public opposition of their inter-caste relationship. The journalistic accounts vary in their details.[21] The *Times of India* News Service claimed that the couple was killed after they eloped and were then tracked down by a manhunt, followed by a public hanging which was witnessed by 200 odd villagers. A CNN report stated that the couple themselves, being aware of the impossibility of their inter-caste marital future, committed suicide together. An Associated Press release claimed that the families discovered the lovers during a rendezvous, and in a united gesture to protect the community peace of the village, made a mutual decision to hang the couple.

Though it was unclear whether the witnesses to the hanging numbered hundreds or only a few close relatives, no one came forth to officially testify. At first glance, this incident may be seen as "shocking" and "brutal" (as described by CNN and the BBC respectively). However, to view the actions of the lover's families as barbaric and as "bizarre" would be to grant their actions clemency, as it would follow from the reasoning that such events are rare (and so the motivations that inform them must also be unusual). In examining the details surrounding the event, certain issues come into focus: Why did both the parents agree to pronounce such violent censure on their offspring? If we believe that, indeed, both families *did* publicly hang Sonu and Vishal, then how did they overcome caste barriers to ensure their death, given that the couple's crime emanated from their attempt to transgress caste? And what is one to make of the Uttar Pradesh Home Minister Ram Nath Mishra's claim that, "There is no tension in the village because both families had given their consent for this. There is peace in the village. Had it been one family that had killed both people there could have been tension."[22] Is this perhaps the very reason that not even one person from the village came forth to testify against the act? Since the parents are assumed to be responsible for killing them, why did they bother to "honor" their deaths, marking them with traditional cremations?

To restrict the import of these questions and limit them to the "incident," is to efface the recurrent and hence more general terms of an interrogation. This incident is not unique because the changing headlines of daily newspapers will testify to its endless repetitions – with new actors, and in different settings – establishing it as a *pattern* and not just a one-time occurrence.

It is my argument that the circumstances which lead up to such violence are a product of the deep-seated tensions accruing from a specific organization of civil society in India. Consequently, in post-Independence national memory, conflicts do not constitute singularly aberrational events in the nation's life, but are often an extension or amplification of violence present in the everyday. The individual acts of corporeal violence are as much a response to the daily violence of the state in fixing social meaning as they are to conflict resolution by other means. The lowest denominator of the national quotidian is expressed by the lives and the individual bodies of the country's citizens, and the regulation of the state is frequently expressed through violence performed by the state's citizens.

The history of democratic forms includes the deployment of physical bodies as communications with the state. Technologies of witness usually foreground processes of relay through literal and visual media. For those outside the ambit of these opportunities, often excluded from the use of such media, corporeality has also served as a technology of witness. In order to understand how literacies of corporeality communicate with the state, attention needs to be directed to the specific "embodiments" of communiqué.

Sacrificial violence has become constitutive to the logic of communitarian strife in India in large measure due to the peculiar legal organization of civil communities. Violent events, which animate the "daily-life" of the state, embody the fundamental antagonisms precipitated by the liberal state's avowal of secularism through awarding citizens cultural rights, and practicing socialism as redistributive justice guaranteed through communitarian policies. To situate the roles of socialism (as redistributive justice) and secularism (as identity) in the state's promise to guaranteeing freedom, the relations between possession, privacy, and community in post-Independence India afford some clues to the reasons for the recurrence of violence in Indian democratic performance.

Possession, privacy and community

The issue of "individual being" is related to the relations between possession, privacy, and community which together produce modern subjects. While addressing the question of agency, Joan W. Scott writes:

> Subjects are produced through multiple identifications, some of which become politically salient for a time in certain contexts, and that the project of history is not to reify identity but to understand its production as an ongoing process of differentiation, relentless in its repetition but also, and this seems to me the important political point, subject to redefinition, resistance, and change.[23]

Following Joan Scott's assertion, the current inquiry into process which started at the level of the corporeal, can now trace a rhetorical process – that of the politicization of individual acts of violence – to its implications on these questions: of possession (and autonomy), community formation (and deviance or non-compliance with the mores of the community at large), and privacy. The points of convergence and divergence between the individual private, the collective private, and the "unclaimed" public sphere (marked by everyone/no-one relations of possession)[24] are brought into relief in the deaths of Sonu and Vishal. Their Jat and Brahmin caste-communities staked the ultimate claim of possession on their beings, resulting in their deaths. The very conception that they were legal "adults," privately free to marry each other under the law, was disavowed with the utmost brutality.[25] Their deaths challenge innocent descriptions of what constitutes the "private" sphere in India. In their case, a "Collective Private" was brought to bear with much more significance on the fates of Sonu and Vishal than did their own willed autonomies. And in a final ironic spin on what *might* be considered "private," the "public sphere" of a law-abiding civil society was obscured by the mass refusal to testify against what would be deemed a "personal" (as in family) matter. The individual biographies of Sonu and Vishal were eventually forged through the collective logics of their communities.

To term the "assertions" of community as resurgences of "primordial" identity is to gesture in the direction of essences. Without positing a straight-forward genealogy between historically substantiated "essences" of community identity, there may be a way to trace the repetition of rhetorical strategies that repeatedly construct communities in a seemingly unchanged way. The refusal to perceive the present as simply deriving from deep-seated fault lines in the past yields a line of questioning that demands an optics of contingency. Thus, in all these violent deaths, the performances constitute the communities from *within* the context of these events. Events, such as the deaths of Sonu and Vishal, serve as *the* sites for a contemporary performance of communitarian identity, but the primary feature of these enactments is that they derive meaning not only through interpretation of *individual bodies*, but through literacies of corporeality.

To understand where the precedents of such performances lie, it becomes necessary to ask if older understandings of privacy and community have formed the historical precedent for the current literacies of corporeality in the post-Independence era

The public body

Varying conceptions of the private and the pubic have generated debates amongst scholars of South Asia, notably between historian, Gyan Prakash and political theorist, Partha Chatterjee. Prakash challenges Partha Chatterjee's widely accepted thesis that "anti-colonial nationalism con-structed an image of the nation's 'inner' sphere of spirituality and culture that *rejected* the 'outer' sphere of modern science, technology and mate-rialism associated with the west."[26] In identifying an explanation for the effectiveness of anti-colonial discourse, Prakash draws attention to dis-courses on the body, which were current during colonial times. He writes,

"The language of Hindu ascesis always functioned under the pressure from corruption and error, always embattled and struggling to restore the body to its original Hindu-national condition. It was through such pulsating deployment of Hindu signs that elite nationalism acted on the medicalized, corrupt and enfeebled body to render it healthy and Hindu-national."[27] The super-imposition of "Hindu" onto "national" may historically echo with the claims of contemporary Hindutva movements, and thus has a protractedly difficult history. But for the purposes of understanding the relations between privacy, community, and bodily possession, its pertinence rests on the centrality of the physical body to a national project.

Prakash argues that during the colonial period, several Indian intellectuals redirected the discourse surrounding the body and sexuality into nationalist rhetoric by basing it on the idea of a healthy body-politic. Construing this health as essential to the struggles against colonialism, such an assumption borrowed heavily from a patriarchal economy (which valued the "strength" of semen alone) as well as from public-health discourses emanating in *colonial governmentality*.[28] It also derived from the primacy of social relations, which sought to subsume individual interest in the service of collective interests.

By reviewing discourses around what historically constituted the private sphere, it is clear that the limits of an individual body cannot be easily supposed to have been individually possessed. In pre-colonial India, Prakash demonstrates that there is no uncomplicated association between individual bodies and individual autonomy. Instead, the limits of the "private" and the "intimate" are expressed in *domestic* terms (of intimate "social-relations"). Thus, if possession (mine/not mine) denotes the situating of identity, then the body too is imbricated in this discourse of possession, which is defined in the act of "staking claim." In *Colonizing the Body*, historian David Arnold draws attention to how anti-colonial sentiment gained impetus from the conflicts over staking claim of material bodies.[29]

Just as the nationalist cause was greatly aided by the primacy of social relations, which sought to subsume individual interest in the service of collective interests, the colonial state too attempted to discipline individuals to manage the body politic. Political theorist Sudipta Kaviraj corroborates the fact that "Colonial rule introduced the conception of disciplining everyday conduct to give shape and form to the body politic [...]. [Colonial] governing conventions were internalized by the Indian middle class, for whom control of everyday uses of space was an indispensable part of the establishment of social sovereignty."[30]

By describing the colonial governance of public space in Calcutta, Kaviraj illustrates the manner in which the *"Ghaire/Baire"* (Bengali terms for "Inside the house/Outside it")[31] distinction is mapped over the distinctions between the public and the private.[32] He argues that the limits of the "private" and the "intimate" are expressed in *domestic* terms (of intimate "social-relations").

In light of this understanding of the body, Sonu and Vishal's deaths afford an examination into how in contemporary India communities stake claims over individuals and reveals the crucial aspect of publicity in democracy. The broader Indian situation and this incident in particular offer provocative ways to think about the larger issues involved in mapping the claims of minoritarian and communitarian politics vis-à-vis the liberal state.

Democracy's aftermath

The ghastly deaths of Sonu and Vishal indicate a failure of "deliberative democracy" as well as the failure of the state to perform its primary function: the protection of individual rights. The especial failure of the state in this instance could be related to its deferral to a communitarian arbitration of retaliatory justice. Despite the media's referencing of the local police action (stymied by a lack of forthcoming witnesses), this event was marked by minimal interference from the liberal state, despite the fact that the Constitutional rights of Sonu and Vishal were brutally violated by their communities. Given that their deaths served to make an ostensible "peace in the village," even the State Minister's comments seemed to pre-empt state "interference" as the ultimate concession was made to community sovereignty.

Events such as the deaths of Sonu and Vishal, which animate the "daily-life" of the state, embody the fundamental antagonisms precipitated by the liberal state's quest to guarantee its citizens particular cultural rights. If we take Chantal Mouffe's characterization of liberalism as "the rule of law, the defense of human rights, and the respect of individual liberty" and that of democracy as the ideologies of "equality, identity [shared] between governing and governed and popular sovereignty,"[33] it will soon become evident how the form of liberal democracy in India differs from its modes in the West.

The specific performance of the Indian state in liberal-democratic terms is predicated on acceding to varying communitarian logics. Clearly, the deaths of Sonu and Vishal demonstrate the resilience of caste differentiation as a powerful principle in organizing community: one that takes precedence even over the contract of citizenship in safeguarding individual rights in liberal democracy. The incident complicates the presumption that contemporary civil society is a social order forever changed by the entrenchment of modern institutions of the state. On reviewing the Indian state's form of liberal democracy, which predicates civil law through an institutionalization of difference (by affirming caste reservation and personal laws that are responsive to religious difference), the question arises if in fact the violent performance of difference is in essence a *product* of the *state's very structuring of difference*.

In upholding the figure of the citizen as a constitutional subject, the state's civil mediations between communities and individuals do not merely derive from an imported idea of *the citizen*, but also derive from an extension of traditional identity applied to individuals. The legal structures of the country reinscribe the religious distinctions between communities by applying specific "community-laws" to administer civil behaviors (marriage, property, inheritance, and divorce). This is most apparent in the cases of marriage, and inheritance, which fortify the bases for the markedly discernible boundaries between communities.

For instance, Article 25 of the Indian Constitution treats "Buddhists, Sikhs and Jains as 'Hindus' for legal purposes" (in effect treating these religions as off-shoots of Hinduism).[34] In addition to the effacement of religious diversity (which needs to be related to the real impossibility of legislating the infinite religious varieties present in the country), the different rights accorded to individuals within these frameworks creates a situation of incommensurability, eventually undermining the "neutral" power of the state to function at the center of competing discourses. The Indian nation-state's legal and political discourses firmly entrench a communitarian bias in providing an official space of sanctioned forms of identity from which claims are then made.

The deaths of Sonu and Vishal exemplify a form of social interaction between different social groups, or what Mouffe terms "the nature of the political."[35] Whether the differences are based in hate or history (and they are often based on a fatal combination of the two), the task of arbitrating justice is often usurped by communities in an immediate exchange, who then choose to implement their own norms, instead of ceding such conflicts to the state. It is this tension between liberal democracy in the abstract, and the distance of this abstraction from a substantive practice (by the state's citizens) that would seem to lead up to the tragedies of communitarian violence.

In fact, it is only through the lens of a liberal democratic perception of justice that such a communitarian presumption of "justice" can be seen as unfair. In a sense, the Indian state's form of liberal democracy is unique because it establishes a priori, the authority of communitarian logic in state-mechanisms by apprehending civil law through communitarian logic. The reasons for such a structuring of civil society in India emanate from a profound particularity of varying definitions of public and private in a local context. This paradox is central to Indian democracy and perhaps the "explosion" of communities is inherently related to such a peculiar legacy.

Enacting the state

In exploring the ways in which communities construct themselves as political actors, Veena Das examines the discourse on cultural rights, the control over memory, and the community's "creation" of heroism in individual deaths. She notes:

If we look at the new political actors emerging on the public scene [...] we see that these actors, until their emergence, led anonymous lives, that they then usurped the domain of visibility momentarily (as a consequence of certain critical events), and that they then disappeared. This does not mean that their emergence is not important, it means that we should expect different kinds of political actors to emerge and disappear.[36]

In the events described here, we see the enactment of politics which, inspired by competing differences, are ultimately transacted through modes of violence. A shared identity is based on a common vision for the future and, in contemporary India, equally frequently on the unfortunate perception of an embodied risk in the present that threatens the future from being realized. The resentment of the LTTE found its loudest articulation in the sounds accompanying Rajiv Gandhi's exploding body. Goswami's attempt at a self-immolation embodied the threats faced by his community. The range of violence perpetrated on riot-victims demonstrates deep-seated community anxieties at a micro-level and index various versions of the state's complicity with majority-community sentiment. These "critical events" in Das's terms can be read as attempts to ensure forms of cultural survival in accordance with the state's own structuring of caste and religion and its arbitration (or lack thereof) of conflict. The state's original bestowal of recognition on the lines of identities premised on caste and religion (as done through the Mandal Commission Report and in Personal laws) is what makes this "dialogue" of staking claims (even through the damages wrought by community violence) between the state and its citizens legible and it is what gives meaning to the *literacies of corporeality* generated by political actors.

Keeping in mind the victims produced in the violence of such narratives of community, one could well ask what would happen if individual assertions – outside of community dictates – found their purest enunciation in the word of the state, that is, the "Fundamental Rights" enshrined in the Constitution? The empirical turns of recent history seem to assert that ideas of community in India are pre-eminently grounded in religious, caste, and linguistic differences, as if they were fixed categories. However, the discursive instantiations of corporeality can be seen as such moments enabling this "historical fixity," which in their aftermath, produce symbolic reclamations. Dead bodies of political leaders (Rajiv and Indira Gandhi), burnt bodies of possible martyrs (Rajiv Goswami and Roop Kanwar), and anonymous bodies of 'over-populating' undisciplined citizens (riot victims) have enacted the semantics and narratives of India's recent history, demonstrating that political communities are only ever formed by repeated performance.

168   *Violence Performed*

Notes

1. Veena Das, *Critical Events: An Anthropological Perspective on Contemporary India* (Delhi and New York: Oxford University Press, 1995), 55.
2. Das, 131.
3. Elaine Scary, *The Body In Pain* (New York: Oxford University Press, 1985), 14.
4. Stuart Corbridge and John Harriss, *Reinventing India* (Malden, MA, and Cambridge: Polity Press, 2000), 21.
5. Brian Massumi, *The Politics Of Everyday Fear* (Minneapolis: University of Minnesota Press, 1993), ix.
6. To see the primary text of the Report as well as subsequent commentaries on it, see K. N. Rao and S. S. Ahluwalia, *Mandal Report X-Rayed* (New Delhi: Eastern Books, 1990); Shriram Maheshwari, *Mandal Commission Revisited : Reservation Bureaucracy in India* (New Delhi: Jawahar Publishers & Distributors, 1995); and S. B. Kolpe, *Mandal Commission's Report: A Charter of Rights of 50 cr. OBCs* (Mumbai: Maharashtra Rajya OBC Sanghatana, 1997).
7. Though he did survive this attempt, and later took up active politics, he tragically passed away at the young age of 33 due to health problems.
8. Altruistic suicide is characterized as premodern suicide in Kenneth Thompson's *Readings from Emile Durkheim* (London: Routledge, 2004), 98.
9. It bears mentioning that Indira Gandhi's 1971 electoral campaign, based on the *"Garibi Hatao"* or "remove poverty" campaign, was already associated with her stunning victories at the polls. She actively pursued a pro-poor public image, even if the Congress policies may not have followed that direction.
10. The infamous case of Roop Kanwar, the much debated "sati" in Rajasthan in the 1980s, offers a gendered version of the same phenomena. As the family asserted their pride in their sacrificing and honorable daughter-in-law, the perceived discussions between "our tradition" and "your state" were essentially located in the ashes of her being. Tradition converted her corporeality to divinity – as based on her loyal and courageous act of following her husband into the burning funeral pyre – Roop Kanwar was deified into a goddess; as deserving of worship as the figures of Kali or Sita.
11. Giorgio Agamben, *The Coming Community,* trans. Michael Hardt (Minneapolis: University of Minnesota Press, 1993), 9.
12. Agamben, 9.
13. Under Indira Gandhi's reign, India was victorious in the 1971 war and her prime ministerial reign was characterized by her famed iron will. Her style of politics guided an aggressive nuclear program and a draconian nineteen-month Emergency, but her direct and uncompromising stance on Sikh separatism finally led to the loss of her life. After Operation Bluestar in June 1984, the hostility that ensued from her militant storming of the Sikh Golden Temple intensified and eventually, within five months of Bluestar, Indira Gandhi was shot dead by her own bodyguards.
14. Press reports detailing this event are collated in K. L. Chanchreek and Saroj Prasad (eds), *Rajiv Gandhi's Assassination: A Blow to Democracy* (Delhi: H.K. Publishers & Distributors, 1991).
15. Giorgio Agamben, *Homo Sacer: Sovereign Power and Bare Life*, trans. Daniel Heller-Roazen (Stanford, CA: Stanford University Press, 1998).
16. *Frontline* was the only print publication which defied an austere and self-imposed media censorship at the time.

17. Brian Massumi, *The Politics Of Everyday Fear* (Minneapolis: University of Minnesota Press, 1993), 19.
18. Urvashi Butalia, *The Other Side of Silence: Voices from the Partition of India* (New Delhi: Penguin, 1998).
19. Francios Ewald, "Insurance and Risk," in Massumi, 224.
20. The full text of the inquiry report can be accessed at: http://www.sabrang.com/srikrish/atr.htm
21. All press releases: 8 August 2001 Posted: 4:40 AM EDT (0840 GMT) by CNN's Andrew Demaria in Hong Kong; BBC, by Adam Mynott in Delhi;AP, TOINS and NDTV at "Teenage lovers Hanged" link where all press releases can be accessed at: http://www.dalitchristians.com/Html/dalitmurder.htm
22. NDTV, "Teenage Lovers Hanged."
23. Joan W. Scott, "Multiculturalism and the Politics of Identity," in John Rajchman (ed.), *The Identity in Question* (New York: Routledge, 1995), 11
24. Though the nature of these events could be addressed through sociological and political theories of the subject, the purpose of this investigative method is to *explore the processes* that constitute subjectivity in these events. I wish to consider an optics that avoids analyses predicated only on current theories of sovereignty and society, as offered within disciplinary frameworks, and so deliberately suspend some possible ways in which these events *could* be interpreted. Thus, I resist discussing these events only through an engagement with omnipresent issues of modernity and tradition; disciplinary mechanisms of the state (in a Foucauldian sense); the philosophical "origins" of individualism and community, etc. Also, this chapter does not provide an alternate totalizing discourse or a meta-interpretive lens (as a substitute for the available theories that could be applied) from which to read "all" political events in post-Independence India.
25. Veena Das says that "It is very difficult to think of a modern state which would give a community the right to decree the death of any of its members, for the state lays sole claim on the rights over the life of its citizens" (p. 13).
26. Gyan Prakash, *Another Reason: Science and the Imagination of Modern India* (Princeton, NJ: Princeton University Press, 1999), 158.
27. Prakash, 151.
28. Prakash, 152.
29. Arnold notes the conflicts that ensued from the colonial governments' health administration policies that sought overall public health which conflicted with the perceptions of Indians. Relatives of plague victims opposed governmental measures to seize the bodies of plague victims from their homes. In David Arnold, *Colonizing the Body: State Medicine and Epidemic Disease in Nineteenth-Century India* (Berkeley: University of California Press, 1993), 140.
30. Sudipta Kaviraj, "Filth and the Public Sphere," *Public Culture* (Durham, NC: Duke University Press), 10:1 (1997): 84.
31. This is similar to the *apnaa/paraya* or mine/not-mine distinction in Hindi.
32. Kaviraj, 93.
33. Chantal Mouffe, *Democratic Paradox* (London ; New York : Verso, 2000), 3.
34. *The Constitution of India* (New Delhi: Government of India, Ministry of Law, Justice and Company Affairs, 1999), 11.
35. Mouffe, 101.
36. Das, 17.

8

Performance, Transitional Justice, and the Law: South Africa's Truth and Reconciliation Commission

Catherine M. Cole

"The trial is pre-eminently a theatrical form," said Susan Sontag.[1] Yet to say that theatre and the law share such affinity is to say nothing new. "From the earliest Greek tragedies to the *Farce de Maistre Pierre Pathelin* to Arthur Miller's *The Crucible* to the contemporary cinematic *Retour de Martin Guerre*," Jody Enders observes, "the exposition and resolution of juridical proceedings have always had latent if not blatantly manifest dramatic value."[2] The uses to which "theatre" and "performance" have been put within the nascent field of transitional justice in the late twentieth and early twenty-first centuries, however, are new. South Africa's Truth and Reconciliation Commission (TRC) embraced performance as a central feature of its operations. While the secondary literature on the TRC is vast, scholars have yet to grapple fully with this unique and defining aspect of the commission – its public, embodied, and performed dimensions.[3] Prior truth commissions were conducted behind closed doors, and they usually became known to the public via the publication of a final report.[4] South Africa's Truth Commission, by way of contrast, transpired in front of live television and radio audiences.

For many, these public hearings *were* the commission. How did the TRC's performative conventions, modes of address, and expressive embodiment shape the experience for both participants and spectators? How is performance being used in the larger field of transitional justice and human rights law? Is performance being embraced by truth commissions as a means to assimilate traumatic history into public memory, and if so, how? Such questions are not only of paramount importance to the world and to the field of international law and human rights, they are also questions that our field – theatre and performance studies – is uniquely equipped to answer. Yet to date, we have not ventured into this field nearly to the degree that one might expect given the prevalence of "performance" within the discourse and operations of transitional justice. My objective in the brief span of this chapter is not to provide answers, but rather a more careful parsing of the necessary questions to be asked if theatre and performance studies scholars are to contribute to the field of transitional justice. I will do so with particular focus on South Africa's Truth and Reconciliation Commission.

Transitional justice

Representing a convergence of the human rights movement and international law, transitional justice is a field of activity that, according to Louis Bickford, focuses on "how societies address legacies of past human rights abuses, mass atrocity, and other forms of severe social trauma, including genocide or civil war, in order to build a more democratic, just, or peaceful future."[5] Transitional justice not only formulates a relationship between political transition and the law, it also addresses the unique challenges that a history of state-sponsored atrocity presents to the law. Crimes of unprecedented nature and magnitude have required new legal principles. The concept "crimes against humanity," for instance, was first articulated in the charter of the International Military Tribunal that authorized the Nuremberg trials.[6]

Without a doubt, Nuremberg is the epicenter for both transitional justice and international criminal justice, "the precedent upon which all ensuing developments are based," in the words of Stephen Landsman.[7] Yet transitional justice departs from international law by suggesting that traditional jurisprudence has limited value when faced with crimes against humanity. Hannah Arendt – one of the first and most articulate critics of the uses to which law was put in the aftermath of the Holocaust – saw the trial of Adolph Eichmann in 1960 as "one example among many to demonstrate the inadequacy of the prevailing legal system and of current juridical concepts to deal with the facts of administrative massacres organized by the state apparatus."[8] She argued that Eichmann's alleged crimes were not extensions of murder, but rather an entirely new crime: "Politically and legally [...] these were 'crimes' different not only in degree of seriousness but in essence."[9] Such crimes, the argument goes, demand not only that the judicial process determines individual responsibility and metes out appropriate punishment, but also that the law must serve didactic functions extending far beyond the courtroom.

When addressing crimes against humanity, the law is often called upon: (a) to place those crimes into the public record; (b) to convey to the public both the documentary facts of what happened and the historical context of those crimes; and finally (c) to perform the restoration of the rule of law. As Lawrence Douglas says of Nuremberg: "The trial was understood as an exercise in the reconstitution of the law, an act staged not simply to punish extreme crimes but to demonstrate visibly the power of the law to submit the most horrific outrages to its sober ministrations. In this regard, the trial was to serve as a spectacle of legality, making visible both the crimes of the Germans and the sweeping neutral authority of the rule of law."[10]

Douglas's use of theatrical language in the above quote – "act," "stage," and "spectacle" – is no accident. Such language is ubiquitous in the discourse of transitional justice. Political scientist Deborah Posel speaks of South Africa's Truth Commission as "theater," and literature scholar Shoshana Felman writes of the Holocaust trials as "theatres of justice."[11] And who can forget Arendt's extended use of theatrical metaphors to critique the Eichmann trial?[12] Though she noted that the judges eschewed "anything theatrical" in their conduct, the trial itself happened on a stage, the remodeled Beit Ha'am municipal theatre.[13] Arendt saw theatre in opposition to the law: "No matter how consistently the judges shunned the limelight, there they were, seated on top of the raised platform, facing the audience as from the stage in a play."[14] The trial had a villain, Eichmann, and an audience, which Arendt contends stood in for the world, and a stage manager, Ben Gurion. Yet as theatrical as the trial itself was, "it was precisely the play aspect of the trial that collapsed under the weight of the hair-raising atrocities," says Arendt.[15] While on one hand Arendt's critique of the trial is that true justice is incompatible with the theatre's limelight, for justice "demands seclusion," ultimately her critique leads us in the direction of saying that the problem with the Eichmann trial may also have been that it was bad theatre.[16] The scope of the play was too large – the "huge panorama of Jewish sufferings" – and the action (the Holocaust) too horrific, unimaginable, colossal, and unspeakable.[17] Furthermore, as a play, the Eichmann trial lacked suspense, for the outcome was inevitable from the beginning: Eichmann would be convicted and executed.

International law in the wake of Nuremberg has traditionally assumed that "both didactic and individual goals can be met in the same proceeding," says Landsman.[18] Yet it is precisely this elision of personal responsibility and didacticism – and the theatrical devices used to accomplish such didactic ends – that makes critics such as Hannah Arendt most acutely uncomfortable. "The purpose of a trial is to render justice and nothing else," she argues, elaborating that even the noblest of ulterior motives for making a historical record of Nazi crimes "can only detract from the law's main business: to weigh the charges brought against the accused, to render judgment, and to mete out due punishment."[19] The need to memorialize, dramatize, and document crimes against humanity is enormous and of paramount importance. But trials, according to the field of transitional justice, may not be the best forums to accomplish these goals. Furthermore, when the law is asked to serve such purposes, and especially when it deploys theatrical devices to do so, the result may be, paradoxically, a corruption of justice rather than the very necessary restoration of the rule of law. If the desired goal is to establish consensus and memorialize "controversial, complex events," says Martha Minow, "trials are not ideal. Even if they were adequate to the task, the theatrical devices and orchestration required threaten the norms of law that are a crucial part of the lesson, at least in societies committed to the rule of law."[20] Mark Osiel contends that "[w]hat makes for a good 'morality play' tends not to make for a fair trial. And if it is the simplifications of melodrama that are needed to influence collective memory, then the production had best be staged somewhere other than in a court of law."[21] These criticisms by Arendt, Minow, and Osiel suggest that although the law may be, as Sontag contended, pre-eminently a theatrical form, theatre and the law may be ultimately incompatible when addressing crimes against humanity.

To use the trials of particular individuals as a vehicle for documenting and promulgating a larger history is morally and legally problematic, for it saddles a particular defendant with responsibility for crimes beyond those he or she committed. Likewise, trials about crimes against humanity can be particularly debilitating and disempowering for victims. As with all trials, the courtroom casts victims in a passive rather than active role, allowing them to speak only when spoken to by an agent of the court, and even then to speak only on certain terms and topics, subject to cross-examination that may be of an adversarial nature.[22] In addition, the principles of evidence and truth operative in the court are often woefully inadequate to grapple with the psychological complexity of trauma, especially trauma perpetrated on a massive scale.[23] The role of the victim in prosecuting crimes against humanity is vexing.[24] At the Nuremberg trials, which were largely based upon Nazi documentation, the victims were generally absent altogether. The Eichmann trial tried to redress this absence by bringing a parade of victims to testify. Yet much of their testimony had no direct relevance to the particular case of Adolph Eichmann. The result was a corruption of justice, for one perpetrator was made to stand in for a whole state apparatus of genocide.

Thus in the face of the twentieth and twenty-first century's breathtaking capacity for genocide, state-sponsored torture, and systemic violations of human rights, the inherited mechanisms for restoring the rule of law have proven inadequate. Crimes against humanity require new means of redress, a mechanism that records hidden histories of atrocity, didactically promotes collective memory, and gives victims a place of respect, dignity, and agency in the process. Such purposes are not well served by traditional jurisprudence. What forum other than trials could serve such complex needs? Enter the "truth commission," a new genre of international law and a defining form of the field of transitional justice.[25] Designed as an alternative to trials, a truth commission is "a body charged with the duty of uncovering the truth about certain historical events rather than prosecuting specific defendants."[26] According to Priscilla Hayner's exhaustive and invaluable comparative study of 21 truth commissions, Uganda's 1974 Commission of Inquiry into the Disappearance of People was the first such commission, and South Africa's TRC, which began in 1995, was one of the largest, most complex, sophisticated, well funded, and successful.[27]

Do truth commissions better serve the didactic objectives that had so bedeviled the Nuremberg and Eichmann trials? Do they more successfully and humanely grant victims voice and agency, record past atrocities, and promulgate collective memory? These questions, while of profound significance, exceed the scope of this chapter. My aim here is rather to focus on an aspect of transitional justice that most pertains to the field of theatre and performance studies. Willingly or not, deliberately or not, the Nuremberg and Eichmann trials were drawn into the domain of performance. Like a moth to light, the attraction proved fatal – at least according to some critics. Does performance hold the same force of attraction for truth commissions? And if so, is this attraction as morally or legally problematic as it was in human rights jurisprudence? Or are performance and theatre embraced in truth commissions because they are seen as effective and appropriate means to accomplish particular ends? As a point of entry, we must ask how performance is being used within truth commissions. To do so, I turn now to the case of South Africa's TRC.

South Africa's Truth and Reconciliation Commission

South Africa's TRC was a product of a negotiated settlement by which South Africa transitioned from apartheid to nonracial democracy.[28] The commission began its work in late 1995, and concluded with the publication of the final volumes of its seven-volume report in 2003. South Africa's was the seventeenth truth commission in the world, according to Hayner, and it departed from precedent in two significant ways: (1) it was empowered to grant conditional amnesty to those who gave full disclosure about gross violations of human rights perpetrated for political motives, and (2) the TRC was the most public and publicized truth commission the world had seen, then or now.[29] Not only were hearings performed before spectators, they also transpired on stages – the raised platforms of town halls and churches throughout the country where the TRC toured like a traveling road show, beginning on 15 April 1996. The audience extended far beyond those spectators who attended in person. TRC Deputy Chairperson Alex Boraine wrote:

> Never in my wildest imaginings did I think that the media would retain its insatiable interest in the Commission throughout its life. Not a day passed when we were not reported on radio. We were very seldom absent from the major television evening news broadcasts, and we were, if not on the front page, on the inside pages of every newspaper throughout the two and a half years of our work. [...] Unlike many other truth commissions, this one was center stage, and the media coverage, particularly radio, enabled the poor, the illiterate, and people living in rural areas to participate in its work so that it was truly a national experience rather than restricted to a small handful of selected commissioners.[30]

To many people, as already observed, both inside and outside South Africa, the public hearings were the Truth Commission, so successful were they at making its work visible and accessible. However, we must recognize that public hearings actually represented only a small proportion of the TRC's activities. Less than 10 percent of the 21,000 victims who gave statements were chosen for public hearings.[31] Hearings – with their focus on embodied testimony – were of greater importance to achieving the commission's didactic aims, I would argue, than was the TRC's final report.[32] Aside from the fact that this tome is, in the words of Deborah Posel, "fragmented, uneven, and at times inaccurate" as well as "formal, voluminous, weighty, considered, authoritative," the report is prohibitively expensive, costing more than 1500 Rand (approximately US $210), and has had extremely limited distribution.[33] For these reasons, archivist Piers Pigou contends that "[t]he vast majority of South Africans, including those who directly engaged the commission, have therefore never seen what the commission actually found and why. There are no apparent plans to rectify this situation."[34]

Thus most South Africans experienced the commission not through its final report but through its hearings, and specifically through the media representations of these.[35] As Ron Krabill argues, "South African mass media have served as both essential actors in the TRC drama, as well as the stage on which much of the drama has been performed."[36] There were two main types of public hearings: (1) Human Rights Violation (HRV) Committee hearings where victims told their "narrative truths" about their experiences, and (2) Amnesty Committee hearings where perpetrators came forward in the hopes of being granted amnesty.[37] In order to be granted amnesty, perpetrators had to give full disclosure about the gross violations of the human rights they had committed, and prove that their deeds were both politically motivated and proportionate. Within the structure of the TRC these two committees were kept institutionally separate, and their rules of decorum and procedure as well as performance conventions were entirely distinct.

Generally speaking, the HRV hearings were more improvisational and victim centered. Their function was explicitly didactic in terms of creating collective public memory. Victims chosen to have a public hearing were given on average about 30 minutes to speak, and they spoke unimpeded and uninterrupted. They could speak in any one of South Africa's 11 official languages, with simultaneous interpretation over headphones available for all participants. Victims were not subjected to cross-examination, although commissioners could and often did ask questions for clarification at the end of their testimony. The aim of these hearings was to establish, in the words of the TRC's constitutional mandate, "the truth in relation to past events as well as the motives for and circumstances in which gross violations of human rights have occurred, and to make the findings known in order to prevent a repetition of such acts in [the] future."[38] The audience was not, one might argue, primarily the commissioners assembled at a table adjacent to and sometimes in front of the victims, but rather the public, represented by spectators who attended the hearings live and, by extension, those who witnessed the proceedings via the media.

The HRV committee could only recommend reparations for victims, thus it had no authority beyond being able to determine which individuals of the 21,000 who gave statements would: (a) be given a public hearing, and (b) be named as a "victim" of gross violations of human rights in the final report and hence be eligible to receive whatever reparations the government – not the commission – ultimately provided. The HRV hearings were highly performative events in terms of their theatrical and dramatic emotional displays, improvisational storytelling, singing, weeping, and ritualistic lighting of candles.[39] But the actual administrative power of the HRV wing of the TRC – its ability to do, to act, to perform in a way that would change the lives of those who testified – was limited.

The judges who presided over the Amnesty Committee hearings, on the other hand, were empowered with the capacity to grant amnesty. So in these hearings something was at stake judicially, and profoundly so. At the end of the process, the fundamental legal status of the perpetrator could be changed forever. If granted amnesty, a perpetrator was free from civil and criminal prosecution for the rest of his or her life. On the other hand, these hearings were generally not as theatrical as the HRV hearings in terms of emotional expressiveness (with notable exceptions). Amnesty hearings were much more constrained by courtroom protocol, with lawyers and advocates making presentations to judges. The commission did not require perpetrators to express contrition or remorse. Rather, the priority of the TRC was truth, the production of information, the full disclosure of deeds: who did what to whom, who gave the orders, and where the bodies were buried. There was no incentive or encouragement for those who appeared before the Amnesty committee to "perform," in the sense of projecting any particular demeanor, emotion, or attitude. In this regard, the South African TRC was quite different from Sierra Leone's truth commission, which during its proceedings in 2003 pressured perpetrators to publicly apologize to the community. As Tim Kelsall has shown, this led to much public discussion about how genuinely contrite particular perpetrators appeared; that is, how "good" were the perpetrators' performances of remorse.[40]

The South African TRC's empowering mandate, the Promotion of National Unity and Reconciliation Act (Act no. 34 of 1995), was complex and sophisticated and, I would argue, well designed in many regards. The mandate "carefully balanced powers," as Hayner says.[41] While the TRC's design was distinct from the paradigmatic trials of the Holocaust – those of Nuremberg and Eichmann – we can also read the TRC's two core committees as being genealogical descendents of these famous trials. The Amnesty Committee hearings shared the perpetrator-centered character of the Nuremberg trials, and the HRV committee hearings descended from the victim-centered impulse within the Eichmann trial. Thus the TRC combined both approaches into one commission, but kept them institutionally distinct. The division between the HRV and Amnesty Committees circumvented many of the most egregious problems that had riddled the Nuremberg and Eichmann trials. Victim testimony was unmoored from prosecution and thus unfettered by the protocols and epistemologies of a court of law. Perpetrators in the Amnesty hearings, drawn by the "carrot" of amnesty, came forward of their own volition to confess their crimes. This was quite different from the defensive position in which criminal proceedings would have placed them. What this meant for the TRC was that the public heard stories about murder, torture, and other gross violations of human rights from the mouths of perpetrators themselves, who spoke of their own free will.

The TRC hearings fully embraced the theatricality and innate performativity that Arendt rejected in the Eichmann trial. For many, these public hearings – with their embodied expressions, their weeping, their silences, their demonstrations of "wet bag" torture techniques, their confrontations between former torturers and those they tortured, the wails, and the moments that transcended language – most defined the Commission. Though not framed as theatre, the TRC was often explicitly described as such both by participants and the media. Some used theatrical metaphors to speak in laudatory terms about the TRC process. Commissioner Pumla Gobodo-Madikizela, for instance, touted the public hearings because they put the victims "center stage" within a state-sponsored investigative process.[42] Deputy Chairperson Boraine praised the public hearings as both "dramas" and "rituals." He vividly recalled the first East London hearings:

> At last the curtain was raised. The drama which was to unfold during the next two and half years had witnessed its first scene. The ritual, which was what the public hearings were, which promised truth, healing, and reconciliation to a deeply divided and traumatized people, began with a story.[43]

Boraine views the TRC as both dramatic and cathartic, walking the line between theatre and ritual that performance studies has also long transgressed.[44] "Here we are, the actors on the stage," said George Bizos, a lawyer who appeared frequently before the TRC. He added: "It's unusual to have a judicial proceeding on a stage."[45] The theatrical qualities of the TRC, the way it literally staged truth and reconciliation, were occasionally cause for great distress. The first Amnesty hearings, for instance, were delayed for hours because the judges worried about the symbolism of having perpetrators sit on the same raised platform as the judges. They also fretted about where the victims should sit: Should they be on the stage or down among the audience? Should they face the commissioners, or outward towards spectators?

"The judges are used to such matters being resolved by the architecture of the courtroom," said journalist Antjie Krog. "Now they have an ordinary hall, and it seems from the human rights hearings that you make a Statement with your seating arrangement" – as any theatre director could have told them.[46]

Not all participants in the TRC embraced its theatrical manifestation. At the East London TRC hearings, Chairperson Archbishop Desmond Tutu struggled to subdue unruly spectators by admonishing: "We have been given a very important task: this is not a show what we are doing. We are trying to get medicines to heal up our wounds."[47] Tutu substituted a medical metaphor – the healing of wounds – for a theatrical one, a "show." As Tutu explained, the entire success of the commission and, indeed, the new democratically elected, non-racial South African government, rested on the public perception of the TRC's even-handedness and effectiveness. Spontaneous expressions of emotion from the audience, he said, would undermine the legitimacy of the commission by turning it into a "show." Using his characteristic mix of South African languages, Tutu elaborated (in Xhosa): "Do not make us a laughing stock, because people will say, 'Because these things [i.e., state-sponsored commissions] are now being run by blacks, now everything is turned into a bioscope.' [In English:] I will not tolerate that please."[48] Code-switching in the masterful way that was so characteristic of Tutu when he presided over the Commission, he made his message clear: the TRC was not entertainment, it was not a movie, it was not a show.

While theatrical metaphors surrounding the TRC were ubiquitous, these metaphors were deployed for a wide range of purposes and meanings. Such metaphors carried a tone of derision as often as they did praise. Thus, we see Boraine embrace the TRC as a ritual and a drama, while Tutu guarded against the proceedings becoming a bioscope. Another prominent South African invoked theatrical metaphors as part of his spectacular refusal to appear before the Commission at all. The former Prime Minister and State President of South Africa, P. W. Botha, rejected the Commission by calling it a "circus."[49] A cartoon published in the *Natal Witness* elaborated on Botha's metaphor and depicted the ambiguity of the TRC as a genre. Chairperson Tutu appears as a lion tamer in a circus. Whip in hand, Tutu shouts into a cavernous pen where the wild beast of the apartheid past, the "Big Crocodile" ("Die Groot Krokodil") – Botha himself – resides. "I don't care what you think the TRC is – You come out here and perform!" Tutu declaims.[50] The artist conveys the underlying imperative to perform that drove the Commission. Even if no one could agree on exactly what genre of performance the Commission was – circus, ritual, drama, bioscope, show – participants had to perform, and they had to perform in a certain way. The cartoonist succinctly captures the multifaceted meanings of "performance" that are operative both in the TRC and in my study of the Commission. These meanings include the following: to accomplish an act, to make a

public presentation, to use embodiment as a central instrument of communication, and to simulate or represent (i.e., to "act"). As Strine, Long and
Hopkins have argued, performance is an essentially contested concept,[51] and
I do not propose here to reiterate or resolve that contestation. Rather, like
the cartoonist Stidy, I deploy these multivalent meanings and embrace the
heuristic potential of simultaneity. Often the most dynamic moments during the TRC proceedings represent a convergence of one or more of these
definitions of performance.

When, early on in the hearings, witness Nomande Calata broke into a loud
wail during her testimony, this disconcerting cry became an emblematic
moment in public memory. Antjie Krog's *Country of My Skull*, an often-
cited memoir of Krog's experience covering the TRC as a journalist, quotes
a Professor Kondlo as allegedly saying of Calata's wail: "For me, this crying is the beginning of the Truth Commission – the signature tune, the
definitive moment, the ultimate sound of what the process is about. She
was wearing this vivid orange-red dress, and she threw herself backward and
that sound [...] that sound [...] it will haunt me for ever and ever."[52] The
importance of this sound – a wail that transcended language and, in doing
so, captured something elemental about the experience of gross violations of
human rights – indicates the degree to which embodied expression was central to the TRC process, even though this aspect of the Commission is rarely,
if ever, given notice in the TRC's final report or in the burgeoning secondary
literature on the TRC.[53] The research of Kay McCormick and Mary Bock from
the University of Cape Town is an exception. They use discourse analysis to
transcribe the embodied, performed meanings expressed by witnesses, and
analyze extra-linguistic communication through breath, cadence, pauses,
eye contact, and gestures.[54] Their research relies on unedited video footage
of hearings that to date is one of the only parts of the TRC archive that
is in the public domain. The National Archives of South Africa in Pretoria
has a collection of over 10,446 hours of audio recordings (60 minutes each)
and 7101 video tapes (each two hours long) of the TRC public hearings. The
video collection alone is so enormous that if one researcher worked eight
hours a day for seven days a week, it would take over four and a half years to
view the collection in its entirety. Yet hardly any researchers are using this
valuable resource at all.

In a recent article on the rise of the terms "performance" and "performativity" in cultural analysis, Julia Walker argues that the metaphor of
performance has come to the fore in critical theory in order to address
"the problematic role of individual agency."[55] She argues that performance,
both as embodied practice and as metaphor, gives access to vocal and
pantomimic/kinetic signification, in addition to textual or verbal signification. Because performance resonates simultaneously in several different
registers – including reason, emotion, and experience – it is more capacious than the culture-as-text metaphor in cultural analysis. The public and

embodied signification expressed through the performative aspects of the HRV Committee hearings of South Africa's TRC potently resonated in the affective and experiential registers of human experience. These aspects of communication – in combination with the actual words spoken – were an essential vehicle for communicating the density of the profound experiences of human suffering that the TRC brought to light.

The TRC was devised to express events and experiences that ultimately are unspeakable. This is one of its core paradoxes. Gross violations of human rights unmake the world,[56] and the TRC was dedicated to remaking that world in a way that honored human rights. The performative dimensions of public hearings allowed the TRC to express the inexpressible, and to human-ize people's experiences of extreme dehumanization. In my forthcoming book *Stages of Transition*, I argue that the TRC's live hearings were affective, and consequently they were effective in facilitating, however imperfectly, a transition from a racist, totalitarian state to a non-racial democracy.[57] The dramatic, unruly, ephemeral, embodied, and performed aspects of live hear-ings potently expressed both the power of the TRC as well as its severe limits in truly grappling with the magnitude of the violations of human rights in South Africa's past.

The theatrical, emotional, subjective truth promulgated by these hearings provided agency to people who had been voiceless in the official documents and discourse of the South African nation. Live testimony given at TRC hear-ings also provided a corrective to the narrow epistemologies of truth-telling operative in venues like the Nuremberg and Eichmann trials. As Deborah Posel says of the TRC:

> The theater of these public hearings produced – necessitated – very dif-ferent genres of truth-telling from those of the more scientific efforts at fact-finding. The hearings gave space for many people to tell their own stories, versions of events that often conflicted with others told in the same forum or which, on closer inspection, were internally incon-sistent. Yet none of this seemed to detract from the truth-telling. Truth lay in the emotional power of individual stories and the capacity of the hearings to uncover seemingly pristine, uncorrupted narratives of past brutalization.[58]

While the compellingly theatrical and emotional nature of the public hearings appeared to provide unmediated access to authentic truth, in actu-ality these public hearings were highly mediated. The Commission itself served as "casting director," determining which victims would receive pub-lic hearing. In addition, the media selected which portions of each day-long hearing would be broadcast on television and radio, or splashed across the newspaper headlines.[59] But no matter how mediated or orchestrated, the projection of these hearings in the public eye provided a palpable connection with victims – victims of all races – and with the larger history that their individual stories, when told in succession, represented. Posel contends that the public hearings made for "good theatre, but bad history," in as much as the conceptual premises of the Commission and its report (such as its wobbly definitions of truth, simplistic bifurcation of victims and perpetrators, lack of quality control in data collection, and absence of explanatory narrative) were deeply flawed.[60] Yet, I would argue that the "good theatre" of the TRC does not necessarily have to lead to bad history. That may have been what the final report produced, which Posel contends is a "disconnected compilation of discrete chunks of information, with little effort at a synthetic unified analysis."[61] But the good theatre of the live hearings with their compelling, densely meaningful embodied testimonies, if analyzed rigorously with methodological creativity and attention to linguistic and performed nuance, can potentially produce good history. The archive is rich and yet largely neglected. As David William Cohen and E. S. Atieno Odhiambo have argued, records of commissions and trials, especially in Africa, have a contingent power "to open views of otherwise thinly documented, poorly understood, or weakly represented worlds."[62] Cohen and Odhiambo's model for reading African trials and commissions suggests that scholars must read these records "against the grain," as it were, of their own internal logic. The massive work of South Africa's Truth and Reconciliation Commission is worth a much more empirically rich analysis than we have seen so far. Most studies have evaluated the successes or failures of the TRC: whether it, in fact, produced truth, or reconciliation, or good history, or bad social science. Theatre and performance studies scholars can contribute close readings of testimony in all its performative richness, readings that open up precisely those contingent powers to which Cohen and Odhiambo refer. To date, the archive produced by the TRC has not been subjected to such analysis. Yet this sad state of affairs does not have to continue. In the remaining part of this chapter, I wish to offer a glimpse of the kinds of analysis and history that attention to performance might yield.

The case of the Guguletu Seven

On 3 March 1986, seven alleged agents of the military wing of the African National Congress (ANC) were killed in a violent shoot-out in the township of Guguletu outside Cape Town, South Africa.[63] The cause of the incident was investigated during two inquests by the apartheid state in 1986 and 1989, as well as in a criminal trial in 1987. Yet these investigations failed to determine who was responsible. Much about the case, known as that of the "Guguletu Seven," remained shrouded in mystery.[64] There were many inconsistencies that the authorities did not pursue, that still haunted the survivors: Why had all seven men been shot dead? Why was it necessary to kill all of them? Why were there no injuries among the police? What accounted for the discrepancies between the official police version of the story and that of eyewitnesses who reported that some of the men had been shot execution style, in the back, while their hands were raised in the air? With so many unresolved questions, the mothers of the Guguletu Seven came forward in 1995, responding to the call for truth-telling from the TRC. They wanted answers. Yet instead, the TRC asked them many, many questions. One of the mothers, Mrs. Ngewu, said during the first day of public testimony: "We cannot see that drama again. We [...] cannot relife [*sic*] this whole experience."[65] And yet that is exactly what the TRC's HRV Committee hearings asked her and the other mothers to do – to relive the drama of their trauma again, and to do so in front of an audience.

The case of the Guguletu Seven was unusual in many regards. It was considered a "window case" for the Western Cape. The dedicated TRC investigative team had managed to seize documents and link up evidence in a way that exposed to an unprecedented degree the corrupt inner workings of the apartheid state at a high level. Unlike many TRC cases, this one threaded through the hearings of both the HRV and Amnesty Committees. Several mothers of young men who were killed testified in the HRV hearings, and two of the perpetrators came forward to seek amnesty. The Guguletu Seven hearings also witnessed one of the few moments of genuine dramatic action in the TRC, when the Commission moved beyond the familiar mode of storytelling, and even beyond the occasional moment of demonstration, by showing, for instance, how someone was tortured with a wet bag as did amnesty applicant Jeffrey Benzien in a now infamous incident.

The first Guguletu Seven hearing was held on 23 April 1996 in the Western Cape township of Heideveld. Four mothers of the deceased testified: Cynthia Ngewu, mother of Christopher Piet; Irene Mxinwa, mother of Simon Mxinwa; Eunice Miya, mother of Jabulani Miya; and Elsie Konile, mother of Zabonke John Konile. Each woman approached the public forum with a distinctive tone, style, and set of stated objectives. Mrs Ngewu was the most forthright. Her narrative of the events surrounding the murder followed chronological order, and her demands and expectations for the TRC were enumerated in clear and specific ways. Mrs Konile's testimony, on the other hand, was far more fragmentary. She described her discovery of her son's death as a period of moving in and out of consciousness. She dreamed of a goat (a bad omen) when a young man came to lead her to the "houses belonging to the comrades." She was seized by tremors as the man took her to Cape Town to see her son, and she was taken to a hospital and given pills and injections. Mrs Konile eventually discovered that the place of the comrades was, in fact, the mortuary, and there she saw her disfigured son with one of his eyes out, covered in blood. She identified Zabonke by his feet. Afterwards, she lost consciousness again. "When I woke up, I felt like I was just getting out of bed. And there was a continuous cry that I could hear. It felt like I was going down-down-down. When

I looked, I was wet-wet-wet – I was wet all over the place." Mrs Konile's testimony poignantly expresses the disorientation and physical toll of trauma. Dream and reality merge. The passage of time is sometimes quite finite – a few hours – and at other times infinite.

The testimony given by each of the mothers was distinctive, and yet all shared certain recurring themes: (1) contentious encounters between survivors and police; (2) trauma that exacted both a physical and psychological toll on survivors; (3) the media as a locus of trauma, for television and newspaper coverage of the Guguletu Seven case was a prime means by which the women learned of their sons' deaths; and (4) official state inquests and police interrogations related to the case compounded the women's trauma and suffering. The TRC itself risked reproducing past trauma, for it too was both an official state inquiry and a highly televised media event. What was eventually discovered through the TRC process was that the police – and specifically a secret unit of the security police known as "Vlakplaas" – had entrapped and staged the execution of the seven young men. At the time of the murders, the police shot and edited a video at the crime scene. When this video was played ten years later during the TRC hearings, the spectacle of violence intensified and refracted. Media spectacle of violence provoked live spectacle. The result was dramatic and explosive.

Mrs Ngewu closed her testimony in April 1996 by demanding that "these boers must be put in front of us, in front of this Commission." Her wish was granted. During subsequent HRV hearings in November 1996, police officers responsible for the entrapment and killing of the Guguletu Seven were subpoenaed to appear before the Commission for something known as a "Section 29" hearing, referring to the section of the law that gave the TRC formidable, if ultimately under-utilized, powers of subpoena.[66] The hearing was technically held within the HRV Committee, but instead of focusing exclusively on victims, the Committee heard testimony from the police who had been subpoenaed and were suspected of perpetrating the murders. This particular hearing was thus more reminiscent of a court of law than typical HRV or Amnesty hearings in as much as the accused perpetrators came to the hearing involuntarily, and the perpetrators and victims testified in the same arena.

At the hearing, the police – along with public spectators, commissioners, and the mothers of the deceased – sat watching the videotape that had been shot shortly after the murders of the Guguletu Seven. Originally created as police propaganda, this video was part of a larger police effort to convince superiors of the unit's effectiveness in stopping so-called communist subversives and to justify its new budget requests. Portions of the video were also released to the media and broadcast on national television at the time of the killings as evidence to the public of the veracity of National Party claims about ANC violence and insurrection.[67] The video was a prop in the apartheid government's expansive war on terror. It contained gruesome footage, including close-up images of the dead, one of whom the police were pulling with a rope.

Viewed ten years later within the context of the TRC hearings in 1996, this same video became part of a very different sort of enactment of governmental authority and identity: a commission that sought to legitimize the new democratic state by putting the atrocities of the old apartheid government on public and publicized view. State-making through spectacle is a global and trans-historical phenomenon, but what is striking about the Guguletu Seven case is its complexity. A video created by the police as a vehicle for the apartheid state to justify its oppression and killing of young black men later became evidence of police (and the state's) culpability.[68] The police officers subpoenaed at the TRC hearings sat like Claudius and Gertrude watching the play-within-the-play of Hamlet. They posed as innocent witnesses watching a video depicting the aftermath of a brutal murder that they had, in fact, committed. The mothers of the Guguletu Seven watched, like the brooding Hamlet, haunted by the specter of the dead. They watched both the representation of the murders (the video) and the police officers, whom they knew (but could not yet prove) to be the murderers.

The spectacle of this particular TRC hearing did not end here. During the video screening one of the mothers hurled a single shoe across the room. This projectile decisively struck two of the nine police officers and completely disrupted the hearing. This dramatic action shifted focus from the gruesome details of the video and the impassivity of the perpetrators to the presence of the mothers of the dead, who sat several rows behind them. The *Cape Argus* reported that the dramatic moment in the hearings came when the video showed the police putting a rope around the dead body of Christopher Piet. "One of the mothers jumped to her feet, shouting hysterically, 'Why a rope? Why a rope?' "[69] The women began, one after the other, to wail and attract attention. Professional "debriefers" – part of the TRC staff – flocked around the women, offering handkerchiefs and physical support. Slowly the mothers were ushered out the door. The subpoenaed police officers walked out of the auditorium and reportedly took shelter in the press room. The TRC transcripts record advocate Dumisa Ntsebeza saying, "Can we stop there – stop the video – we'll take a break now, if you would just stop the video right at this moment." Perhaps betraying the transcriber's desperation, the record then reads in capital letters: "PEOPLE ARE HYSTERICAL – CRYING AND SCREAMING."[70]

The TRC hearings followed explicit rules of decorum and rationality. Even though it was not a court, the TRC borrowed heavily from juridical protocol. Commissioners would begin each session with ground rules that attempted to ensure that the subject of the hearings could be discussed and viewed in a rational manner. The Commission took great pains to prepare the mothers of the Guguletu Seven for the psychological experience of viewing this video. They even offered the women the option of not viewing it, but the mothers insisted they wanted to see and know what had happened to their loved ones. Whatever the motivation of the unidentified mother who tossed the shoe during the hearing on 28 November 1996, the effect of this missile was a complete disruption of the session. Whether planned or spontaneous, the women's responses as they stood, threw their arms over their heads, and wailed in grief effectively upstaged the video. The women asserted themselves as active agents of performed spectacle rather than passive consumers of the video spectacle.

When the session resumed, Ntsebeza commented that he regretted the theatrics of the women and that commissioners had been concerned about screening the video. Before the session, the women "were told that it might not be in their interest to see the video, precisely because it might lead to this sort of scenes [*sic*] that we have seen which we regret as a Commission," he said.[71] He then delivered an extensive exhortation, using the TRC's juridical vocabulary of truth-finding to address the crime of the thrown shoe. Regarding "an allegation that a shoe or a similar like missile was thrown at Mr. Kleyn and Mr. Knipe presumably by one of the witnesses," Ntsebeza said:

> I did not see it, I don't know who saw it, clearly the gentleman also didn't see who it was, because obviously the missile came from behind their backs. I don't want to make a finding about this matter. I need merely say if it happened, it is something we take a very dim view of, we understand emotions of people who come to this proceedings [*sic*] carrying the sort of trauma that is commensurate with a loss that they obviously have. But it must be emphasized – it must be emphasized very strongly and in the strongest terms that the Commission has a task to perform which must be performed in circumstances where it's [*sic*] integrity and reputation will not be undermined. The entire aim and the broader aims of the Commission is to achieve not only the exposure of truth, but reconciliation. And the test, the acid test is going to be moments like this – at moments where no version of the truth as we are trying to achieve it, should be suppressed because witnesses are not made comfortable to testify.

Ntsebeza's concern that the TRC proceedings be conducted fairly and safely for all participants was imperative. Indeed, it was his duty as acting chairperson of the Commission that day to ensure such fairness and to ensure that the TRC performed restorative justice, not acts of retribution. Yet in light of the way in which rationalization fueled the apartheid government's organized murder and torture of black people – a rationalization so evident in the Guguletu Seven case itself – perhaps something other than a calm, rational viewing of videotapes of murder was necessary in post-apartheid rituals of truth and reconciliation. The Guguletu mothers provided a visceral, emotional, and embodied response that officially had no place in this TRC hearing. Why was their response regrettable, and the calm, impassive viewing of a video of murder by the murderers somehow not regrettable theatrics? The police officers' behavior was far more of a "performance" in as much as their demeanor at the hearing was a masquerade, a disguise. They performed as passive spectators of the atrocity, when in fact they had actually been its perpetrators.

Ntsebeza stated that the Commission not only "has a task to perform," but this task must be performed in a certain way. In order to maintain the TRC's integrity and reputation, spectators and participants must also perform, and they too must perform in a certain way. Ntsebeza's exhortation recalls the cartoon from Stidy in the Natal Witness. While South Africa's Truth and Reconciliation Commission was designed to contain and make manageable the effects of atrocity, the magnitude of atrocity constantly exceeded its bounds. The Guguletu Seven hearings contained one moment when a state-sponsored spectacle was derailed and transformed by the volatility of the experiences it was designed to manage. Ntsebeza was right to identify this moment as an acid test of the truth "as we are trying to achieve it."[72]

We can read this incident as being ghosted by other spectacular disruptions at the Holocaust trials that are the TRC's genealogical heritage. The screening of the gruesome video of the Guguletu Seven crime scene evokes the screening of the film *Nazi Concentration Camps* at the Nuremberg trials. "By providing a visual register of extreme atrocity," says Lawrence Douglas, "the film crossed a threshold of representation from which there was no turning back."[73] Likewise, the mothers' emotional outburst evokes the collapse of writer Yehiel Dinur on the stand at the Eichmann trial.[74]

There are other ghosts as well: those of the Guguletu Seven deceased. Ntsebeza refers to the shoe thrown at the police officers as a "missile," an interestingly martial reference, and he notes that part of what was so appalling about this incident is the fact that this missile was thrown from behind. We might read this moment as a symbolic dramatization of reversal: several of the Guguletu Seven themselves were shot from behind, execution style, even with their hands in the air in a posture of surrender – with no chance to see the perpetrator, no chance to respond, no chance to negotiate.

What particularly interest me about what happened during this hearing are the ways in which the procedures of the Commission broke down. The TRC did not perform – or to be more precise, it did not perform according to script. Such occasions reveal the uncertainties and indeterminacies that were as much a part of the TRC's production of truth and knowledge as were the sweeping narratives it generated in its final report. I argue that the realm of performance – with its ambiguity, embodiment, and potent ability to convey the affective dimensions of human experience, as well as its dramatic action – was as efficacious at producing truth as were the words spoken by those who appeared before the Commission, words that were then mediated by translation, transcription, and publication on the World Wide Web.

Although the Guguletu mothers' expression of rage was suppressed at the hearings, the Commission's format did allow the mothers themselves to demand that the entire command structure of the metropolitan police force responsible for the murders of their sons be summoned to answer their crimes in public. And the live nature of the public hearings allowed the women agency of expression even in ways that the Commission censured. This incident from the Guguletu Seven hearings demonstrates that while only some truths could be contained within the Commission's mandate and procedures, other truths constantly erupted in the live, embodied experience of public hearings. Yes, everyone had to perform, but the structure and format of live hearings also allowed room for those moments when individual agents took charge in unscripted and unexpected ways. In such moments, I argue, the TRC performed truth most potently. Such moments revealed that the truth could not be tightly packaged, it could not be fully regulated by protocol, and it could not – and perhaps should not – be calmly and passively viewed on television.

Conclusion

The theatrical and performative nature of the TRC is, indeed, axiomatic, and has been noted by others. However, I argue that scholars have barely begun analyzing what the performative nature of the South African TRC might tell us about the impact, outcomes, and daily practices of this unique model of transitional justice, a model that has been widely celebrated and replicated throughout the world. While many performance and theatre scholars have been drawn to write about South Africa's TRC, most have focused on theatrical or aesthetic representations of the Commission rather than on the Commission itself as performance.[75] While such work is indeed valuable, I would argue that our field has a unique opportunity at this moment to make a critical intervention in human rights discourse. The word "performance" is often on the lips of scholars who write about truth commissions, international human rights law, and crimes against humanity. Yet performance and theatre studies scholars have not yet been central interlocutors in this discourse. This is a loss for all.

There is a great range of research on public testimony from the TRC that remains to be done, and an even greater amount of research that must be done on the role of theatre and performance in the nascent and burgeoning field of transitional justice. In the case of South Africa, I contend that the "completeness" of the vision of the apartheid past, which was mandated in the Commission's authorizing act, can be discerned as vividly through in-depth analysis of performed testimony as it can through macro-narratives that calculate in quantitative terms apartheid's national dimensions, as the Commission's final report attempted to do. Furthermore, I would argue that the integrity of victims – which was supposed to be paramount within the TRC's "victim-centered" approach to restorative justice – requires this level of attention to testimony. Whatever the victims' intentions or expectations were of the Commission, their words, gestures, cadence, intonation, and embodied expressions are now in the public domain, and this material deserves to be closely analyzed. The picture that emerges from such analysis is complex and contradictory, full of details that both confirm and resist the dominant narratives of the past and of the TRC's own mission. We also see how individuals performed within the Commission the particular truths that they were trying to achieve. In the disjunctions between participants' performances of truth they wished to perform and the Commission's public iteration of the truths it wished to perform, we come closest to perceiving the complexity of the knowledge the TRC brought forth.

Notes

1. Susan Sontag, *Against Interpretation* (New York: Farrar, Straus & Giroux, 1966), 126. This research has been made possible by generous funding from the National Humanities Center, the National Endowment for the Humanities, the University of California Regents Humanities Fellowship, and University of California-Santa Barbara's Interdisciplinary Humanities Center and Academic Senate. Earlier versions of this essay were presented in seminars at Northwestern University, Yale University, the University of Michigan, Performance Studies International, and the American Society for Theatre Research. For their consultation and responses to the work, I would particularly like to thank Kwame Braun, Catherine Burns, Keith Breckenridge, Leo Cabranes-Grant, James Campbell, David William Cohen, Michael Levine, Stephan Miescher, Joseph Roach, Diana Taylor, Richard Wilson, and William B. Worthen, as well as the editors and three anonymous readers from *Theatre Journal*. I am also grateful for Bianca Murillo's diligent research assistance.
2. Jody Enders, *Rhetoric and the Origins of Medieval Drama* (Ithaca, NY: Cornell University Press, 1992), 2.
3. The "select bibliography" of the unpublished "Guide to Archival Sources Relating to the Truth and Reconciliation Commission," compiled in 2004 by the South African History Archive and Historical Papers at the University of the Witwatersrand (Johannesburg), lists over 90 books and monographs published on the TRC only eight years after the commission began its work.
4. See, for instance, *Nunca Más: The Report of the Argentine National Commission on the Disappeared* (New York: Farrar, Straus & Giroux, in association with Index on Censorship, 1986).
5. Louis Bickford, "Transitional Justice," in Dinah L. Shelton (ed.), *The Encyclopedia of Genocide and Crimes against Humanity* (Farmington Hills, MI: Macmillan Reference, 2004), 1045.
6. See Lawrence Douglas, *The Memory of Judgment: Making Law and History in the Trials of the Holocaust* (New Haven: Yale University Press, 2001), 38–64.
7. Stephen Landsman, "Those Who Remember the Past May Not Be Condemned to Repeat It," *Michigan Law Review*, 100:6 (2002): 1571.
8. Hannah Arendt, *Eichmann in Jerusalem: A Report on the Banality of Evil* (New York: Penguin Books, 1963), 294.

9. Arendt, 267.
10. Douglas, 41.
11. Deborah Posel, "Truth? The View from South Africa's Truth and Reconciliation Commission," in Deborah Posel et al. (eds), *Keywords: Truth* (New York: Other Press, 2004), 1–25; Shoshana Felman, *The Juridical Unconscious: Trials and Traumas in the Twentieth Century* (Cambridge, MA: Harvard University Press, 2002), 106–30.
12. Arendt.
13. Arendt, 4. See also Douglas, 98. It is not uncommon for transitional justice to be enacted in theatrical spaces. While the title of the "Commission to Clarify Past Human Rights Violations and Acts of Violence that Have Caused the Guatemalan People to Suffer" would have been a challenge to market as "theatre," the emotional presentation of its final report did transpire in the National Theater in Guatemala City before an audience of thousands. See Priscilla Hayner, *Unspeakable Truths: Confronting State Terror and Atrocity* (New York: Routledge, 2001), 48.
14. Arendt, 6.
15. Arendt, 6.
16. Arendt, 6.
17. Arendt, 8.
18. Landsman, 1571.
19. Arendt, 253.
20. Martha Minow, *Between Vengeance and Forgiveness: Facing History after Genocide and Mass Violence* (Boston, MA: Beacon Press, 1998), 47.
21. Mark Osiel, *Mass Atrocity, Collective Memory, and the Law* (New Brunswick: Transaction Publishers, 1997), 59.
22. See Marie-Bénédicte Dembour and Emily Haslam, "Silencing Hearings? Victim-Witnesses at War Crimes Trials," *European Journal of International Law*, 15:1 (2004): 151–77. See also Peter Brooks and Paul Gerwitz (eds), *Law's Stories: Narrative and Rhetoric in the Law* (New Haven: Yale University Press, 1996).
23. Felman, *The Juridical Unconscious*. See also Judith Lewis Herman, *Trauma and Recovery: The Aftermath of Violence – from Domestic Abuse to Political Terror* (New York: Basic Books, 1997); Shoshana Felman and Dori Laub (eds), *Testimony: Crises of Witnessing in Literature, Psychoanalysis, and History* (New York: Routledge, 1991).
24. For a moving depiction of the potentially offensive way in which victims are positioned and treated in the International Criminal Tribunal for the former Yugoslavia, see Dembour and Haslam, "Silencing Hearings?"
25. For an overview of transitional justice, see Neil J. Kritz (ed.), *Transitional Justice: How Emerging Democracies Reckon with Former Regimes*, 3 vols. (Washington, DC: US Institute of Peace Press, 1995).
26. Landsman, 1589.
27. Hayner, 40–5, 305–36. Ambivalence about the outcomes of the TRC is evident in such works as Christopher James Colvin, "Performing the Signs of Injury: Critical Perspectives on Traumatic Storytelling after Apartheid" (PhD diss., University of Virginia, 2004); Brandon Hamber et al., "Submission to the Truth and Reconciliation Commission: Survivors' Perceptions of the Truth and Reconciliation Commission and Suggestions for the Final Report" (compilation, Centre for the Study of Violence and Reconciliation, and The Khulumani Support Group, 1998); Terry Bell, in collaboration with Dumisa Buhle Ntsebeza, *Unfinished Business: South Africa Apartheid and Truth* (Cape Town: RedWorks, 2001). Yet the most extensive empirical survey conducted on the reception of the TRC in South

Africa suggests that the process is viewed by the majority of South Africans – and especially non-whites – as having been successful in achieving its goals. See James L. Gibson, *Overcoming Apartheid: Can Truth Reconcile a Divided Nation?* (New York: Russell Sage Foundation, 2004); James L. Gibson, "The Truth about Truth and Reconciliation in South Africa," *International Political Science Review*, 26:4 (2005).

28. An excellent treatment of South Africa's transitional period is Allister Haddon Sparks, *Tomorrow Is Another Country: The Inside Story of South Africa's Road to Change* (Chicago: University of Chicago Press, 1996).

29. See Hayner, *Unspeakable Truths*, 40–5. Hayner writes: "The South African Truth and Reconciliation Commission succeeded in bringing this subject to the center of international attention, especially through its public hearings of both victims and perpetrators outlining horrific details of past crimes. Although quite a few of such truth commissions existed prior to the South African body, most did not hold hearings in public, and none of the others included such a compelling (if also ethically problematic) offer of individualized amnesty, which succeeded in enticing many South African wrongdoers to confess their crimes in front of television cameras" (p. 5).

30. Alex Boraine, *A Country Unmasked: Inside South Africa's Truth and Reconciliation Commission* (New York: Oxford University Press, 2000), 89.

31. An examination of the process and rationale by which cases were chosen for public hearings is too complex to cover here. This subject will be addressed in my book, *Stages of Transition: Performing South Africa's Truth Commission* (Bloomington: Indiana University Press, forthcoming).

32. Truth and Reconciliation Commission, *Truth and Reconciliation Commission of South Africa Report*, vols. 1–5 (London: MacMillan Reference, 1999); vols. 6–7 (Cape Town: Juta, 2002).

33. Posel, "Truth?", 16; Deborah Posel, "The TRC Report: What Kind of History? What Kind of Truth?" in Deborah Posel and Graeme Simpson (eds), *Commissioning the Past: Understanding South Africa's Truth and Reconciliation Commission* (Johannesburg: Witwatersrand University Press, 2002), 147; Piers Pigou, "There Are More Truths to be Uncovered before We Can Achieve Reconciliation," *Sunday Independent*, 23 April 2006, 9.

34. Pigou, 9.

35. Although the TRC was both nationally and internationally a media event of the century, there has been astonishingly little scholarship on this aspect of the commission with the exception of Edward Bird and Zureida Garda, "Reporting the Truth Commission: Analysis of Media Coverage of the Truth and Reconciliation Commission of South Africa," *Gazette*, 59:4/5 (1997): 331–42; Ron Krabill, "Symbiosis: Mass Media and the Truth and Reconciliation Commission of South Africa," *Media, Culture and Society*, 23 (2001): 567–85; and Annelies Verdoolaege, "Media Representations of the South African Truth and Reconciliation Commission and their Commitment to Reconciliation," *Journal of African Cultural Studies*, 17:2 (2005): 181–99. However, there appears to be no research analyzing just how widely and via what medium (radio, print, or television) most South Africans experienced the Commission. Former director of communications for the TRC, John Allen, confirms Boraine's view that the conventional wisdom is that radio reached the highest numbers of people, yet there has been a total absence of scholarship on this medium. John Allen, email message to the author, 2 December 2006.

36. Krabill, 568.

37. For an explanation of the different epistemological assumptions about truth that guided the Commission and the writing of its final report, see Posel, "The TRC Report," 150–7.

38. Promotion of National Unity and Reconciliation Act, Act no. 34, 1995. *Government Gazette*, 26 July 1995, 16579.

39. Belinda Bozzoli, "Public Ritual and Private Transition: The Truth Commission in Alexandra Township, South Africa 1996," *African Studies Quarterly*, 57:2 (1998): 167–95.

40. Tim Kelsall, "Truth, Lies, Ritual: Preliminary Reflections on the Truth and Reconciliation Commission in Sierra Leone," *Human Rights Quarterly*, 27 (2005): 361–91.

41. Hayner, 41.

42. Pumla Gobodo-Madikizela, "Prefatory Remarks, and Response to Long Night's Journey into Day," Brandeis Initiative in Intercommunal Coexistence, 27 March 2001: http://www.brandeis.edu/ethics/publications/biic_student_writing/sa/pumla_remarks.pdf (Accessed 20 April 2005.)

43. Boraine, 99.

44. Richard Schechner, *Between Theatre and Anthropology* (Philadelphia: University of Pennsylvania Press, 1985).

45. Frances Reid and Deborah Hoffmann, *Long Night's Journey into Day: South Africa's Search for Truth and Reconciliation*, VHS (San Francisco: California Newsreel, 2000). Transcript available at California Newsreel: http://www.newsreel.org/transcripts/longnight.htm (Accessed 20 April 2005.)

46. Antjie Krog, *Country of My Skull: Guilt, Sorrow, and the Limits of Forgiveness in the New South Africa* (New York: Three Rivers Press, 2000), 78.

47. TRC online testimony available at: http://www.doj.gov.za/trc. See East London section, Port Elizabeth hearings, 21 May 1996. Emphasis added.

48. TRC online testimony. Note that the transcriptions available on the TRC's web site are mostly in English. Transcriptions were created hastily at the time of the Commission and speakers who switched between languages, as Tutu habitually did, often did not have the fullness of their testimony transcribed. See Z. Bock et al., "An Analysis of What Has Been 'Lost' in the Interpretation and Transcription Process of Selected TRC Testimonies" (paper presented at the Intercultural Communication Conference, Stellenbosch University, 6–8 September 2004). The quotation listed here from Desmond Tutu is based on the TRC web-site material, augmented with corrections based on the author's viewing of unedited video from the hearings.

49. "PW Botha to be Prosecuted," *Mail and Guardian Archive*: *Online Edition*, 7 January 1998.

50. Wilhelm Verwoerd and Mahlubi Mabizela, *Truths Drawn in Jest: Commentary on the TRC through Cartoons* (Cape Town: David Philip Publishers, 2000), 54.

51. Mary Strine, Beverly Long and Mary Francis Hopkins, "Research in Interpretation and Performance Studies: Trends, Issues, Priorities," in Gerald Phillips and Julia Wood (eds), *Speech Communication: Essays to Commemorate the Seventy-Fifth Anniversary of the Speech Communication Association* (Carbondale: Southern Illinois University Press, 1990), 181–204.

52. Krog, 57. Krog's invention of a fictionalized Professor Kondlo has received much criticism of late. See Chris Mann, "Letter to the Editor: Watson Deserves an Answer," *Mail and Guardian Archive*: Online Edition, 25 November 2005. Krog's

book, for all its limitations, is more effective than any other secondary sources in terms of conveying the performative dimensions and dynamics of the TRC live hearings.

53. The bibliography on South Africa's TRC is extensive. For an overview, see the following review essays: Olayiwola Abegunrin, "Truth and Reconciliation," *African Studies Review*, 45:3 (2002): 31–4; and Richard Dale, "The Politics of the Rainbow Nation: Truth, Legitimacy, and Memory in South Africa," *African Studies Review*, 45:3 (2002): 39–44. Key secondary sources include: Kader Asmal, Louise Asmal and Ronald Suresh Roberts, *Reconciliation through Truth: A Reckoning of Apartheid's Criminal Governance* (Cape Town: David Philip Publishers, in association with Mayibue Books, University of the Western Cape, 1996); Wilmot Godfrey James and Linda van de Vijver (eds), *After the TRC: Reflections on Truth and Reconciliation in South Africa* (Athens: Ohio University Press, 2001); Deborah Posel and Graeme Simpson (eds), *Commissioning the Past: Understanding South Africa's Truth and Reconciliation Commission* (Johannesburg: Witwatersrand University Press, 2002); Fiona C. Ross, *Bearing Witness: Women and the Truth and Reconciliation Commission in South Africa* (Sterling, VA: Pluto Press, 2003); Charles Villa-Vicencio and Erik Doxtader, *The Provocations of Amnesty: Memory, Justice, and Impunity* (Trenton, NJ: Africa World Press, 2003); Charles Villa-Vicencio and Wilhelm Verwoerd, *Looking Back, Reaching Forward: Reflections on the Truth and Reconciliation Commission of South Africa* (New York: St. Martin's Press, 2000); Richard Wilson, *The Politics of Truth and Reconciliation in South Africa: Legitimizing the Post-Apartheid State* (Cambridge: Cambridge University Press, 2001).
54. See Kay McCormick and Mary Bock, "Negotiating the Public/Personal Interface: An Analysis of Testimonies on Human Rights Violations" (unpublished paper, n.d.).
55. Julia A. Walker, "Why Performance? Why Now? Textuality and the Rearticulation of Human Presence," *The Yale Journal of Criticism*, 16:1 (2003):149.
56. See Elaine Scarry, *The Body in Pain: The Making and Unmaking of the World* (New York: Oxford University Press, 1985).
57. Cole, *Stages of Transition*.
58. Posel, "Truth?" 17.
59. Bird and Garda.
60. Posel, "Truth?" 17–18.
61. Posel, "Truth?" 19.
62. David William Cohen and E. S. Atieno Odhiambo, *The Risks of Knowledge: Investigations into the Death of the Hon. Minister John Robert Ouko in Kenya, 1990* (Athens: Ohio University Press 2004), 19.
63. The "Guguletu Seven" were Zabonke John Konile, Godfrey Jabulani Miya, Zanisile Zenith Mjobo, Mandla Simon Mxinwa, Christopher Piet, Zola Alfred Swelani, and Themba Mlifi Zanisile.
64. For an excellent video documentary on this case, see Lindy Wilson, *The Guguletu Seven* (South Africa: Ring Records, 2001). The lead investigator in the case has written a memoir about his experiences; see Zenzile Khoisan, *Jakaranda Time: An Investigator's View of South Africa's Truth and Reconciliation Commission* (Cape Town: Garib Communications, 2001). The Guguletu Seven are also featured in the documentary by Reid and Hoffmann, *Long Night's Journey into Day*.

65. For TRC online transcripts, see http://www.doj.gov.za/trc. The first day of testimony on the case of the Guguletu Seven can be found on the TRC website in the Cape Town section, Heideveld hearing, 23 April 1996, case of Christopher Piet, see: http://www.doj.gov.za/trc/hrvtrans/heide/ct00100. htm (Accessed 25 April 2005.)

66. The Guguletu Seven case was one of ten "window cases" identified by the Truth Commission as representative of broader patterns of abuse. This hearing in November 1996 was also referred to as an "event hearing," for the focus was on specific events in which gross violations of human rights occurred, rather than on the individual experiences of victims (TRC, *Truth and Reconciliation Commission of South Africa Report*, 1: 147–8).

67. Khoisan, 73.

68. It was partly through these video images that counsel finally proved that the police had planted evidence at the crime scene; see Wilson.

69. Joseph Aranes and John Yeld, "Top Cops Storm Out of Hearing into Guguletu 7," *Cape Argus*, 28 November 1996: http://www.iol.co.za (Accessed 4 November 2004).

70. "Day 3: Introduction and Showing of Video," TRC, *Human Rights Violation Committee Transcripts*, Cape Town, Pollsmoor, 28 November 1996: http://www.doj. gov.za/trc/hrvtrans/polls/video.htm (Accessed 29 June 2005).

71. "Day 3."

72. "Day 3."

73. Douglas, 29.

74. Douglas, 145–9.

75. See, for instance Hazel Estella Barnes, "Theatre for Reconciliation: Desire and South African Students," *Theatre Journal*, 49:1 (1997): 41–52; Yvette Coetzee, "Visibly Invisible: How Shifting the Conventions of the Traditionally Visible Puppeteer Allows for more Dimensions in both the Puppeteer-Puppet Relationship and the Creation of Theatrical Meaning in Ubu and the Truth Commission," *South African Theatre Journal*, 12:1/2 (1998): 35–51; Loren Kruger, "Making Sense of Sensation: Enlightenment, Embodiment, and the End(s) of Modern Drama," *Modern Drama*, 43 (2000): 543–66; Yvette Hutchinson, "Truth or Bust: Consensualizing a Historic Narrative or Provoking through Theatre. The Place of the Personal Narrative in the Truth and Reconciliation Commission," *Contemporary Theatre Review*, 15:3 (2005): 354–62; Stephanie Marlin-Curiel, "Performing Memory, Rehearsing Reconciliation: The Art of Truth in the New South Africa" (PhD diss., New York University, 2001); Kimberly Wedeven Segall, "Over My Dead Body: Trauma and Unreconciled Truths in South African Performance" (PhD diss., Northwestern University, 2001). Generally, performance studies and theatre scholars write on theatrical or aesthetic adaptations of truth commission material, such as the famous work by Jane Taylor, William Kentridge and Handspring Puppet Company, *Ubu and the Truth Commission* (Cape Town: University of Cape Town Press, 1998).

9

"To Lie Down to Death for Days": The Turkish Hunger Strike, 2000–2003

Patrick Anderson

Violence remains the founding language of social representation.
Allen Feldman[1]

The prison must be an exhaustive disciplinary apparatus.
Michel Foucault[2]

Ebru Dinçer sits at a flimsy table, wearing a nondescript beige uniform in front of a nondescript concrete-block wall. As film-maker Metin Yegin interviews her, she stares back at him with the intensity of one who has a short time to live. Her face is sunken, worn thin from months of starvation, and scarred from the many surgeries required to heal her chemically burned skin. Her eyes are her largest features, and when they occasionally drift to the floor or to the ceiling, they are filled with a despair that will never find its way into language; she is despondent, but intense. She speaks clearly, and responds to Yegin's questions with impressionistic precision in describing life since her recent release from prison: "Life now is like a dream. Like a fantasy. It's like a non-existent thing. [. . .] It has lost its reality. Nothing to add. No, there is nothing else I want to add."[3]

Dinçer was one of 816 Turkish political prisoners who, on 20 October 2000, began what would become the longest and most deadly hunger strike in modern history – and was still, as of summer 2003, in process.[4] Of the strikers' original demands, the most central is the termination of all construction on, and transfer of prisoners to, so-called "F-Type" prisons, based on Euro-American designs intended to isolate prisoners, to limit (and largely to deny) their rights to interact with one another, and ultimately to make it easier for prison torture to go unrecorded, unchecked, and unpunished. In the three years since the strike's commencement, more than 2000 people have been on strike at various times and for various intervals; by the middle of 2003, 107 strikers had died.[5] The impasse between prisoners and both prison and governmental officials continues; despite the efforts of Human

Rights organizations, legal advocacy groups, foreign committees, and others, it shows no signs of traditional resolution.

The staging of the hunger strike coincides with two larger scale affairs that have dominated domestic and foreign policy in Turkey in the last several years: continuing attempts to obtain member status in the European Union; and renewed Western (imperialist) interest in the Middle East, especially after 11 September 2001. Turkey was officially named a candidate for EU membership in 1999 despite promises from several nations to veto any final accession.[6] This step renewed Turkish governmental interest in continuing to adapt official policy to European standards – particularly policy concerning capital punishment, prison conditions, human rights protection, and economic development – and simultaneously renewed critics' charges against Turkey. As Turkey continues to attempt to color itself European, those unquestionably "inside" the Union (e.g., Greece, Germany, France) continue to position Turks as Other, and Turkey as Outside. At the same time, Turkey represents a pivotal ally for US and UK military forces (and others) who have sought easy access to majority-Muslim nations including Iraq, Iran, and Syria. At once solicited by Western superpowers for right of entry to its land between Europe and the Middle East, *and* maligned by European Unionization officials who criticize its policy and sneer at its agrarian culture, Turkey is uniquely situated – or rather, imagined – as ally-other.

Of course, the dynamics that go by the terms *inside/outside*, and that so dramatically mark Turkey's role in contemporary world politics, are hardly as simplistic as such a characterization may make them seem. Typically figured (as the subtitle for a recent book puts it) "between two worlds"[7] – a positioning that will forever refuse to recognize Turkey as its own "world" – Turkey is generally treated as the *effect* of a staid borderland geography. In the logic of imperialism and Unionization, Turkey marks that place where the potential for European identification (the referent of EU membership) is unsettled by the proximity of a disparate set of states (Iraq, Iran, Syria, et al.) generally imagined as the Other against which Europe defines itself. Against this geographical background – conceived, as geography often is, as stable, natural, or fixed[8] – Turkey represents both the promise of Western expansion and the ultimate threat to the integrity of Western identification. In the words of Valery Giscard d'Estaing, former French President and chair of the EU Constitutional Convention, Turkey's membership would "open the door" to West Asian and North African Arab states and would thus "be the end of the European Union."[9]

Given this context in which Turkey is asked to perform the role of Ally when it comes to colonizing the Middle East and Other when it comes to identification with Europe, the current hunger strike articulates on a domestic scale the anxieties and the violence associated with the production of a state whose global affiliations are in flux. I will argue in this chapter that the hunger strike reframes that question of affiliation in terms of political subjectivity, staging a challenge to more conventional relationships between state and subject. Indeed, the Turkish hunger strike – itself bound up in a complicated reversal of violence performed on individual subjects by the state – demonstrates that the political subjectivity (which we might recognize as a consortium of "freedoms": of representation, of expression, of association) so desperately sought by the strike's constituency is underpinned by subjugation to more dire forms of institutional and ideological power.

This chapter is intended, then, to explore both the role of the F-Type prison in contemporary Turkish politics, and the conditions of the prisoners' hunger strike: its aims and demands, the experiences of its practitioners and casualties, its representation in state-sanctioned and other media, and its effects. I trace the context of the strike's beginnings and the history of the demands named by its constituency. I explore the progression of the strike and the implications of its various main events. I ask how and why this strike stages and interrogates the relationship between political subjectivity and the integrity of the Turkish state. And, finally, I consider the potential of hunger striking as a viable, sustainable form of political action, and the production of political subjects through the performance of the hunger strike.

The contours of my argument follow four specific suggestions for how the Turkish hunger strike functions as performance to refigure the relationship between state and subject, to facilitate the deployment of new kinds of political subjectivity, and to redraw the meaning of Turkey as a borderland state. First: the community of strikers has expanded to include not only political prisoners in a number of prisons throughout Turkey, but also activists outside the prison altogether, particularly those in the Kuçuk Armutlu neighborhood of Istanbul. The manner in which the strike has been staged *across prison walls* resymbolizes the significance of the boundary represented materially and discursively by those walls – as a producer of political alliance rather than a block to it. This is not to say, of course, that the techniques of surveillance, discipline, and containment performed within Turkish prisons are empty of the meanings and effects that distinguish them as specifically carceral; but, rather, that this particular strike has begun to open up the possibility for identification and alliance – questioning, that is, what precisely "Turkishness" means in the context of state penal policy – across the very walls that are traditionally conceived as intense divisions between those who are held within and those who are "free" to move outside penal institutions.

Second, in part because such intense lines of affiliation have developed across prison walls, the hunger strike has effected the signification of the prison, and in particular the F-Type prison, as the context and the condition for contemporary Turkish state-production. That is, due largely to the specific performance of this hunger strike (and the rhetoric surrounding its performance), the prison has become both a specific site for the production of "Turkishness" *and* a metaphor for the production of state power in Turkey more generally. This metaphor takes the form, for example, of specific statements aimed to solicit support for the strike ("The entire Turkey is like the F-Type prisons," wrote Turkish human rights workers after military attacks on Kuçuk Armutlu were reported[10]) and of sophisticated attempts by prisoners and activists alike to show that the effects of the prison crisis extend well beyond any penal policy that may be linguistically amended or erased. The hunger strike, devoted explicitly to a list of demands, is more than simply a question of prison design, of increasingly restrictive legislation, or of the impunity afforded to the practitioners of prison torture. At question is the very nature of Turkish political subjectivity as defined, delimited, and compelled by the state.

Third – again, in part as a result of efforts to force into crisis the difference between being inside and being outside the prison – the hunger strikes have effected a mode of resistance for which *coalition* is the primary unit of political action and signification. This is exceptional in the history of hunger striking, which traditionally promotes the names, and not the numbers, of hunger strikers as the standout representatives of the strikers' struggles. Bobby Sands, Mahatma Gandhi, Cesar Chavez: these names do not merely denote oft-cited events in the history of hunger striking; as symbols, they embody those events, and so bear the burdens of both representing the issues at the heart of past strikes and commemorating those strikes. In Turkey, though press releases name those strikers who have died, there is no single figurehead whose image or name signifies the movement as a whole. More significant for this strike are the immense numbers of those who have been on strike and those who have died; this shift in attention repositions the effects of resistance and the desired outcome of the strikes in terms of a political subjectivity squarely situated within the context of group affiliation. Rather than striving for a kind of martyrdom in individualization, for the recognition of the individual subject as the principle unit and site of political signification, the multiple solidarities epitomized by the Turkish strikers produce a subject deeply connected to her/his political community despite the potential for what is essentially a highly individualized consequence of striking: death.

Finally, and also exceptionally, the Turkish strikes have redefined the practice of hunger striking: from what is traditionally conceived as a necessarily short-term, last-ditch effort to effect change, to a budding sustainable mode of political action whose organizing center is the potential death of individual subjects. That is, the Turkish hunger strike is not merely *notable* for its unusual duration; rather, it recasts hunger striking itself as a form of protest whose extended performance could alter both the meaning of the protest itself, and the notions of "success" and "failure" that often limit how the protest's larger impacts may be understood. As I discuss below, attempts to read the effects of the Turkish hunger strike strictly in terms of whether or not its specific demands are met ignore its many implications not only for the conditions of prisons themselves, but for the very heart of the production of Turkish subjects.

In calling the hunger strike a performance, I mean, first, to draw attention to the theatricality of the strike; and, second, to suggest that the strike speaks to several classic conundrums for research on performance and specifically for the field of performance studies. Rhetoric surrounding the strike has indicated the prison as a kind of "stage," set in particular for spectacularizing prisoners' complaints about treatment and attempts at reform.[11] Simply put, the prison represents an arena in which the state asserts its power both to define the limits of "upright citizenship" and to confine those subjects who deviate from that designation; this casts the prison as a critical site for the production and maintenance of the integrity of the state. The prison is also a stage on which those subjects marked as "delinquent" – a term I investigate in greater depth below – perform compliance with, or resistance to, the discipline exerted by representative agents of the state.

Conceptualizing the prison – specifically the F-Type prison – as a stage also implies that hunger striking is an embodied cultural practice, whose display before national and international audiences occasions the re-symbolization of participating subjects. This casts the hunger strike as both performance and performative. For if the hunger strikers are rehearsing a set of gestures legible to their audiences – specifically: the refusal to eat, and the expression of central demands – they are also attempting to produce a new, viable model for political alliance and political subjectivity. This exemplifies the complicated means by which performance can resist social and ideological forces. The hunger strike has infused the very notion of resistance with an extreme form of embodied practice, made all the more vivid by the ever-increasing possibility that its practitioners will at some point die. Such embodiment heaves hunger strikers – to use Peggy Phelan's phrase – into a "maniacally charged present," significant not only for the tremendous intensity of its actions but also for its very real – potentially lethal, potentially revolutionary – final effects.[12]

Moreover, the "liveness" so celebrated by Phelan and others as ontologically linked to the value of performance as a cultural form becomes especially charged in the example of the Turkish hunger strikers, for whom "life" and "to live" become radically redefined in the context of the inevitability and (for the larger cause) the necessity of death as a political mode. Hunger striking pushes the question of "liveness" to its extreme degree of cultural significance, suggesting that even in its most mediated incarnations – journalistic exposés, autopsies, official statements from Ankara – the performance of the hunger strike is exceptionally "live," almost vibrating with intensity before the gazes of those who behold it. At any moment, another hunger striker may die as a result of her/his performance, adding to the already high number of casualties. The "maniacally charged present" in which the hunger strike is staged, then, is a matter of life and death. The strikers' seizure and radical embrace of the political significance of death and its potential for resistance, are at the center and at the limit of the power of the hunger strike to recodify and reorganize state-sanctioned regimes of violence and oppression, represented fully by the F-Type penal system: at the center because they are the foundation on which the strike has been staged and sustained, at the limit because death, the potential final outcome, is simultaneously their final political action.

To call 20 October 2000 the *beginning* of the current Turkish hunger strike is, however technically accurate, a misnomer. First, the strike itself follows several previous hunger strikes in Turkish prisons, in particular one in 1996 that claimed 12 lives.[13] Rhetoric used both to announce the current strike and to name its demands not only refers back to these previous strikes, but names their participants – especially those killed in military operations intended to break up the strikes, and those suffering from medical problems as a result of the strikes – as part of its own constituency:

> Our friends named below were imprisoned alive at different times. The state was responsible for the security of these friends' lives. As is well known, the acceptability and legality of the prisoners' status [as imprisoned] is based on the state's guarantee of the security of their lives. These friends were massacred by the state. We want our friends back.
>
> [. . .]
>
> All our friends who have suffered from permanent illness, whose health problems have continued since the 1996 Death Fast, who were injured during state operations and denied medical treatment, must be immediately released.[14]

Second, the strikers' demands circulate within a larger set of concerns about the relationship between the Turkish state and its many marginalized subjects: Kurds, Islamists, dissident writers and journalists, human rights workers, and members of outlawed leftist political parties.[15] Since the inception of the Turkish state in the 1920s, this relationship has been contentious at best – and warlike at worst. Thousands of Kurds, for example, have been incarcerated after being charged – many fewer have actually been convicted – with membership in one of the country's many leftist political groups, several of which have historically fought for an independent Kurdish state. Islamists have struggled to find a way in to Turkey's political scene, reformed as strictly secular by Kemal Ätatürk's government in the 1920s – and are often imprisoned because of those struggles. Writers and journalists are regularly sentenced for publishing texts critical of the Turkish government, explicitly covering the "Kurdish problem," or any of a number of other restricted journalistic and literary activities.[16] These struggles have defined the context and the conditions of political resistance in Turkey since its establishment as a Republic in 1923.

In this context, the commencement of the current hunger strike was announced in a manifesto released to the press on 18 October 2000 by the Devrimci Halk Kurtulus Partisi-Cephesi (DHKP-C) and the Turkiye Komunist Partisi/Marksist-Leninist (TKP/M-L),[17] and on 20 December, strikers began their fasts. The most immediate concern of the prisoners was the government's plans to move political prisoners (who make up 10,000–12,000 of the 72,000 people incarcerated in Turkish prisons[18]) into the new F-Type prisons. Traditional Turkish prisons have been built in dormitory-style, consisting of large bunkrooms (holding up to 100 people) in which prisoners were able to interact with one another regularly.[19] Concerned that certain political groups use these prisons as "indoctrination and recruitment centers," and as part of its intense push for EU membership, the Turkish government decided to reform its penal system by adopting a US/European-based design for the F-Type prison in order to hold incarcerated people in isolated cells and to limit (if not deny outright) time allotted for socialization.[20] This decision followed attempts in the 1980s to develop what was then called a "Special-Type Prison" – most notoriously, Istanbul's Kartal Prison, which has served, more or less, as an immense holding cell for political prisoners *awaiting trial* for offenses under the draconian Anti-Terrorism Laws enacted in 1991. Kartal represented a jarring shift in the penal system, from the old-style dormitory prisons to a large institution broken up into single or small-group cells where prisoners are being held (according to most reports) all day, every day with little to no interaction (or even visual contact) with other human beings. By 1999, of the roughly 300 prisoners held at Kartal, only one had been convicted of any crime. The Kartal Special-Type prison and its conditions of incarceration both rehearsed and set the stage for the implementation of the F-Type prison regime.[21]

The F-Type prisons subject political prisoners to isolation, and to vulnerability to ill-treatment and torture. The prisons are divided into 109 cells holding up to three prisoners each, and 59 solitary cells. Typically, cells in F-Type prisons are windowless, are relatively small, and include the most basic furnishings: a bed, a shower, a toilet, and often a desk. Each cell opens into a small high-walled courtyard of 16–50 square meters. Prisoners have reported that those doors, as well as the doors leading to the wards' corridors, remain locked for most of the day – typically the entire day. According to the Ministry of Justice, the prisons also include "lawyers reception rooms, a library and an extensive reading room for personnel and prisoners, and indoor multipurpose area for social, cultural, and sports activities, [...] and working sites for various purposes (so prisoners can learn a trade)."[22] These spaces, compulsory for meeting the Minimum Prison Requirements of the

United Nations and Council of Europe, are, however, typically closed to prisoners, who report that their access to libraries, education, and socialization programs is random and sparse to nonexistent. These conditions render the F-Type prison both an isolationist prison *regime* as well as the specific architectural space in which that regime may assert itself. In other words, the prisons represent both an ideological force that restricts contact and communication between political prisoners, and the most intense site of that ideological constraint.

Plans to develop the F-Type prisons were announced in 1989. Following that announcement, and throughout the 1990s, several strikes were staged to oppose the transfer of political prisoners to these new prisons.[23] The transfers were explicitly justified in Article 16 of the Anti-Terrorism Law. This article serves as the legal basis for discriminating between political prisoners and other prisoners, for isolating political prisoners in small cells, and for thoroughly restricting communication among prisoners and between prisoners and their families. The beginning of the article reads:

Sentences of those convicted of offences within the scope of the provisions of this law [i.e., the Anti-Terrorism Law] shall be served in special penal establishments, constructed according to a system of one-person and three-person cells. No open visits shall be permitted in such establishments. Communication between inmates and with other convicts shall be prevented.[24]

Designating between "acts of terror" and other crimes and (more importantly) between terrorists and other criminals has been a major project for Turkish legislators since the early 1980s. In 1983, the new Official Prison Regulations were adopted; Articles 78/3 and 78/4 of those regulations designated the categories of "political prisoner," "terrorist," and "anarchist," and rendered legal the confinement of prisoners designated as such in special prisons.[25] Strikes against such legislation intensified after the adoption of the Anti-Terrorism legislation in 1991; the 1996 strike was the first to specify the F-Type prison as its central complaint. Despite an official announcement that all plans to build the new prisons would be put on hold to allow time for reconsideration, construction continued, and by mid-2000, four F-Type prisons had been opened.[26] Further, the Turkish government undertook military operations in the prisons in 1995, 1996, 1999, and 2000 in order to end the strikes and forcibly to extract selected prisoners from the large wards and relocate them in the new prisons.[27] Under the Anti-Terrorism Laws, political prisoners were classified as "terrorists," enabling the government to exact even more specialized discipline and punishment including, again, solitary confinement and the restriction of many, if not all, daily activities (including socialization, exercise, and formal education).

The terms of the Anti-Terrorism Law gathered in their expansive sweep a range of activities that together construct an idiom of Terrorism synonymous with dissent, with discourse, and indeed with difference. Article 8 reads:

> No one shall, by any means or with any intention or idea, make written and oral propaganda or hold assemblies, demonstrations and manifestations against the indivisible integrity of the state of the Turkish Republic with its land and nation.[28]

In addition to outlawing a wide (and vague) range of activities, this article draws attention to the fundamental relationship between the state and its various geographies – the land it occupies, or aims to occupy; social service infrastructures and the territories they transverse; and the architectures of the government's many institutions, including the prison. Article 8 is not merely a crackdown on resistance "propaganda" or protest; it renders many oppositional voices "terrorist" simply by virtue of their dissenting political claims. If officials have been centrally concerned with attempts by Kurdish political groups in the east attempting to develop and enforce independent rule (an immediately recognizable "assembly" aimed to divide the land Turkey claims as its own),[29] they have also been fully aware that prisons built to house such dissidents are themselves vital architectures of the "indivisible integrity of the state." Legislation such as Article 8, then, enables the official classification of hunger strikers – whose demands include massive prison reform – as terrorists, simply by virtue of their demonstrating against state penal policies. As I describe above, this legislation has a genealogy, including, for example, the Turkish Constitution, the most recent version of which was written during three years after a military coup in 1980 in order to "restore" democratic control of the government. The preamble to the Constitution makes it clear that:

> No protection shall be given to thoughts or opinions that run counter to Turkish national interests, the fundamental principle of the existence of the indivisibility of the Turkish state and territory, the historical and moral values of Turkishness, or the nationalism, principles, reforms, and modernism of Ätatürk, and that as required by the principle of secularism there shall be absolutely no interference of sacred religious feeling in the affairs of state and politics.[30]

Kemal Ätatürk was the first president appointed after the fall of the Ottoman Empire in the early 1920s. Deeply invested in "dragging Turkey into the West," his government initiated a series of policies in the 1920s aimed to outlaw traditional practices associated with former notions of Ottoman identity – including such everyday actions as wearing the fez, and such abstract practices as organizing as a "minority" group. Further, in restricting citizens' travel and controlling communications and mass media, these policies had the effect of isolating Turkey from neighboring states and effectively locking down Turkey's national borders. At the same time, Ätatürk's government opened the door to Western imperialism by welcoming – and, in some ways, legally enforcing – influence from Western nations, including the United States. Sweeping changes mesmerized the national populace, which was nonetheless suspicious of the rapidly changing Turkish scene. This "modernism," however, made Ätatürk something of a national hero – not merely a president whose leadership was pronounced in its sweeping reform of the former Ottoman Empire, but himself an intensely loaded metaphor for bold economic and cultural "progress." To this day, Ätatürk's name appears throughout national legislation as a sign of Turkey's desire to implement many Western policies, to open the door to identification with the West, and to deny the more pluralist nature of Ottoman culture; that is, the "principles" of Ätatürk – typically called "Kemalism" – are based on the coherence of a congealed, conventional Turkish national identity based on the denial of difference.[31]

In what is perhaps the most remarkable example of this kind of legislated denial, the Political Parties Law includes an article "preventing the creation of minorities": "Political parties cannot put forward that minorities exist in the Turkish Republic based on national, religious, confessional, racial, or language differences."[32] This third piece of legislation is not simply an institutional denial of difference among the Turkish populace, but the ideological arm of a nationalist project of racial and ethnic cleansing; as many of those incarcerated in Turkey's prisons have learned, it has the force not only to designate certain activities as "terrorist," but to activate various state-sanctioned techniques of punishment. Kevin Robins artfully and compellingly dispatches cultural theory to this scene of nationalized homogeneity, suggesting that the denial of internal difference is a kind of "repression of identity." The repressed, he argues, is "returning":

> There is still iron in the soul of the [Turkish] state; seemingly devoid of political imagination or nuance, it continues to pursue the uncompromisingly adamantine principles. [. . .] The "Other" Turkey is making its declaration of independence, making its reality felt, manifesting the complexity of its social being.[33]

In the context of these articles of legislation, the "Other Turkey" has little to hope for by way of political recognition or representation. The forced patterns of identification initiated by such legislation, and the widespread arrest and incarceration of political prisoners that it has sanctioned, are suggestive of Foucault's argument in *Discipline and Punish*: that the juridical is subordinate to the punitive in the service of the state[34] – and that the prison is not merely a disciplinary edifice that houses criminal offenders, but a "penitentiary *technique*" that produces marginalized subjects. He writes:

> The penitentiary technique bears not on the relation between author and crime, but on the criminal's *affinity* with his [sic] crime. [...] The correlative of penal justice may well be the offender, but the correlative of the penitentiary apparatus is someone other; this is the delinquent, a biographical unity, a kernel of danger, representing a type of anomaly. [...] The penitentiary technique and the delinquent are in a sense twin brothers. [...] It is now this delinquency, this anomaly, this deviation, this potential danger, this illness, this form of existence, that must be taken into account when the codes are rewritten. Delinquency is the vengeance of the prison on justice.[35]

Political prisoners are "delinquents" par excellence; they represent the most dangerous threat against – and their incarceration is the *sine qua non* of – the integrity of the state. Incarceration enables, through its many forms of discipline, the production of "docile bodies" whose political subjectivities are defined and determined by their subservience to the state.[36] The ultimate aim of the penitentiary technique, that is, includes not simply the "rehabilitation" or "reform" of the incarcerated, but the specific production of "delinquents" whose subjugation becomes the condition of their very existence. Such subjugation is likewise the condition of state integrity; Foucault argues:

> The carceral network does not cast the unassimilable into a confused hell; there is no outside. It takes back with one hand what it seems to exclude with the other. It saves everything, including what it punishes. It is unwilling to waste even what it has decided to disqualify. [...] The delinquent is not outside the law; he [sic] is, from the very outset, in the law, at the very heart of the law.[37]

The state defines itself, that is, with these delinquents at its very core; it asserts itself in designating delinquents as such; it produces itself precisely through the exclusion-by-incarceration exemplified by the F-Type regime. "Rehabilitated" comes to mean "docile."

Below, I focus more directly on the hunger strike's gendered makeup and implications; however, it is important to take notice here of Foucault's complicated omission of gender from his analysis. This omission is represented linguistically by his constant referral to prisoners as "he," narratively by his exclusive use of historical examples of imprisoned men, and generally by the lack of any legible recognition that his theory may be differently inflected across lines of gender identifications. And yet, it seems that gender, though unexplored by Foucault, is at the heart of his argument: "docility," after all, is not so far from "domesticity," both of which are hallmarks of patri-archal ideological frameworks that situate feminized subjects, spaces, and behaviors on the far-right side of the pairing "active/passive." This pairing, often treated as synonymous with the binaries "male/female" and "mascu-line/feminine," also describes the performance and the anticipated effects of the hunger strike, variously portrayed as political *action, passive* resistance, or an *act* of terror. Though I will return to these issues below, it is critical to note at the outset that Foucault's insistent concentration on male sub-jects is thus deeply infused with gendered implications: his "docile" bodies are essentially feminized bodies; his "law" is essentially a masculine regime whose power, in part, is derived from its highly specialized, and specifically gendered, tactics of discipline and constraint.

Foucault's focus on *affinity* suggests that in the logic of the prison, a "crime" is not, as conventionally conceived, just an act derived from a moment in which a person chooses badly; prisoners are "delinquents" inso-far as their crimes become a part of them, an extension of their characters – indeed, central to their characters. Imprisoned, a criminal *is* her/his crime. This logic becomes an articulation of state-sanctioned racism in the case of Turkey, where the legislation I discuss above works to marginalize, if not prohibit outright, political activity as traditionally practiced by most Turkish Kurds (who maintain that sharp racial and ethnic differences divide them from non-Kurdish Turks), Islamists (for whom "no interference of sacred religious feeling" has little practical meaning), and others. Turkey's attempts at various points in its 75-year national history to prohibit pub-lic use (in education, publication, and radio transmission) of the Kurdish and Arabic languages only extends this agenda of erasing difference in the interest of protecting a streamlined, purified Turkish national identity.[38] If, in Foucault's theory of the penitentiary technique, the "criminal" *becomes* her/his "crime," many Turkish political prisoners become delinquents pre-cisely when and where they identify as (racially, ethnically, religiously) different.

The articles of legislation described above, along with various portions of the Penal Code similarly intended to suppress political dissidence,[39] are the courts' main legal justification for handing down sentences for the great majority of Turkey's political prisoners. They represent core values of the systems of political representation and domestic policy-making – systems that reproduce and institutionalize the "vengeance of the prison on jus-tice" – against which the current hunger strike has been organized. The construction of F-Type prisons is, in other words, the central issue behind which the hunger strikers have organized – but not the only issue. For if the prison *qua* prison is a material *function* of the state's attempts to expel certain groups from its populace, it is also (to use Foucault's word) a *tactic* used to *produce* state power in immediate acts of ideological and material violence. The prison may be the most concentrated architecture of that production – the site and the sign of the state becoming itself in discrete enactments of violence and in the institutionalization of violence-as-law.

This renders the construction of F-Type prisons the most coherent rallying-point for the hunger strikers. Resplendent in its own very real dangers (extended solitary confinement, torture), the F-Type prison is also emblematic of the state's reproduction and defense of its veracity and of its power; and as such, it sets the terms – if not the limits – of resistance. As the hunger strikers know all too well, these two operations are correlative. Implicitly drawing attention to the prison-as-tactic, the strikers list eight other demands in their manifesto. These include the immediate and unconditional repeal of the Anti-Terror Law, independent monitoring of prisons and treatment of prisoners, release of prisoners suffering from permanent injury and/or illness as a result of previous hunger strikes, and trials for prison staff and others accused of torturing incarcerated people during the past several decades.[40] Couched in the language of an astonishingly violent genealogy of carceral practice and juridical policy, these demands clarify the meaning of the prison in terms of an experiential economy not just of subordination and resistance, but of nationalism, state-making, and the production of "docile" *and* resistant political subjectivities.

The meaning of resistance within that economy becomes especially potent when the terms of its performance potentially occasion the death of its practitioners. Hunger striking, that is, explicitly ups the stakes of political action. In his intensely intimate ethnography of political violence in Northern Ireland, Allen Feldman pays special attention to the 1981 Blanketmen hunger strike in Belfast. Probably the most famous strike of its kind in modern history, the Blanketmen hunger strike lasted for 217 days and resulted in the deaths of ten incarcerated members of the IRA (including MP Bobby Sands).[41] Feldman says of the strike:

> The performance of the hunger strike would stage the abuse and violence of the Other in the eviscerated flesh of the dying protester. The penal imperative to incorporate the panoptic presence of the Other as a form of compliance and subjugation would itself be subjected to deflating mimesis and a final ironic reversal. [...] No other action more eloquently demonstrated the condition and image of the human body infested with the state apparatus.[42]

The hunger strike's eloquence lies in its potential: (1) to throw into crisis the binary passive/active in terms of violence performed, and (2) to undermine the conception, suggested in Ankara's official responses to the strike, of individual and state as absolutely discrete entities at odds with one another – a conception that facilitates the dumbing-down of questions about political terror into the language of cause and effect. In other words, hunger striking rebuffs a particular notion of domination and resistance that positions subjects simply as the *victims* of state power, and simultaneously stages the seizure, re-symbolization, and enactment – one might say the *ingestion* – of modes of violence typically performed by the state; this is the "condition and image of the human body infested with the state apparatus," the function of the hunger striker as representative of, deeply entrenched in, and radically resistant to the complicated machinery of state violence. The body of the hunger striker, in other words, asserts itself as a body – as a visceral representative for state-produced "delinquency" – by performing its own gradual decline, through self-consumption, to death. And so that body becomes not only the object of state punishment and torture, but simultaneously an agent imminently responsible for performing violence upon itself. In the case of the Turkish hunger strikers, that is, embodiment becomes not only a mode of resistance, but also a seizure of state power – especially the state's power to enact violence upon its subjects. This kind of embodiment, further, represents the dying symbol, ever larger in its political and cultural effects, of what has otherwise been denied to the incarcerated: political agency. And hence:

> Starvation of the flesh in the hunger striker was the inverting and bitter interiorization of the power of the state. Hunger striking to the death used the body of the prisoner to recodify and to transfer state power from one topos to another.[43]

The most dramatic example of the state's attempt to short-circuit such transfer of power in Turkey took the form of an intense military operation on 19 December 2000, two months after prisoners began their strike. Government officials were no doubt concerned that the rhetoric of the strikers' demands, and the inevitable images of dead strikers' bodies being carried out of Turkish prisons, would be a PR disaster for a country attempting to mend its reputation as backward, pre-modern, and barbaric. And so Operation Return to Life – perhaps the most ironically named military exercise in modern history – was staged in 20 prisons across the country. Narratives of what exactly happened on that day, including the few official statements from Ankara, confirm that at 4:30 a.m., as people in those 20 prisons were sleeping, military commandos dressed in war fatigues and carrying automatic weapons, semi-automatic weapons, tear gas, flame throwers, and various chemical agents broke through the prisons' roofs. Prisoners were terrified as they were torn from their slumber, herded or chased through different sections of the prisons, gassed, set afire, shot, and/or chemically burned. Thirty were killed[44]; many more were seriously wounded. And more than 1000 were dragged out of the prisons and transferred to cells in the four completed F-Type prisons.[45]

Of course, reports about the operation from different sources tell radically opposed stories about conditions inside the prisons and about those responsible for the 30 deaths and the very many more serious injuries. Governmental sources argued that they were regaining control of certain parts of the dormitory prisons to which they had had no access for nearly a decade.[46] These sources also claimed that the prisoners set themselves and their comrades on fire, and used small arms weapons to fire on soldiers, prison staff, and their fellow prisoners. Investigations after the operation – and statements from various prisoners – have suggested that these claims have mostly been fabricated.[47] But because many of the prisoners were transferred to F-Type isolation cells, where they are not permitted to communicate with one another, with legal advocates, and in many cases with their families, very few narratives from prisoners have been publicized. All of those that have been released indicate that the soldiers' attack on the prisoners was direct and relentless:

> Before we put our clothes on, they opened fire at us and we took cover. We soaked all the towels we could find. Then they started to dig into the ceiling from various places. [...] "We came here to kill you," they were saying. [...] After opening several holes in the ceiling, the bombardment with gas bombs began. We were choking and trying to gasp for air.[48]

> They told us that they wanted us to capitulate. But we said no. Our friends brought those of us who were on hunger strike to death to the back of the dormitory in an effort to save [us]. The soldiers sprayed a chemical agent that we couldn't recognize on [some of us]. Six of our friends were burned in front of us. I was also on hunger strike to death. Ozlem, one of my friends who was not on hunger strike to death lay on my body in order to save me. But they sprayed that chemical agent on her and burned her. She died as a result.[49]

> The bombardment continued non-stop for 9 hours. And we could not leave the dormitory. Because we had no place to go. We could not even raise our heads. We could not stand. [...] Most of those who survived, except for a couple of our friends, have been burnt. 6 of us were burnt completely and even their corpses could not be identified.[50]

These narratives illustrate one of the most central – and complicated – "rules" of hunger striking: that hunger strikers must not die directly at the hands of police, military, or prison officials, and yet that their deaths stand as representatives of the terror of the state. This "rule" symbolizes the hunger striking subject's seizure of one of the mechanisms of state power: discrete acts of violence performed on the bodies of prisoners by state and prison officials. Further, in the case of Turkey, it has become imperative that every striker, and every striker's death, should stand as equally significant representatives of the strike's mission; no single striker (or category of strikers, or type of striker) exceeds the significance of the group as a whole. Again, writing about Northern Ireland, Allen Feldman says of the Blanketmen strike: "The individual hunger striker was a representative of their collective condition. [H]e would sacrifice his individuality at the same time that he committed the most individual of acts."[51]

Feldman's argument here is that hunger striking positions "the most individual of acts," dying, as representative not of the individual striker's needs and desires, but of the group as a whole. In such an economy of representation – an economy some may call "martyrdom" – a single hunger striker infuses her/his rapidly disappearing, eviscerated flesh with meanings that renounce the individual as the base unit of political action and signification; the body of the dying hunger striker becomes fully saturated with significance, particularly at the moment when the "sacrifice" is complete. Feldman continues:

> At the moment of the striking Blanketman's disappearance, he was to attain his highest condition of visibility. But despite this parasitic swarming over his body, the hunger striker became an isolate, dominated by the solitary ordeal of his dying body, witnessing the submersion of his politically constructed self by the decimations of biological process. [...] In reaching the edge, death, pain, blindness, and coma moved the body beyond all cultural/political constructs. As he entered into the time of his death, the hunger striker realized the politics of silence as the termination of his long passage through the joined labyrinths of the prison and his body.[52]

In the case of Northern Ireland, the bodies of the dying hunger strikers were "swarmed over" with their representation of both their fellow prisoners and the republican movement more generally; similarly, the Turkish hunger strikers stand out as political symbols for both the country's many political prisoners, and the maligned communities outside the prisons: the "Other Turkey." But whereas in Northern Ireland the post-mortem unit of signification became the name of the strikers themselves – Bobby Sands, as a figure, became an almost literary production of the strike's eschatological function[53] – in Turkey, the "isolate" defies any expectation of being the most eminent (and most cited) unit of political significance. This is no doubt a strategic use of the hunger strikers' capacity to embody the demands of the larger movement; it mirrors, that is, the strikers' opposition to isolation in the penitentiaries. Conscious of the extreme significance of the strikers and in particular of their strong lines of affiliation, government officials knew that they needed to bring an end to the strikes in order to seize control of the prisoners' pithy productions of political meaning. The violence performed on 19 December 2000 – staged after the strikes had been in process for several months – was to be the definitive end to the hunger strike, and (through the transfer of political prisoners to isolation cells) the containment of any potential future action. Individualizing the prisoners' experience in the prisons – "breaking up" the communities that had formed, disallowing "contact and communication" between individual prisoners – was the method by which the government hoped to confound the prisoners so fully that political action, if it were able to continue at all, would have to be radically reconfigured.

Operation Return to Life further asserted the government's active role in "weeding out" political prisoners and squelching the dissent so powerfully articulated by the hunger strikes. This process of isolation and elimination was explicitly demonstrated with transfers of strikers from prison to prison: the violent, rushed transport of prisoners from the dormitory cells to the F-Type prisons. These transfers were themselves discrete performances; scenes from Metin Yegin's films show prisoners – many of whom are covered in scars, wounds, and burns – being moved in and out of police cars, vans, and ambulances. As the prison officials exerted the power to confine, to silence, and to convey the inmates between spaces, the inmates themselves seized the opportunity to shout their testimonials during every single exposure to the "outside" world: "Dozens were killed! [...] Look at my burns! [...] Six of my friends are dead!"[54]

But given these increasingly intense conditions of isolation, how have the strikers' lines of affiliations developed, and what traditional lines of division and difference have they transgressed? First, the strikers have gone to greats lengths to show that the effects of the F-Type regime extend to all prisoners, even those (e.g., female prisoners) who will never be incarcerated in the F-Type prisons themselves. The excerpts of narratives about Operation Return to Life I have selected for inclusion above are drawn from interviews with incarcerated women, typically underemphasized in literature and critical theory about prison torture. In her compelling ethnography of women's prisons, Jill McCorkle argues that "the history of the prison is almost exclusively the history of men's institutions," and that:

> The omission of gender [in the vast majority of prison literature] is particularly troubling in the case of Foucault, whose historical analyses of a variety of eighteenth-century disciplinary institutions have earned a prominent place in contemporary discussions of punishment and modern power. [...] Bodies are not only gendered by specific practices and regimes intended to code them as masculine or feminine. Bodies are actively gendered within institutions whose stated mission is directed to other goals.[55]

McCorkle's argument is useful for considering the case of the Turkish penal system, which (like all penal systems purporting to conform to the UN's international prison standards) distinguishes between men's and women's institutions or men's and women's wards within a single prison. Women transferred from the dormitory-style prisons during operations such as Return to Life are moved to "remand" prisons, many of which have been restructured to divide large groups into smaller, often solitary, cells resembling those in the Kartal Special Prison. According to Turkey's Human Rights Association, there are no women housed in F-Type prisons.[56] What does it mean that many women strike explicitly against the construction of prisons holding men? How are their lines of affiliation drawn and maintained? Further, is women's self-starvation a different kind of political activity than men's, with different forms, meanings, and effects? And does the death of a striking woman have political currency distinct from that of a striking man?

Reports from 19 December 2000 from men and women suggest that the amount of brutality used to transfer prisoners to the F-Type prisons was extreme: verbal abuse, beatings, rape. Images from the operation, along with prisoners' narratives and autopsy reports, show that many men and women suffered burns (from chemical agents dropped through the roofs) over large portions of their bodies, along with many other injuries and abuses. In both the men's wards and the women's wards, prisoners described mobilizing themselves and their comrades strategically in order to save, at all costs, the lives of the hunger strikers among them.[57] Men and women strikers were similarly resilient after the operation, recommitting themselves to the strike and its demands. The coalition of strikers, indeed, seems to have strengthened after 19 December 2000, becoming larger, more resolute, and (as I discuss below), more refined.

If these similarities suggest that the Turkish government is indiscriminate, in terms of gender, in its extreme use of violent techniques to attempt to bring an end to the hunger strike, they also imply that the solidarity of the strikers has been developed, in part, by minimizing emphasis on gendered differences between men's and women's experiences of the prison system specifically, and of the political climate in Turkey more generally. Simply put, in order to justify their demands, and in order to organize effectively around those demands, the strikers have stressed the great similarities among experiences within the penitentiary system; and the government, for its part, has exacted similar forms of violence among both women's and men's prison communities. Or, to cast this issue in a different light: in order to strengthen the community participating in the strikes, hunger strikers have developed intense lines of identification and affiliation *across* gender differences. Indeed, what is remarkable about the Turkish hunger strike is its development of a large base of strikers and supporters across several traditional divisions (between many Kurds and Islamists, for example; and between working-class or "peasant" populations and strikers whose backgrounds indicate greater degrees of privilege[58]).

At the same time, gendered differences between strikers have not gone fully unremarked by reporters and others interested in the cultural and political ramifications of this unprecedented protest. Among many medical professionals asked both to investigate the physical effects of the extended hunger strike and to develop methods of intervention in order to force an end to its enactment, Clinical Psychiatrist Sahika Huksel argues that among the strikers she has observed:

> There's an anorectic aspect to [the strike], especially among the women. They have this morbid fascination with watching their bodies deteriorate. And just as with normal anorexics, they reach a point where they cannot think straight, where they literally cannot see how bad off they are.[59]

In her strange fusion of two different practices, Huksel implies that the fasting woman is a recognizable form (the anorectic) within a larger representational domain, and that her function as such affects the meaning of her re-articulation within the context of a hunger strike. That is, for Huksel, Anorexia Nervosa seems to be the prior term against which women on hunger strike are defined, one that specifies the meaning of the self-starving woman in terms of an historically articulated vernacular of diminished mental capacities, hysteria, and a "morbid fascination with watching" their gradual emaciation. Cast as genealogically and symbolically linked to anorectics, in Huksel's observation women on hunger strike symbolize the particular terms of their specific strike, and simultaneously gesture towards more general concerns surrounding gender inequality and the oppression of women.

Similarly, in her study of the meaning of self-starvation in various kinds of literary texts, Maud Ellmann argues that self-starvation has become a "symptom of the discontents of womankind"; she writes of "the" self-starving woman: "Her emaciated form belongs to a collective economy of images, symbolizing not only her own malaise but that of the community at large."[60] For Ellmann, images of women on hunger strike are always already marked in relation to the "malaise" associated with patriarchal oppression, signifying not only the individual striker, but also the larger community ("women") of which she is a part. Huksel's comment on the "anorectic aspect" of hunger striking directs our attention to a community she calls "normal anorexics" – which we may take to mean those starving women whose political aims may be implied in their self-starvation, but who remain silent on the explicit expression of their political discontents. Hunger strikers, in Huksel's imaginary, are "normal" only in terms of the physical effects of long-term self-starvation – emaciation, dementia, blindness, and others – and presumably "abnormal" in terms not only of the context of their fasts, but also in their explicit acknowledgement of the concerns that have driven them to strike. In Ellmann's terms, Huksel's comment suggests that the striking women are "symptomatic" in the same ways that anorectics are, but deviant in their use of self-starvation as an explicitly political practice, perhaps ever more deviant for their affiliation with striking men.

If these arguments seem to emphasize a critical distinction between men and women on strike, they are also specific to observations made by or on behalf of audiences of the strike – especially *Western* audiences. These arguments are not, that is, explicitly articulated meanings produced locally by the strikers themselves, whose protests are defined by the lines of identification drawn across traditional boundaries. This is not to say that gender differences have no relevance for the hunger strikers; it does suggest, conceivably, a mode of conceptualizing gender differences as a bridge to identification rather than a barrier to it. Difference has indeed become a register for affiliation more generally for the strikers and their supporters.Perhaps most significantly, lines of identification have been consciously fostered and intensified across prison boundaries, from the prison cells themselves to areas outside the prison altogether. Within weeks of 19 December 2000, more than 2000 people had joined the strike, which had spread outside the bounds of the prison and into Kuçuk Armutlu, a so-called shantytown on the northern ("European") edge of Istanbul. Kuçuk Armutlu is primarily populated by DHKP-C members and sympathizers; many of its residents are former political prisoners and/or relatives of current prisoners – but many are not.[61] The strike's expansion beyond the walls of the prisons is significant not only because it meant that more people were striking, but also because the lines of political affiliation transgressed the otherwise impenetrable steel-and-stone walls of penal institutions and the specific penitentiary technique of the F-Type regime. The mode and the manner of containing "delinquency" were suddenly thrown into crisis; and the government responded with several military operations outside the prisons similar to the one staged on 19 December 2000. Rather than ending the strike, it seems that Operation Return to Life galvanized a movement that had been characterized (in national and international media and by Turkish governmental officials) as hovering on the brink of "failure."

By October 2001, one year and 40 deaths after the beginning of the strike, four houses in Kuçuk Armutlu had been designated "houses of resistance." Together, the houses held 19 strikers, 11 of whom were women and eight of whom were men; 29 strikers had already died in Kuçuk Armutlu.[62] These strikers, along with the prisoners, adopted a meticulous set of distinctions between a hunger strike and a death fast. While the death fasts were intended to advance through the late stages of self-starvation quickly, hunger strikes were to be sustained over lengthy periods of time. Those on hunger strike were keeping themselves alive as long as possible, relying on the political currency of gradual emaciation and increasing physical weakness. Those on death fast were moving speedily and relentlessly towards the inevitable, relying for political effect on the moment of death and the slow transfer of corpses from bed to grave.

In practice, the difference between being on hunger strike and being on death fast comes down to a drink. Most strikers prepare their bodies for extended hunger strikes with a "feast-and-famine" regimen (eating huge quantities of food and then starving for days at a time) for weeks or months before the strike begins. Death fasters then drink only salt water for the duration of their strike. In contrast, Onder Ozkalipci (forensic doctor at the Human Rights Foundation of Turkey) describes the drink used by hunger strikers:

> They take a lot of liquids because that slows down the muscular atrophy. They've discovered that potassium chloride is better than sodium chloride and that crude sugar is better than refined. In 1996, the strikers took only one spoonful of salt and sugar a day, and their daily weight loss was about 400 grams. This group, by taking a lot more salt and sugar, has brought that way down.[63]

The strikers have designed, tested, and implemented a regimented mode of self-starvation that both short-circuits the government's attempts to paint them as violent terrorists, and has transformed them into highly-charged, gravely marked political subjects who manage, at the scale of the individual body, the physical effects of state violence. As I discuss above, hunger striking represents a reformulation of the binary active/passive in terms of political action and, more generally, political subjectivity. Colloquially understood as practitioners of *passive* resistance – "peaceful" protesters vividly performing the domination of the individual subject by the state – hunger strikers are also depicted (by the Turkish government now, as by the British government in 1981) as terrorists, actively engaged in discrete acts of political terror and rabidly invested in dismantling state apparati. Recognizing the dynamic tension between these two interpretations, Allen Feldman argues that hunger striking presents such a conundrum for the state precisely because it "fus[es] the subject and object of violent enactment into a single body."[64]

Moreover, in honing the performance – we could almost call it a *skill*, and certainly a *tactic* – of hunger striking, prisoners and Kuçuk Armutlu residents aim to develop a sustainable mode of resistance based on the practice of self-starvation. This is a radical turn in the history of hunger strikes, which typically have a relatively short life span precisely because either strikers die after no more than six months, or mediators broker a resolution to the crisis that eventuated the hunger strike in the first place. Indeed, before the recent strikes in Turkey, the very idea of a hunger strike was predicated upon its unsustainability as political practice. The inevitable death of its practitioners is fantasized at the core of the hunger strike; each death of a hunger striker marks the end of a hunger strike. It is this death that has historically given hunger striking its power as political protest and its force in effecting change. But in the Turkish case, government officials have made it painstakingly clear that they will not capitulate on the question of the F-Type prisons; strike leaders have similarly refused to call off the strike until demands are met.

The events of 19 December were the government's attempt to declare the hunger strike a failure, both in terms of its original demands and in terms of its sustainability as political action. That is, the government hoped on that day to *end* the hunger strike, to place a final and definitive stamp of refusal on the list of its demands, and to dissuade prisoners from planning any future strikes. This was to be accomplished, first, by terrorizing prisoners into compliance, and second, by transferring many prisoners to isolation cells, where they would have no further contact with their comrades. The fact that the operation occasioned both the *expansion* of the strike and the *refinement* of the strike's regimen befuddled government officials, who had been holding out hope that the strike would simply fade away into a productive amnesia that would simultaneously reinstall the strikers into a collective vernacular of Terrorism. Officials had also hoped to force strikers into their roles as compliant prisoners ("docile bodies"); if it is ever to become European, Turkey needs its prisons and its delinquents – and it needs them quiet, clean, and orderly, in Foucault's words, "at the heart of the law." In beginning their own arm of the hunger strike, the residents of Kuçuk Armutlu radically unsettled the state's attempts to contain and defuse the strike, and shortchanged governmental attempts to equate hunger striking with terrorist activity.

Despite all of this, however, the rhetoric of failure underlies most mainstream media representations of the strike and governmental responses to the strikers' demands. Consider, for example, the following section from journalist Scott Anderson's exposé on the strike featured in the *New York Times Magazine*:

[The hunger strike seems like] an act of desperation, a weapon of last resort for the powerless, but the reality is a bit more complex. Politically motivated hunger strikes tend to occur in a very specific kind of society and at a very specific time: namely, in places with a long history of official repression, but where that repression has gradually begun to loosen. If it is the institutionalized nature of abuse that fuels the strikers to such extreme action, it is the cracks of liberalization that lead them to believe that such a course might shame the government into change – and often they are right. [...] What is remarkable about the Turkish hunger strike, by contrast, is both the apparent smallness of the issue that sparked it and that it continues despite all evidence that it is and will remain a failure.[65]

Anderson's article was published just a few weeks after 11 September 2001 in an issue of the *New York Times Magazine* whose other lead article, called

"Jihad's Women," spectacularized and demonized anti-American sentiment among Middle Eastern Muslim women. The article on the hunger strike was centered on Anderson's obsession with Fatma Sener – "an extraordinarily beautiful woman, with an infectious smile and penetrating brown eyes" – who was striking in one of the houses of resistance in Kuçuk Armutlu.[66] At the end of the piece, Anderson reveals that during his research, he was working with Fatma's father to convince her to stop striking. He confesses, moreover, his desire to "proselytize" the entire group of Kuçuk Armutlu strikers: "I will do this by bluntly telling them what the justice minister so bluntly told me: that there will be no concessions, that there is no hope" – a confession that reveals his conflation of "hope" with governmental concessions, that delimits "success" to legislative change, and that exposes his fundamental disregard for the strike's potential to foster and to produce on its own terms a new kind of political subject and coalition.[67] He is, of course, unable to convince any of the strikers to quit; one responds:

> by looking at me with a slight tilt of the head, the way a mother does when consoling a child. "You shouldn't take this so hard," she says softly. "This is a war, and there is nothing you can do about it." She gives me a smile that at one time must have been very pretty. "Be calm," she whispers.[68]

Note how in writing about the strike, Anderson focuses on the women involved: "extraordinarily beautiful," "infectious," "penetrating," "a mother," as if he, the Westerner who has arrived to clarify the strike for the participants, to make its dangers and its meanings known to those who are performing it, is being overtaken – sexually, maternally – by the women he interviews. This distinctly gendered approach to the strike mirrors, in many ways, Western colonial encounters with what becomes, in representation and in colonization, the "feminized East." Anderson argues in the long citation above that hunger striking is made possible by "cracks of liberalization" in an institutional history of violence – as if hunger striking could be the penultimate chapter in a myth called Liberal Democracy's Conquests, as if hunger striking were a symptom of pre-modern, "primitive" statehood. Even more astonishing is the author's reference to the "smallness of the issue" of prison torture and a long history of political violence. But most disturbing is Anderson's notion that the strike "is and will remain a failure," a statement that implies that "success" could only be read in terms of the government's agreement to the strikers' demands.

The effects of a practice as complex as hunger striking cannot be summarized with the conventional vocabulary of "success" and/or "failure." The Turkish hunger strike in particular may have little hope of provoking officials to reform Turkey's penal system; as I have argued throughout this chapter; however, the strike is significant in other ways – for example, in the reversals it stages in the production and representation of political violence. I would suggest, further, that the Turkish hunger strike – in, for instance, its production of nontraditional coalitions inside the prisons, and in its expansion to areas outside the prisons – performs a sophisticated iteration of Foucault's critique of the penitentiary technique. For the strikers, that is, the space of the prison-as-tactic extends far beyond any single prison's walls, just as the reality of prison conditions intensifies within the carceral buildings themselves. In short, the Turkish hunger strike embodies Foucault's charge: "the political, ethical, social, philosophical problem of our day is not to try to liberate the individual from the state, and from the state's institutions, but to liberate us both from the state and from the type of individualization which is linked to the state."[69]

In the case of the Turkish hunger strikes, is emaciation emancipation? Certainly the strikers have not eventuated radical shifts in Turkey's penal policies, and in the production and enforcement of a naturalized, Europeanized sense and state of "Turkishness." Certainly, too, the Turkish government has succeeded neither in fully subjugating strikers to the bounds of that nationalist identity, nor in silencing the isolated strikers or their protests. In performing and sustaining such an unresolved, agonistic struggle between state and subject, the hunger strikers insist upon the recognition and articulation of the intricate complexities both of subject-production within a nation that has worked to erase difference and to discipline patterns of identification that diverge from anticipated norms, and of state-production within a world that continues to situate Turkey at, and as, the border of the West.

During six days in October 1927, Turkish President and national hero Kemal Ätatürk gave an extended address to representatives of his Republican Party. That address, which outlined the manner in which Turkey had gained independence in 1923 and presented a roadmap for the secularization and "modernization" of Turkey in the following several decades, has since become an official nationalist text as important to Turkish policy-makers as the country's constantly changing Constitution. On 20 October 1927, Ätatürk ended his address with an appeal to young Turkish citizens:

> Turkish Youth! Your first duty is ever to preserve and defend National Independence, the Turkish Republic. That is the only basis of your existence and your future. This basis contains your most precious treasure. [...] If one day you are compelled to defend your independence and the Republic, then, in order to fulfill your duty, you will have to look beyond the possibilities and conditions in which you might find yourself. It may be that these conditions and possibilities are altogether unfavorable. [...] Assuming, in order to look still darker possibilities in the face, that those who hold the power of Government within the country have fallen into

error, that they are fools or traitors, yes, even that these leading persons may identify their personal interests with the enemy's political goals, it might happen that the nation came into complete privation, into the most extreme distress; that it found itself in a condition of ruin and complete exhaustion. Even under those circumstances, O Turkish child of future generations! It is your duty to save the independence, the Turkish Republic. The strength that you will need for this is mighty in the noble blood which flows in your veins.[70]

Fusing the "only basis of your existence" with the preservation and maintenance of the Turkish State, Ätatürk suggested that Turkish citizenship was not merely one of the many productions of the State, but the very site of its ontology. The rhetorical force of Ätatürk's declaration of "your first duty" was to link, essentially, state formation and political subjectivity – to argue, that is, that the very political subjectivity that was the condition for "independence" was itself subordinate to the broader charge of defending the integrity of the state. The force of that charge, moreover, was tied directly to the bodies of Turkish subjects, those expendable sites in which "noble blood" – the stuff and the sign of an ascendant national kinship binding contemporary Turk to ancient Ottoman king to the project of nationalist passion – would forever flow. If not a particularly unusual articulation of republican citizenship, it does remind us that modern Turkey was formed out of a desire to deny what Kevin Robins has called the "cosmopolitanism [and] pluralism of identity" represented by the Ottoman Empire, and to replace it with a homogenous, Westernized "nation without minorities."[71]

But what is the state of Ätatürk's "independence"? And who, today, is defending it? The hunger strikers? Government officials attempting to subdue their strike? In the context of Turkey's current attempts – so similar to the Kemalists' work in the 1920s – to accede to the European Union, to prove beyond the shadow of many other nations' doubts that it is essentially Western, modern, democratic, and sound, might not the hunger strikers be fully (and ironically) taking up Ätatürk's charge to lay their bodies on the line for the larger cause of global independence? Or is the meaning of "independence" here too bound up with the notion of the active, self-determining individual so precious to Western liberalism?

In rebutting human rights organizations' critiques of the F-Type prison, (now former) Justice Minister Hikmet Sami Turk explains that:

the old prisons with communal dormitories had become "training camps" for the left-wing groups; that the new prisons conformed with UN and Council of Europe guidelines; and that, far from crushing the prisoners, the F-Types would allow them to develop their identity away from the ideological constraints of their colleagues.[72]

In the same interview, Turk suggests that the F-Type prisons "are state policy, and that [will] not change." Fusing state policy with the prisons, and arguing that it is through the function of the prison that prisoners "develop their identity," Turk implies that the isolationist tactics practiced and policed by the F-Type prisons are intended to produce a particular type of political subject, one who submits to enactments of state power and eschews association with the "ideological constraints" of resistance. The Justice Ministry hopes to eventuate that production with both a penitentiary regime of isolation and through the specific architecture of the F-Type prison.

With the F-Type prisons, the Turkish state strives to cleave group opposition into fragments and then incarcerate those fragments in rehabilitation (read: torture) cells that function not to restore prisoners into a social order, but to de-socialize them entirely. In other words, there's nothing "rehabilitating" about solitary confinement. Indeed, the isolation cells of the F-Type prisons exemplify what Orlando Patterson has called "social death": alienation from all other human attachments, subjection to random and unrecorded acts of torture, and general dishonor as inhuman.[73] Social death is registered on the level of personal interaction and experience, and also institutionally, as a systemic function. As I argue above, the F-Type prisons represent the site and the sign of that function, providing not merely the space for its most explicit enactment, but the model for its dispersal over space and time.

I have refrained in this chapter from writing about any of the 107 deaths resulting from the current hunger strike – in part because death implies a sort of completion, an end to one individual's participation in the strikes, an instance of resolution. To focus too closely on actual deaths would imply some closure to the meaning and to the effects of a hunger strike that has so far refused to allow death to diminish its resolve. Such resilience – not in spite of the many deaths, but because of them – represents a direct challenge to the social death described in Patterson's study and ratified by a prison system designed to terrorize through direct and indirect performances of violence and through systemic processes of isolation and torture. Against all hope of traditional "success," the Turkish hunger strike stages a refusal to submit to that system, just as it has effected the demise of 107 strikers – a seeming paradox reflected in the declaration at the end of the original manifesto: "LONG LIVE OUR DEATH FAST RESISTANCE!" Such a complicated declaration, one that situates "death" as the theme and the mode of a resistance that can "live" over an extended period of time, epitomizes the Turkish hunger striker's attempts – and varied, underemphasized successes – to build a coalition aimed explicitly to confound nationalist attempts to produce "docile" subjects whose relationship to the state is defined by subjugation, streamlined identification, and unwavering participation in the project of purification that such a relationship requires.

220   *Violence Performed*

Notes

1. Alan Feldman, Formations of Violence: *The Narrative of the Body and Political Terror in Northern Ireland* (Chicago: University of Chicago Press, 1991).
2. Michel Foucault, *Discipline and Punish*, trans. Alan Sheridan (New York: Vintage, 1977).
3. Metin Yegin, *F: To Lie Down to Death for Days* (Film 2001).
4. TAYAD Komite Nederland. 2001a. "Documentation on the Death Fast in Turkey" (Rotterdam: TAYAD printers), 2.
5. Hüsnü Öndül, "Isolation, Death Fasts, and Women Deaths," Human Rights Association of Turkey. Press Release, 27 Aug. 2002; Dogan Tilic, "Turkish Hunger Strike Takes 107th Life," *EFE News Service*, 13 Jan. 2003.
6. BBC News, "Analysis: Can Turkey Fit In?" *BBC News Online*, 26 Jan. 2000a; BBC News, "EU Urges Turkey to Reform," *BBC News Online*, 9 March 2000b. Greek, German, and French representatives, among others, have expressed strong resistance to Turkey's accession to the EU.
7. Stephen Kinzer's *Crescent and Star* explicitly represents Turkey as inescapably bound up in a borderlands geography that compromises the country's ability to develop a sustainable form of liberal democracy. Kinzer, *Crescent and Star: Turkey Between Two Worlds* (New York: Farrar, Straus, and Giroux, 2001).
8. For a detailed articulation of this problem, see Neil Smith and Cindi Katz, "Grounding Metaphor: Towards a Spatialized Politics," in Michael Keith and Steve Pile (eds), *Place and the Politics of Identity* (New York: Routledge, 1993), 67–83.
9. Batuk Gathani, "Giscard Warns Against E.U. Membership for Turkey," *The Hindu*, 10 Nov. 2002.
10. TAYAD, "Death Fast in Turkey," 12.
11. Turkish Justice Ministry, "High Security F-Type Prisons," (2003). Available at: http://www.adalet.gov.tr/cte/english/events/f-type.htm
12. Peggy Phelan, *Unmarked: The Politics of Performance* (New York: Routledge, 1993).
13. Joe Beynon, "Hunger Strikes in Turkish Prisons," *The Lancet*, 348:9029 (1996): 737; Peggy Green, "Turkish Jails, Hunger Strikes, and the European Drive for Prison Reform," *Punishment and Society*, 4:1 (2002): 97–101.
14. DHKC London Information Bureau, "Political Prisoners in Turkey to Begin Hunger Strike on October 20," Press Release, 18 Oct. 2000.
15. Cemile Cakir and Frank Neisser, "'Courage and Determination' Fuel Prison Hunger Strike," *Worker's World*, 21 Dec. 2000: 2.
16. See, for example: Kinzer (2001); Kevin Robins, "Interrupting Identities: Turkey/Europe," in Stuart Hall and Paul du Gay (eds), *Questions of Cultural Identity* (London: SAGE, 1996), 61–86; and Elsa Le Pennec and Sally Eberhardt, *The F-Type Prison Crisis and the Repression of Human Rights Defenders in Turkey*, Report for the Euro-Mediterranean Human Rights Network, The Kurdish Human Rights Project, and the World Organisation Against Torture (2001).
17. These names, in English, mean *Revolutionary People's Liberation Party/Front* and *Turkish Communist Party (Marxist-Leninist)*, respectively.
18. Cakir and Neisser (2000), 2. After a conditional amnesty intended to relieve drastic overcrowding in Turkey's prisons, political prisoners made up roughly 6000 of the remaining 60,000 prisoners (Öndül [2002]).
19. Le Pennec and Eberhardt, 9–10.
20. Turkish penal officials are also regularly sent to the United States for training in techniques of prisoner discipline and punishment (Cakir and Neisser [2000]).

21. Human Rights Watch, "Small Group Isolation in Turkish Prisons: An Avoidable Disaster," Human Rights Watch Briefing Paper, 24 May 2000.
22. Turkish Justice Ministry, "High Security F-Type Prisons".
23. Human Rights Watch (2000); Beynon (1996); Le Pennec and Eberhardt (2002); and Cemile Cakir and Frank Neisser, "Massacre Can't Stop Resistance," *Worker's World*, 11 Jan. 2001.
24. Human Rights Watch, "List of Turkish Laws Violating Human Rights," Press Release, Feb. 1998.
25. Human Rights Watch (1998).
26. BBC, "No Compromise to Turkey Prisoners," *BBC News Online*, 3 April 2002.
27. Le Pennec and Eberhardt (2002).
28. Human Rights Watch (1998).
29. See Kinzer (2001), and Robins (1996).
30. Preamble to the Constitution of the Turkish Republic, Paragraph 5.
31. Kinzer (2001); Robins (1996); and Jon Gorvett, "Ankara Seethes as Parliamentary Reforms Fail to Open Door to EU," *Washington Report on Middle Eastern Affairs*, 21:9 (2002): 31–2.
32. Human Rights Watch (1998).
33. Robins, 72.
34. "Although it is true that prison punishes delinquency, delinquency is for the most part produced in and by an incarceration which, ultimately, prison perpetuates in its turn" (Foucault, 301).
35. Foucault, 253–5.
36. See Foucault, 135–69.
37. Foucault, 301.
38. See Human Rights Watch, "Human Rights Watch World Report 1989," (1989). Available at: http://www.hrw.org/reports/1989/WR89/index.htm; Human Rights Watch, "Turkey Violates Rights of Free Expression," Press Release, 15 April 1999a.
39. The particularly offensive Penal codes are Number 159, which outlaws "insult[ing] or ridicul[ing] the moral personality of Turkishness"; and Number 312, which outlaws "openly prais[ing] an action considered criminal."
40. DHKC (2000).
41. See Padraig O'Malley, *Biting at the Grave: The Irish Hunger Strikes and the Politics of Despair* (Boston, MA: Beacon Press, 1990); and David Beresford, *Ten Men Dead: The Story of the 1981 Irish Hunger Strike* (New York: Atlantic Monthly Press, 1989).
42. Feldman, 236.
43. Feldman, 237.
44. The number of casualties reported ranged from 28 to 32, depending on source. The BBC reported 30 prisoner casualties and 2 military casualties, "Shadow Hangs over Turkish Jails," *BBC News Online*, 10 Jan. 2001a.
45. BBC News, "Turkey Halts Prison Plans," *BBC News Online*, 15 July 2001b.
46. BBC News (2001a).
47. See TAYAD (2001a) and Cakir and Neisser (2001).
48. TAYAD (2001a), 14.
49. Cakir and Neisser (2001).
50. TAYAD (2001b), 8.
51. Feldman, 241.
52. Feldman, 251.
53. "Death in the Hunger Strike was conceived as both the literal termination of biological functions and 'the countdown,' the long drawn-out sociobiological

222   *Violence Performed*

death that the endurance of starvation dramatically stretched into an iconic act of historic mediation. [...] Military eschatology and biological eschatology were intertwined" (Feldman [1991], 225, 237).

54. Yegin (2001).
55. Jill A. McCorkel, "Embodied Surveillance and the Gendering of Punishment," *Journal of Contemporary Ethnography*, 32:1 (2003): 44–5.
56. Öndül (2002).
57. Le Pennec and Eberhardt (2002), TAYAD (2000a), TAYAD (2000b).
58. Scott Anderson, "Starving Their Way to Martyrdom," *The New York Times Magazine*, 21 Oct. 2001.
59. Quoted in Anderson, p. 47.
60. Maud Ellman, *The Hunger Artists: Starving, Writing, and Imprisonment* (Cambridge: Harvard University Press, 1993), 2.
61. Ellman, 2.
62. Anderson, 46.
63. Anderson, 46.
64. Feldman, 264.
65. Anderson, 44.
66. Anderson, 43.
67. Anderson, 124.
68. Anderson, 124.
69. Michel Foucault, "The Subject and Power," Afterword in Hubert Dreyfus and Paul Rabinow (eds), *Michel Foucault: Beyond Structuralism and Hermeneutics* (Chicago: University of Chicago Press, 1982), 208–26, at p. 216.
70. Various translations of this speech exist. I have chosen the one preferred by the official Kemal Ätäturk Library.
71. Robins, 69.
72. BBC (2002).
73. Orlando Patterson, *Slavery and Social Death* (Cambridge, MA: Harvard University Press, 1982).

10
Violent Reformations: Image Theatre with Youth in Conflict Regions

Sonja Arsham Kuftinec

The modern state is predicated on the idea of the body politic, characterized by the relationship between individual citizens' bodies and the state's political body. Yet, to achieve control over territory and citizens in an era of increasing globalization and cultural hybridity some contemporary nation-states have engaged in radical population politics, including ethnic cleansing, occupation, and targeted assaults, to reshape the body politic through acts of violence. How might performance intervene in these violent acts, particularly those that incur and maintain divisions among youth in conflict regions? I propose that Theatre can intersect with conflict transformation techniques to reflect and potentially reshape an ethos of violence and sense of social identity grounded in group difference. In addition to creating an aesthetic space in which to reform images of violent authority, theatrical facilitation offers a way to understand how youth can move from positions of identity associated with ethno-religious nation-formation towards relationships more grounded in ethical thought – from ethnic to ethical relationships. This chapter assesses comparative strategies for working with youth in Kabul, Macedonia, and Jerusalem based on over ten years of fieldwork.

In the spring and summer of 2004 I traveled to conflict regions served by Seeds of Peace, a US-based organization with regional affiliates that brings together adolescents from South Asia (India, Pakistan, and Afghanistan), the Balkans, and the Middle East for dialogue and summer camp activities. Formed in 1993 by journalist John Wallach, this kind of organization can be critiqued as yet another example of "benevolent" US political brokering. While certainly part of post-Cold War humanitarian interventions that cast the United States as global facilitator, the dispersed nature of the Seeds of Peace organization also allows for more radical political interventions. In fact, Seeds of Peace dialogue facilitators are not primarily American, but Israeli and Palestinian, many of whom trained with *Neve Shalom/Wahat al-Salaam's* School for Peace, a program rooted in social identity theory. School for Peace encourages consciousness-raising about power relations as a step towards more active political intervention. Thus, though Seeds of Peace as an organization emphasizes inter-personal contact, many of its facilitators challenge this framework with more critically informed practices.[1]

In concert with this more radical facilitation approach, in my work with Seeds of Peace youth I adapted Augusto Boal's Theatre of the Oppressed. Like the School for Peace, Boal grounds his practices in a belief that consciousness-raising must precede and lead towards action. Yet, his techniques differ from facilitated dialogue by emphasizing the language of the body. Boal's exercises thus additionally challenge the use of English in Seeds of Peace as a "neutral" language of negotiation; techniques such as Image Theatre non-verbally concretize abstract ideas and experiences with embodied sculptures. I used Image Theatre to model and explore some of the challenges of reconciliation in these regions, in sites with both intra-national (Afghanistan) and inter-ethnic (Middle East) types of conflict, and opposing paradigms of justice that resist reconciliation.[2]

While Image Theatre in Jerusalem clarified the competing ideological paradigms through which Israeli and Palestinian youth understood the conflict situation in their own communities, images that emerged through this activity in multi-ethnic Kabul centered on undoing a culture of violence and highlighting the commonality of concerns among Afghan youth while also offering strategies for addressing those concerns. Image Theatre with Balkan youth, in a region grappling with both inter-ethnic and intra-national conflict, illuminated how young people constitute their social identities when threatened by acts of violence, and how a less rooted understanding of ethnicity shifted their consciousness and relationships towards each other. In each situation, performance provided a space for reforming relationships and understandings initially forged in acts of violence.

Dramatic changes in Kabul

While ongoing inter-ethnic conflicts inflected relationships among Seeds of Peace youth in the Balkans and Middle East, common hardships united the multi-ethnic youth in Afghanistan: an environment marked by decades of violence and a deep uncertainty about their future. Though the US government support of insurgency in the 1980s and subsequent 2001 invasion certainly contributed to this environment, the youth I worked with in Kabul expressed a deep sense of gratitude for the hope with which Seeds of Peace, a US organization, had provided them. With the guidance of an Afghan-Australian teacher, the group of young men and women I worked with in the spring of 2004 had already established a sense of common purpose and ways of thinking critically about their recent history and their shared future prior to my arrival. I knew all this abstractly, yet I was unprepared for the visceral details of their suffering and their capacity to analyze and envision its transformation.

Weda, one of the young participants, had received her markedly visible scar from a sharp slap on the palm with a splintered ruler while protecting a cowed schoolfriend. Under the Taliban, Weda had secretly continued her studies and could easily transmit the rote answer required by her newly employed teacher. But her still-smarting scar spoke as powerfully as her story about the legacy of brutality in Afghanistan: while the Taliban had been overthrown, a culture of violence remained in Afghanistan, grounded in an obedience that both recognized and constituted authority.

In Kabul, 11 participating youth, aged 16 to 18, created images animating the internal complexity of their society, culture, and history alongside impassioned, and often risky, commitment towards change. Seeds of Peace had provided them with a vision of a future less determined by capricious authority and violence; they felt a duty to enable that future, and enacted their vision through four extremely powerful and dynamic images.

Boal proposes that these kinds of embodied images uncover essential truths about society and culture without resorting to spoken language. This process short-circuits cerebral censorship, silencing the "cops in the head" put into place by experience or social education. While images generated in this way can be concrete, metaphoric, or allegorical, Boal encourages participants to search for poetic rather than literal truths, emphasizing the polysemous nature of images – their multiplicity of meanings. According to Boal, images do not function as charades with a one-to-one correspondence between an idea and its "correct" interpretation as determined by performers. Instead, they offer a screen onto which a participating group can project a variety of ideas and interpretations. Image Theatre thus engages social rather than individual problem-solving. The process works particularly well with a group marked by varying levels of verbal articulation, leveling the playing field so that active participation does not require verbal skill. Boal's proposals were made manifest in Kabul via interactions with, and dynamic discussions prompted by, a series of images the youth entitled "Education Killer."

Given the explicitly *non*-linguistic character of images, it is difficult to capture in words their expressive nuance absent of interpretation. In fact, Boal advises that the facilitator ask for physical observations prior to subjective interpretations, encouraging the "spect-actors" (his preferred term for a participating audience) to really *see* what they are looking at. That said, I will attempt to conjure a semblance of the images presented in Kabul: (1) a girl stands looking at a book while a male youth points what appears to be a weapon at her; (2) a third party pushes away the weapon; (3) a seated boy studies a sheet of paper beside the still standing girl, while a turbaned youth reaches to strike the girl and the boy; (4) all three figures sit together with books, the girl reaching over to point towards a passage in the boy's book.

Taken together the images provoked dynamizations and imagined trans-formations as well as a number of astute physical observations. Many of these observations were based on absence, such as the lack of the girl's burqa and the invisible forces ("foreign influence") that allowed the weapon to be present. Some participants perceived a visible and immanent threat, while others pointed towards the book-reading as an act of resistance. A lively debate ensued about the reality of the image, and whether it existed pri-marily in Kabul at the time, the only city in Afghanistan where rule of law trumped rule of might.

While the first image provided a forum for verbal analysis, the sec-ond prompted physical dynamization and interventions. The spect-actors added to or shifted the image to indicate their ideas for change. Theatre of the Oppressed has impact, Boal insists, only if it becomes a rehearsal for transformation, where participants struggle to enact change while also acknowledging the real forces of oppression. Thus, Mir stood up to assist the anti-gunman while Ahmed leapt up behind the gunman, rendering vis-ible the external support that he believed still existed outside the image's frame. An energetic crowd of spect-actors gathered behind each of the inter-ventions, asserting belief in either the power of collective resistance or of external force. Still, attention remained focused on the gunman; Parnian remained alone holding her book. Finally Weda, attentively observing the chaos from a distance, gathered a handful of pens and folders lying about the floor and bounded up behind Parnian. Mustafa, cautiously watching from another corner of the room, arose and offered a book to the gunman, while Noor proffered imaginary tools for a job insisting, "stop violence first – then jobs, then education." Parnian nodded sagely at the weary but engaged spect-actors. "This shows," she offered, "that it takes a lot of people and effort and time to change a gun to a book."

The third image provoked fewer physical interventions and more debate about why the seated boy didn't resist the threats of the turbaned youth. "He is weak, the Talib are powerful." "He doesn't care." "He thinks about only his future." "During Talib rule, when you saw someone being beaten, no one helped out of fear for themselves." "Because the people were not united!" "Is the population more powerful or is the Taliban?" "People can't oppose the army, even when united – unity plus arms equals power." "But a person needs equipment and books and security to really learn; in the picture the boy has none. He is only studying a sheet of paper, not a real book. His studying was only for the Talib, not for himself." They all agreed that under the Taliban there was "study but no education."

While the discussions provoked by the first three images emerged from multiple interpretations and spect-actor interventions, Boal's techniques work most effectively when an image becomes dynamized – transformed by the spect-actors – suggesting the changeable nature of society and the possibilities for individuals to influence that change. The final image that the Kabul youth created emerged as a proposed "image of the ideal." I asked the group to first note any physical distinctions from the other images: "The studying girl is finally allowed to sit down," noted Weda. "No guns are present." "They are all physically on the same level." The turban has been replaced with a baseball cap." "They are using pens and not guns." I then asked another question that Boal maintains is essential for Image Theatre's effectiveness. "Is it a possible or only a 'magical' future?" I had the group physically place themselves into a spectrum of positions from "possible" to "magical" and speak from these places. An extraordinary conversation emerged about the "reality" of the Taliban. "The switch from turban to pen is real," stated Khabir, "people who were forced by the Talib to wear a turban and enforce rules have switched to the side of education – this happened in my family." Others proposed that the turban was "real" but the change "faked." "People can be judged by what they do when they are in power," insisted Parnian. Mir, resplendent in his shalwar kameez (long shirt and pants often worn by South Asian Muslims) then told a story about his mullah, a man who initially supported the Taliban because he wanted the good Islamic government they professed to bring. Once he witnessed their methods of enforcement and concern for outer manifestations of Islam rather than inner conversion and belief, he worked from within the regime to undermine the state, for example, by not forcing his congregation to sign the Taliban prayer log as a sign of their devotion.

The group sat in silence for a moment, absorbing all they had heard and learned, until an impassioned Weda asked them what they could now do. A number of ideas emerged – working in their schools and families, and especially continuing to learn together – as a group of young men and women seated on the same level, with Weda's provocation like a finger underlining an important passage in a common book. In this environment one is not chastised for cooperative education, for learning with others. The socially mechanized body transforms: from a template for the visible marks of punishment to a screen for dreaming and rehearsing dramatic change.

Workshops in Kabul allowed for multi-ethnic urban youth to begin envisioning the labor required to transform their existential conditions and conditioning forged over two decades of violence. While the Afghan Seeds represented a variety of ethnic backgrounds, they were not in direct conflict with each other, but with a greater cultural ethos of violence. In contrast, a workshop in Ohrid, Macedonia sparked a difficult dialogue among groups whose communities had recently engaged in violent conflict with each other.

Revelations of Image Theatre and identity in Macedonia

Arta, Agon, Vladimir, and Demo were stuck. They sat as though paralyzed, heads bowed to the floor, bodies twisted away from each other, unable

to begin a conversation whose itinerary and destination they felt they
already knew too well. "If we start, we will not be able to finish," Demo
acknowledged to me when I strode across the workshop space in Ohrid. Connected by a shared Seeds of Peace experience but divided by attachments to
various ethnic identities in the regions of Kosovo/a and Macedonia, these
four individuals had become immobilized by the apparent geopolitical and
discursive limits of their identities.[3] They were temporarily frozen within a
conceptual terrain, the borders of which were defined by what was currently
thinkable to them. At the same time, their acts of performance, through
Image Theatre, symbolic representation, and everyday behavior, moved
them literally and figuratively into a more complexly negotiated space, one
in which they reaffirmed and challenged identity- and nation-formation in
the shattered terrain of former Yugoslavia.

Respectively Macedonian Albanian, Kosovar Albanian, Kosovo Serb and
Roma (Gypsy), Arta, Agon, Vladimir and Demo's participation in the workshop, alongside 20 other youth aged 16 to 21, illuminated the formation
of social and individual identity in the south Balkans. The trauma marked
by the silence of this particular small group meeting, a trauma in which
language had little place, alluded to the occasionally violent formations of
these identities. In this situation, theatre can become a site for reflection and
re-imagining, an alternative to the "industry of consciousness" that shapes
young peoples' minds in the Balkans.[4]

The impacts of this industry, reinforced by education, experience, and
family, emerged via formal and informal actions during our three-day workshop in Macedonia. I had designed this convening with my partners Andy
Arsham and Scot McElvany; Scot and I had developed multi-ethnic theatre
projects in the Balkans since 1996 and Andy and I (in addition to being
married) had worked as Seeds of Peace facilitators since 2000. The Ohrid
workshop thus became an opportunity to further coordinate facilitation and
theatre techniques.

I began the workshop with a cultural mapping exercise designed to
heighten awareness of the various networks of relationships and belonging
that modulate social identities in the Balkans. Groups were formed based on
non-national identity (year of attendance at Seeds of Peace camp), external
signs of fashion (facial hair), self-selecting relationships (those in a hotmail
chat room), and month of birth (zodiac sign). I then asked participants to
locate themselves geographically on an imagined map of the region, in the
places where they were born and where they now lived. For the latter position, participants grouped themselves according to "where they felt most
comfortable."

This exercise relied for its effectiveness on various understandings of space
and identity. The participants' ability to conceptualize and situate themselves on the imagined map signified a grasp of emplacement, of location
within a set of agreed upon coordinates. The second position, marking where

participants felt most comfortable, relied upon what cultural geographer Yi Fu Tuan refers to as "sense of place" or "attachment to place."[5] Participants could confidently locate themselves on both a Cartesian grid of location and a phenomenological terrain of belonging. The relationship between these two formations alludes to one of the central tensions in the south Balkans region – nationalism and the geopolitical borders of the state, particularly the tensions between ethnic and imperial derivations of modern nation-states.

In *The New Century*, historian Eric Hobsbawm identifies two distinct definitions of the nation-state. He situates the first politically, as a territorial location in which the people hold sovereign power.[6] "The people," in this definition, do not need to belong to the same ethnic or cultural group, as in (arguably) the ethnically diverse United States. Thus, citizenship rather than common cultural origin emerges as one defining principle of national belonging. The second definition conforms more closely to received understandings of the nation-state, in which a particular, ethnically or culturally homogeneous group of people all live in and control a territorial state. In the purest form of this type of nation-state, members of a shared cultural group all live together in one place and only members of that culture may participate politically.[7]

Following the collapse of the Hapsburg and Ottoman Empires in the early twentieth century, Yugoslavia emerged as a newly constituted nation-state of the first order, though inflected by the dominance of Serbia, particularly in the early formation of the Kingdom of Serb, Croats, and Slovenes (1918–24). Still, the territory of the state embraced the various ethnolinguistic groups of Slovenes, Croats, Serbs, Albanians, and Roma, among others. Following the defeat of the more nationalist Croatian fascist Ustasha and Serb royalist Chetniks in the early 1940s, Tito's partisans formulated a state designed to quell the extremities of ethnic nationalism, offering some autonomy to the various republics, while outwardly condemning signs of nationalism and encouraging the mixing of ethnic groups in the Yugoslav republics. Ethnic national sentiment found expression in sub-cultural groups and in support for regional sporting teams, music, and dance.

The current situation in the south Balkans may be described as a conceptual shift from the first to the second definition of the nation-state, and from sub-cultural to overt political expression of ethnic identification. Given the former state's constitution under Tito, however, and the movement of labor from more homogeneous rural areas to multi-ethnic urban industrial centers, this realignment led to forced "returns," ethnic cleansing, extermination, and collective resistance in order to "re-homogenize" various territories. The current unstable situation in the south Balkans, with national identity rehearsed through the industries of consciousness, origin myths, violent rhetoric, and often brutal expressions of "self-defense," projected itself onto the bodies of our workshop participants, in everyday life as well as in its theatrical representation.

In addition to proposing that embodied images expose essential truths, Boal asserts that our bodies become mechanized by labor and culture. Physical Theatre can de-mechanize the body, enabling more creative thinking by silencing the censoring mechanisms put into place by experience or social education (the consciousness industry). Having warmed up earlier in the morning with de-mechanizing exercises – the cultural mapping exercises just described – the youth participants proved ready to speak with their bodies. From their current positions on the imagined map of relational affinity and attachments to place, I asked everyone, with intentional ambiguity, to "form groups and create an image without speaking." The spontaneous set of images that emerged provided a stark reflection of the conflicting dynamics of national belonging as experienced by the participants.

A "kick-line" of Albanians from Tirana (Albania), Pristinë (Kosova), and Tetova (Macedonia) spanned the room, hands raised in the air and linked together, effectively uniting the Albanians across geo-political borders.[8] Behind this linked chain sat a group of four Macedonian Slavs, physically unconnected, with their backs towards the Albanians. In front of the Albanians a tight circle of Kosovo Serbs and Roma squatted, arms crossed, hands linked together, heads down. Finally, a lone Kosovo Serb stood behind the pan-Albanians, looking through their raised and linked arms towards the circle of Serbs and Roma. This one Kosovo Serb, Jelena, had been physically removed from the group of Serbs and Roma in two ways. Her family had recently lost its home in the March conflicts between Serbs and Albanians in the region, so she had been apart from the other Kosovo Serbs when I had asked them to locate themselves where they currently lived. Jelena had also detached herself symbolically, as the only Kosovo Serb who had traveled to Ohrid through Pristinë with the Albanians. In my writing here about the images, my own interpretations invariably creep in to influence the reader's perception. The participating youth had for reference only the images themselves, which we examined one by one.

Several participants noted that the Albanians were standing, physically connected, their backs towards two of the groups, their line of sight raised above the squatting Serbs and Roma. Self perceptions included interpretations and emotional projections such as "open and inclusive," "representative of the double-headed Albanian eagle" (located in the center of the Albanian flag), and "celebratory." Others saw a "traditional dance" as well as a "chain or fence," insightfully noting, "they are so involved in their celebrations that they can't see how lonely Jelena is." "Yes, she is by herself, away from everyone, and no one can see her because they are all looking only in their own groups." The spect-actors saw the Macedonians as lower, disconnected, "not really expressing anything significant." We concluded our reflections with the tight circle. "Their backs are to everyone else," noted

Agon, followed immediately by the interpretation, "They don't want to cooperate." Many saw the circle as "exclusive" and "suspicious." "They are planning something," Bisej suggested. After these projections, which we as facilitators felt effectively summarized the current situation, we took a break, intending to move on to another exercise that would build on these images. But the Serbs and Roma were distraught. "They didn't get it." Scot overheard one of the Roma kids, Ramiz, mutter, "We are in jail."

The Kosovo Serbs and Roma, who had both been under attack by Kosovar Albanian mobs only a few months previously, felt completely misunderstood. So we returned to their image after establishing a more focused tone in the room and again asked the spect-actors what they saw. The first two observations arose from Macedonian Slavs. "It looks like a protective circle," noted Sneshka after several moments of silent consideration, "they are holding hands and united, but I don't think they are excluding, just protecting." "They look sad," added Maja. Then Arlind, an Albanian chimed in, employing a more accusatory second-person rather than observational third-person construction. "It's good that you're protecting each other, but why are your heads down? You should be optimistic. You are the future!" "Why do you feel like that?" asked a more inviting Arta. Sneshka jumped in with an answer to both, using a third-person construction that implied her ability to empathize with the Slavic and Roma group. "They feel like they have no future in their surroundings." Other Albanians added their more critical perspectives, slipping once again from the more objective third person to exhortative second. "They are gathered, they have fears," acknowledged Gent, "but are too close to each other to cooperate with others. Don't just stay there and moan. You need to cooperate to make the future!" "Our group was open, accepting others," added Iqballe, without noting that the Albanians' backs had been turned to Jelena and the Macedonian Slavs, "there is no space in this group for others." "But there is more national diversity in this circle," chided Sneshka gently, drawing attention to the presence of Roma and Serbs together.

As Boal proposes, the questions and projections onto the image revealed as much about the spect-actors as the participants. The Macedonians seemed more aligned with or at least empathetic towards the Serb/Roma image while the Albanians remained accusatory. At the same time, the image work slowed down the conversation, keeping it focused on reflections and observations of a single site, one that complicated the Kosovar/Serb binary.

Once the image broke up and the group began conversing directly, they quickly moved to a more inflammatory and diffuse debate that, while beginning with a focus on the image, gradually erased the way it complicated binary ethnic divisions. "Why were your heads down?" asked Arta. Tefik, a Roma, responded, "we were showing our feeling of enclosure." Then Dijana leapt to the heart of the matter. "That was the aim of March 17, ethnic cleansing to push the Kosovo Serbs into even smaller areas than they were already. This is all about Serbs and Albanians," she continued, eliminating the Roma who had been physically beside her moments ago. "It is real progress that we are here in this room and can try to understand each other, but you [Albanians] have done something to us that was done to you." Gent chose to deflect this accusation with another. "Why didn't you travel through Prishtinë and Skopje with the rest of us? It showed disloyalty and that you don't trust us. Why don't you feel safe with us?" "My parents lost their home," responded an emotional Jelena (who had now joined this image group), "My friends are in the hospital, two are dead, it was the worst night of my life, worse than in the war. I came on the bus through Prishtina – but it's not freedom of movement if there is not freedom for all Serbs."

Dijana added passionately, "The Albanians are impatient to get rid of UNMIK [United Nations Mission in Kosovo] so that they can finish their takeover of Kosovo. I keep waiting for the Kosovar Albanians to declare independence. They wanted so much for the internationals to help, and to show the world their suffering, and now are killing Serbs and destroying churches, same as what was done to them." "You don't know what we went through," responded Gent, alternately defending what happened and avoiding collective responsibility. "One hundred years of Serb rule. Don't paint us all with the same brush."

From their positions, experience, and consciousness, Dijana and Gent were arguing for competing elements of self-determination that require a unified people to have a unified land over which to rule. Aspects of this self-determination centered on religious symbolism, as well as on the actual and symbolic control of space and culture. Dijana slipped between assumptions of legitimate Serb ownership (the Albanians were waiting to "take over" Kosovo) and more resistant Albanian "independence." Neither Dijana nor Gent seemed able to envision a multi-ethnic state, and despite the presence of Roma, figured the conflict bi-nationally. But it was the Roma who eventually complicated this figuration in another image exercise.

We had earlier asked participants to develop an image of "where they were coming from." We had intended for participants to form physical images, but since we had provided them with markers and paper to document their discussion, they instead created provocative iconographics, further reifying and challenging the nation-state-territory debate. The groups for this activity shifted slightly, reflecting ethnic nationalism, with Jelena joining the Kosovar Serb group and the three Roma electing to work separately. The Macedonian Slavs and pan-state Albanians formed the remaining two groups.

The Macedonian Slavs and Macedonian Albanians had effectively tabled an earlier discussion of self-determination through their initial images. The pan-Albanian kick-line crossed over (and glossed over) geopolitical state borders while ignoring the Slavs in the territory behind them. In the later exercise, the Macedonians created a symbolic image of their country that emphasized their Slavic Orthodox heritage (church, cross), unintentionally but visibly erasing the predominantly Muslim Albanians, who made up 20 percent of the population. In defending this erasure, they noted that the Albanians had chosen to join with their ethnic compatriots from other states, and thus hadn't been available to "remind the Slavic Macedonians of their presence." The Macedonian nation and country emerged in this construction as identical, a unified second-order nation-state.

The Albanians focused their iconography on the double-headed eagle pictured on the Albanian country flag. The group further underlined the elision of ethnicity, territory, and state by beginning their presentation with a revision of the preamble to the US constitution "We the Albanian people, [...]" Spiraling away from the eagle, the Albanians had written a series of words that represented their common roots. The problems with this unified national formation were somewhat sketchily represented, as broken lines at the edge of their poster and the undeveloped statement, "There are also some problems, which we have represented with these broken lines." The problematic construction of an Albanian ethnic group across geopolitical borders thus emerged without articulation at the limits of what the youth could discursively imagine of themselves. Any elements of their identity that did not conform to the image of a spiraling unity, such as the multiplicity of their religious identity, simply did not appear.

The Kosovo Serbs, on the other hand, having represented where they were coming from with the image of a church, quite consciously linked their Orthodox Christian heritage to both their ethnic unity and their territory. The group discussed the long history of the Orthodox church in Kosovo, and how their name for part of the land, Metohija, emerged from the word for "church property." Linking church, state, and nation, the image effectively erased the majority Albanian population from Kosovo and Metohija.

While the Slavic Macedonians had been able to temporarily "forget" the Albanians who resided in Macedonia, and the border-crossing Albanians had imagined a nation absent of territory, the Serbs chose to emphasize their attachment to land, rootedness, and historical claims to the territory that the Albanians now occupied. Ernest Gellner links this attachment to roots and rootedness with Romanticism and the emergence of nationalism via modernity. In *Nationalism*, Gellner argues that agricultural societies require stability, particularly in status positions, and cultural differences to demarcate these positions. In contrast, modern industrial societies require social mobility and shared culture; administrative bureaucracies linguistically codify culture. Romanticism, Gellner argues, disseminated the sentimental attachment to nation via the construction of traditional folk heritage rooted in the past. Hobsbawm argues that national myths create consensus around a regime, and argue for the group's right of precedence over others. "The current political situation is justified by something that has nothing to do with the present but was true six centuries ago [...] it is used to replace everything else that has happened in the intervening period."[9] Thus, the Battle of Kosovo Polje in 1389, in which the Ottoman Turks defeated the Serbs, historically and eternally linked the Serbs to the territory of Kosovo and Metohija.

The spirit of collective belonging that allows for nationalistic expression figured as iconographic, symbolic, and sentimental among the Albanians, correlated to territory and heritage construction among the Kosovo Serbs, and culture and land for the Slavic Macedonians. Though expressed in ways that differentiate them from each other, the concepts of identity, ethnicity, culture, and tradition are based on the same set of ideas – a perceived sense of unity organized around the construction of common heritage. The participants in this Image Theatre could not imagine a culture disconnected from past tradition, a hybrid identity, a homeland without political borders, or a history disconnected from current articulations of the nation. The presentation and performance of the Roma group challenged many of these assumptions, and thus the very foundations of nation-formation.[10]

The three Roma youth, Tefik, Demo and Ramiz, presented "where they were coming from" through the Roma flag, consciously developed at the First World Romani Congress in London in 1971 to symbolically link this diasporic, diffused, and highly mobile community. The flag consists of a yellow, wheel-shaped *chakra* superimposed on horizontal fields of blue (sky) and green (grass). The boys explained that the wheel represented both their traditional mobility and the sun towards which they travel. Betraying the limitations of nation-state thinking, one Albanian participant asked why the Roma weren't "organized enough to have their own country." They responded that they didn't need a geopolitical homeland because "the whole world is our country," and also discussed the historical mistreatment of the Roma – the persecution of the rootless by the rooted. The Roma, as represented by Tefik, Demo, and Ramiz, had no desire for a nation-state, only for individual security and a modicum of cultural preservation. Yet, even this "preservation" allowed for absorption, exchange, and hybridity. The three Roma boys had in fact adopted, displayed, and performed a highly visible global youth culture, clad in low slung, wide-legged jeans, nylon track jackets, large gold jewelry, and delivering an astounding hip hop dance performance that evening. Although from the village of Djilan in Kosovo, these boys considered themselves as being from "the ghetto."

Theorists of nationalism, such as Gellner and Hobsbawm, posit a particular construction of social identity rooted in the conditions of modernity. Their definitions depend, for the most part, on both individual mobility and anonymity and inter-group differentiation. But according to Anthony Giddens, under the social conditions of late modernity or postmodernity, identity must be theorized as a reflexive project, shaped by institutions of education, family, and labor, impacted by both physical and symbolic violence, and sustained through narratives of self that are continually monitored and constantly revised.[11] Adolescents in particular undergo a complex process of developing a sense of self and social identity that tends to move towards a more closed or "achieved" status.[12] In this more complex context, Demo's comment within the silent, traumatized group takes on a different shading: "If we start we will not be able to finish." This comment, rather than referencing only the paralysis of engaging in a perceived zero sum conflict analysis, can be reframed as a marker of the ongoing project of identity formation as reflexive and continuous. "There are more groups within groups, greater divisions within delegations," one participant reflected in an evaluation at the end of the session, suggesting that the workshop had prompted a greater awareness of the complexity of social identity.

Constructing identity in experiences like these may allow the participating youth to form connections rooted in ethical alliances rather than cultural similarities and differences. "I learned something more deep about this conflict between us through the images," one participant noted, "that it wasn't so black and white." The industry of consciousness relies on assumptions of deep and easily defined ethnic differentiation as well as on similar frameworks for perceiving that difference. The presence of a third party, the Roma, ruptured the assumptions of both the Kosovo Serbs and the Albanians. Embodied and symbolic images provided a space to expose and extend the limits of consciousness, to blur and enhance the notion that these youths could only be perceived in black and white.

This kind of blurring, allowing for the formation of ethical relationships in the Balkans, was rendered possible because of a shared history and a complex interrogation of the binary conflict over contested territory through the presence of the Roma. Image Theatre exposed the ways in which individuals attach to social groups and helped, at least temporarily, to realign those relationships. In the even more contested site of Jerusalem, such labors proved more challenging. At the Seeds of Peace Regional Center, Image Theatre manifested the distinct existential frameworks within which participating Palestinian, Jewish Israeli, and Palestinian Israeli youth saw and interpreted their worlds.[13] Yet the opportunity to articulate these competing paradigms allowed for the possibility of more authentic dialogue that recognized how power operates in the region on and through the individual body.

Oppositional images in Jerusalem

Following a year marked by increasing violence in the region and Israel's decision to construct a controversial dividing wall beyond the 1967 Green Line borders, generating interaction between Israeli and Palestinian youth proved particularly challenging. Yet, shortly after the Kabul and Ohrid programs, I led a series of workshops for Palestinian and Israeli graduates of Seeds of Peace.

Boal's techniques are grounded in philosophies asserting that the smallest social unit reflects the larger dynamics of a society. In the summer of 2004, the Palestinian Authority and Israeli governments weren't speaking to each other and neither were some of the Seeds of Peace youth. While the Israeli government built a separating wall, some Palestinian Seeds wrote fiery letters to their Israeli peers and stopped attending year-round facilitation sessions. The Seeds of Peace Jerusalem Center staff (led by both US and regional employees) had thus focused summer activities on work that gradually led up to re-encounters between Israeli and Palestinian youth.

A Spread the Word symposium took place in the middle of the summer following separate "uni-national" meetings of Palestinians living in the Occupied Territories and Israeli citizens – both Jewish and Arab. The Spread the Word program focused on developing community-based activities for three distinct groups: Jewish Israelis, Palestinian citizens of Israel, and Palestinians living in East Jerusalem. Each of these groups of Seeds graduates, aged 15 to 17, would work with an older Seed on projects specific to that group's ethnic community: facilitating encounters between religious and more secular Jewish Israelis, documenting the oral history of Arab villages in Israel, and volunteering at a local Palestinian hospital. Prior to their separate departures, the three groups met together at the Center, connecting for the first time in six months. After some warm-up activities in English[14] designed to resituate the youth as multi-ethnic Seeds of Peace by recalling memories of the camp and what they had learned there,[15] I led an Image Theatre activity.

To reflect the seminar's theme, we asked participants to work in their three sub-groups (Jewish Israeli, Palestinian Israeli, and Palestinian) to each develop an image of their self-defined community's strengths and weaknesses or problems.[16] Each group could create more than one image if members had different ideas about the definition, strengths, and/or problems marking their community. Alternatively, groups could create one image that represented both the strengths and weaknesses of their self-defined community. This variation on Boal's Image Theatre techniques worked towards generating consciousness of what Freire refers to as "limit situations," or obstacles impeding authentic individual growth that, when recognized, can potentially be transformed. A well-articulated image represents both group values and limits, and the images generated by the Seeds participants tangibly reflected both their existential situations and the inter-group dynamics in the room and in the region.

The Israeli group presented two highly concretized images, each combining perceptions of strengths and weaknesses in their community. One focused on religious/inter-generational differences and the other on socio-economic disparities. Since the images specifically referenced an everyday social situation, the spectators' projections proved fairly consistent. At the same time, the Israelis' coding suggested openness to communicating

contradictory world-views within their community – the images presented multiple imaginings of community and proposed that the expressed
socio-cultural differences reflected both community strengths and weakness.
This relative public openness about community differences dovetails with
interpretations growing out of inter-group theory, underlining more radical
facilitation practices in the region.

Facilitation theory focused on inter-group rather than inter-personal relations asserts that identification with a social group and power asymmetry
mark the dynamics of encounter between Israelis and Palestinians.[17] In
encounter groups attuned to these theories, Palestinians tend to present
a more unified political front, where Israelis tend to articulate a sense of
their society as more differentiated – emphasizing inter-personal relations
rather than political concerns.[18] Not surprisingly given these findings, while
the Israelis developed social images reflecting difference and dissent, the
Palestinian Israelis and the Palestinian group each developed politically
informed images with which their entire group concurred. At the same time,
differences in these images signaled some of the complexities of national
identity formation among Palestinians as a whole.

In their very name the Palestinian Israelis suggest a complex and conflicted identity,[19] and this group generated a single image reflecting that
tension. The seven Palestinian Israeli participants stood in an outward-facing
circle with one hand reaching behind them to connect with each other.
The group's expressed intentions and the other groups' projections surfaced
the tendencies enunciated in social identity theory. A Jewish Israeli participant proposed a reading emphasizing inter-personal relations within the
group: "They share something, but they're apart." In contrast, a Palestinian
woman provided a more political historical analysis, linking the group to the
Palestinian nationalist aspects of their identity. "It's the Palestinian diaspora,
but they are still connected to their roots."

Boal insists that the projections and analyses of an image come from
its witnesses; he discourages the image's producers from "explaining" their
choices, arguing that verbal explication causes the images to lose their
potency and potential for multiplicity. I would argue, however, that this
constraint makes sense primarily in the context of homogeneous groups,
where individual projections work together to forge social rather than
idiosyncratic meanings. In more heterogeneous circumstances, particularly
those that bring together groups in conflict with each other, I have found
that letting the image producers articulate the rationale behind their creation can be enlightening, marking key differences in how various groups
decode the same image. Indeed, in this situation the Palestinian Israelis'
collective response spoke to a lack of clarity in negotiating their own existential situation, as well as to the importance of *acting* rather than simply
being together. "We are all somehow connected," they asserted, "but don't
know how to take action together." Freire might suggest that a "generative

theme" that emerged for the Palestinian Israeli group was their diffuse and conflicted identity. That is, their very inability to concretely code their existential situation alluded to their confused position within both Israeli and Palestinian cultures – their limit situation.

In contrast to the Palestinian Israelis' collective sense of stagnation, and to the multiplicity of views expressed by the Jewish Israeli group, the Palestinians' images proved the most unified, and ultimately the most provocative. These images also directly engaged the conflict, and incorporated the most representational complexity, employing both literal and symbolic elements. The Palestinians' images additionally provoked the most dissonance between projection and intention – the most disparity among the world views expressed through analytic decoding – attesting to the existential disparities between this group and the other two. The first image expressed "strength" and consisted of two parts. In the foreground, a man stood above another lying on the ground, his foot firmly planted on the other's chest. The man on the ground had his arm raised with fingers in the shape of a "V." Behind these two men, a woman stood with her arms raised, and two other women clinging to her legs.

The Palestinian group had coded the image to communicate their "strength in resistance" (to an implied but unspoken Israeli oppression) and connection to their land/identity, deliberately conflating the two. The coding thus emphasized social identity as attached to both a particular place and a collective narrative of resistance. Many of the witnesses, both Jewish and Palestinian Israeli, did indeed decode part of the image as signifying Palestinian resistance. Yet the witnessing groups projected varying readings onto the women's image, included symbolic decoding ("a tree to which people cling") and more expressive readings by Palestinian Israelis ("the love of land").

While these projections remained nominal – Palestinian Israelis tended to see a represented object or idea – a Jewish Israeli expressed her decoding as a verb phrase about the actions of the other: "They lower themselves to cling to a symbol." Indeed, in their decodings, Israelis tended to emphasize the symbolic over the "real" referent (which would perhaps have acknowledged a claim to the land in dispute). Thus rather than "a love of the land," perceived by a Palestinian Israeli, an Israeli witness saw "a love of a map of the land."

These comments delineating differential world-views among the Seeds might seem provocative and divisive, yet they were listened to without volatility, perhaps because the process emphasizes that decodings serve as projections rather than authoritative definitions; participants put more energy into interpretation and reflection than argumentation. Within the framework of the workshop, no one group had the power to define anyone else's existential situation. In this atmosphere, the second image of Palestinian "weakness" provoked even more differentiated yet respectfully attended responses.

In the second image, four participants faced each other in a circle while one stood outside with his fist raised. Both witnessing groups, Israelis and Palestinian Israelis, decoded the image as representing an extremist suicide bomber breaking the unity of the more peaceful Palestinian majority. When asked to decode their own image, however, the Palestinians attested that the man with the raised fist represented not a martyr/extremist, but rather a collaborator with the Israelis. A sense of shock ran through the room, both because the Israelis were unaware of the Palestinians' extreme distaste for "collaboration" and because of the (exaggerated) punishment described by the Palestinians: "They are drawn and quartered in the street." To this remark, a witnessing Palestinian Israeli – a close friend to many of the Israelis – muttered, to their horror, "They [the collaborators] deserve it."

This second image most clearly indicated the conflicting world-views in the room, the distinct existential situations of each group. The groups produced not only different images, but radically different interpretive frameworks. I would argue that this shocking, but revelatory, moment would not have occurred in either a non-theatrical facilitation setting or in a more doctrinaire Image Theatre workshop that assumes a relatively homogeneous group of "oppressed" participants. By developing community images within an aesthetic space and analyzing those images through projections as well as verbalizations, the decodings catalyzed a more sustained and mutually engaged discussion. As the young participants stepped back to reflect on the session as a whole, their comments suggested a critical capacity to reflect meta-theatrically on the power politics alluded to in the coded images.

As many of the Israeli participants remarked, unlike the other images that largely reflected the internal dynamics of their communities, the Palestinian images emerged in relation to an external force (the Israelis). One of the Palestinians responded that the group did not feel comfortable revealing internal weaknesses to the community they see as their oppressors. An Israeli added that the images reflected the political reality of power asymmetry in the region: Israelis have the luxury of examining the internal dynamics of their community in a more open public space while Palestinians do not. She also cited the luxury of disengagement, that outside of a violent moment of crisis, it is easier for Israelis to "forget about" the Palestinians, while Palestinian daily existence (freedom of movement and association, access to resources) remains largely defined by Israeli occupation. The Israelis saw their existential situation as one in which they had the privilege to code more complex images of their society. According to Freire, acknowledging that privilege marks the first step towards establishing a space for authentic dialogue. Thus this encounter with Image Theatre within a facilitation setting allowed for mutual modifications of each activity format, and more open conversations about the power politics operating within the group and the region.

The cross-ethnic communication that occurred in the Image Theatre debriefing seemed to have the most impact on the participants. "I've been waiting for this for a year," noted one Palestinian to a Jewish Israeli. As he later explained to me, he had been waiting not for the opportunity for "shared dialogue," but for the revelations of a representational moment that pierced through received understandings. When this puncturing occurs within an inter-ethnic conflict setting it may allow for the conditions of authentic dialogue to emerge – a dialogue that is aware of the fact that distinct existential conditions frame and shape one's point of view.

Towards some conclusions

As I noted at the outset of this chapter, the modern nation-state requires a sense of virtual community, a body politic in which the majority of individuals identify, at least in part, with the nation and with the state. They see themselves as Afghan, Kosovar, Serbian, Israeli, or Palestinian, as well as Muslim, secular Jewish, Orthodox, Roma, or female. Acts of violence can reinforce these ethno-national alignments while at the same time drawing attention to how the discourse of identity is often vigilantly controlled. Theatre of the Oppressed offers a place to reflect on and re-imagine this control, puncture received narratives, and suggest ways to forge social relationships across ethnic and national borders.

In *The Rainbow of Desire*, Boal theorizes about how Theatre of the Oppressed techniques allow for social and individual transformation. He proposes that the aesthetic space offers spect-actors the opportunity to create an imagined yet embodied representation, one that is connected to and yet separate from their existential reality – a place to see themselves seeing, and to envision change.

In Kabul, the aesthetic space offered the opportunity for a group of young men and women to reflect on and reform an image of violence grounded in their experiential understanding. The image-making provided the youth with a way to both create distance from and encode the ethos of violence in which they were steeped. This encoding also allowed for consequent decoding and for the rehearsal of transformation. The images activated the youths' imaginations in ways that supported their ongoing actions to educate themselves in a multi-ethnic, mixed gendered group.

Following the occasion of more recent and immediate ethnic violence, the aesthetic space in Ohrid helped the group to articulate a violent trauma beyond language. While animating how the youth formed attachments to myth and history, the images also ruptured the naturalization of these narratives. Encoded and decoded in the presence of the witnessing Other, the images moved beyond expressive self-definition towards a dialogue that ultimately complicated the group image.

In the even more violently contested site of Jerusalem, the aesthetic space highlighted the existential and epistemological positions of each group, and their ways of constructing knowledge about themselves and others. As in Ohrid, the beginnings of renegotiating social relationships via ethical rather than ethnic orientations emerged through a kind of epistemological rupture, an un-housing of received understandings about social and national identification.

This un-housing of received understandings applies to Theatre of the Oppressed work itself. In the process of transforming relationships in sites of violence, the strategies of Theatre of the Oppressed become themselves transformed, particularly in a situation where Theatre of the Oppressed works with groups in conflict. Both Boal and Freire maintain that in a situation of oppression, where a group of people feel that they do not have the power to transform their existential situation, they must find a way to liberate themselves, thereby liberating their oppressors towards the goal of being fully and authentically human. But in addition to my own limited Image Theatre work, at least one ongoing Theatre of the Oppressed group in Israel has proposed the necessity of creating a "theatre of the oppressor." The goal is to create allies by presenting the spect-actors with a contradiction, a kind of minor trauma.

In order to transform situations of violence, to transform relationships from those based on unquestioned ethnic identifications towards more just, ethical relationships, it may be necessary to introduce this kind of epistemological rupture. The transformation of violent conflict will require a great deal more labor than a few Image Theatre workshops. But a transformed Image Theatre offers a way to manifest perspectival difference, a step towards the possibility of imagining a shared vision of a just future built neither on symbolic, nor on physical violence. In Kabul, Ohrid, and Jerusalem, the transformation of individual bodies in relationship with each other offers one paradigm for resisting political violence in the (post)modern state.

Notes

1. International studies scholar, Ned Lazarus, who co-founded the Seeds of Peace Jerusalem Center with Palestinian Sami al-Jundi, has written eloquently about these complex political dynamics in his unpublished evaluation of the organization, "The Political Economy of Seeds of Peace." Lazarus argues for an assessment of the organization's impact grounded in critical theory, particularly as articulated by the Frankfurt School. At the same time, Lazarus situates his study in the shift from "problem-solving" (a Western rationalist paradigm of conflict management) towards "indigenous methods and the critical holistic approach of 'conflict transformation'" (p. 25).
2. See Augusto Boal, *Games for Actors and Non-Actors*, trans. Adrian Jackson, 2nd edn (London: Routledge, 1992); and Boal, *Rainbow of Desire*, trans. Adrian Jackson (London: Routledge, 1995).

3. The multiple terms of reference for these territories reflects and inscribes the conflicts played out there. While territorial signifiers often retain the same etymological root, these terms shift to fit the linguistic rules of each ethnicity. The Serbian "Kosovo" thus translates to "Kosova" in Albanian, and Albanians refer to the capital city of "Pristinë" while Serbs speak of "Pristina." This terminological shift represents more of an emotional attachment to heritage production and ethno-religious distinction than mere linguistic habit. Kosovo Serbs refer to the territory they inhabit as Kosovo and Metohija, signaling the close relationship between the Orthodox church, the state, and myths of nationhood. Metohija in Serbian refers to "church-owned land." Despite their minority inhabitance, Kosovo remains important to present-day Serbs because of the Battle of Kosovo Polje, which the Serbs lost to the Turks in 1389, introducing 600 years of Ottoman rule, and thus establishing a myth of origin grounded in victimhood. Until recently, Macedonia was officially recognized as the Former Yugoslav Republic of Macedonia due to vigorous lobbying by Greece, fearing that the name "Macedonia" would allow the inhabitants of this region to lay claim to northern Greece. The passion and problematics of these selectively historical claims was sardonically illuminated for me by graffiti scrawled on a Turkish bridge in Skopje, imagining "The Former Turkish Republic of Greece."
4. Hans Enzenbsberger, *The Consciousness Industry: On Literature, Politics and the Media*, selected and with a postscript by Michael Roloff (New York: Seabury, 1974).
5. Yi Fu Tuan, *Space and Place: The Perspective of Experience* (Minnesota: University of Minnesota Press, 1977).
6. Eric Hobsbawm, *The New Century* (London: Little, Brown, 2000).
7. Ernest Gellner, *Nationalism* (New York: New York University Press, 1998), 4.
8. I intentinly use the Albanian inflected signifiers for these regions. I also refer to the Serb-preferred modifier "Kosovo" rather than "Kosovar," a relatively new construction grounded in the ongoing movement towards further ethnic, cultural, and linguistic differentiation from Albanians.
9. Hobsbawm, 26.
10. I would like to acknowledge my colleague Branislav Jakovljevic, for sharing this proposal about the Roma and the conceptual limits of identity.
11. Anthony Giddens, *Modernity and Self-Identity: Self and Society in the Late Modern Age* (Stanford, CA: Stanford University Press, 1991). Sociologist Pierre Bourdieu identifies symbolic violence as that which occurs when one group imposes a set of ideas and symbols on another through, for example, education, enslavement, or colonization. See *Language and Symbolic Power* (Cambridge: Polity, 1985).
12. Jenny Makros and Marita P. McCabe,"Relationships Between Identity and Self-Representations During Adolescence," *Journal of Youth and Adolescence*, 30 (2001): 623–39.
13. The politics of naming collective identities in the region are fraught. Within the framework of this chapter I refer to "Jewish" (secular or religious) Israelis as "Israelis" or "Jewish Israelis," Palestinians living in Israeli as Palestinian Israelis, and Palestinians living in the Occupied Territories as Palestinians.
14. For more on the politics of language, see Rabah Halabi and Michal Zak, "Language as Bridge and as Obstacle," in Rabah Halabi (ed.), *Israeli and Palestinian Identities in Dialogue: The School for Peace Approach*, trans. Deb Reich (New Brunswick, NJ: Rutgers University Press, 2000), 119–40.
15. The connection to a "Seeds of Peace identity" works within theories of task-oriented encounters that emphasize a superordinate identity as a way to reduce

conflict affiliation. See Muzafir Sherif et. al, *Intergroup Conflict and Cooperation: The Robber's Cave Experiment* (Norman, OK: University of Oklahoma Press, 1961).

16. In Freirian practice, after extensive research within the community, the educator generally provides the coded situations. Boal's work differs in encouraging the community to produce its own coded existential situations (within the parameters established by the facilitator). See Paulo Freire, *Pedagogy of the Oppressed* (New York: Continuum, 2003).

17. Thus "community," a term that often connotes a more intimate social group in a US context, was immediately translated by the Seeds youth into a larger ethnic framework.

18. Rabah Halabi and Nava Sonnenschein elaborate on these findings in "Awareness, Identity, and Reality: The School for Peace Approach" (in Halabi, ed). Rela Mazali and Haggith Gor Ziv offer a more critical political reading of the tendency for Israelis to focus on interpersonal dynamics, proposing that as the Israeli state had sponsored most 1990s encounter groups, they tended to favor models that preserved the status quo. According to Ziv and Mazali, these models were "not designed to achieve change but to placate individual distress within existing social structures" (Halabi and Sonnenschein, 23).

19. When he heard Palestinian Israelis referred to as a "bridge" in the region, my colleague Walid, a Palestinian Israeli facilitator and poet, wondered whether they were "a bridge to connect or to be walked on."

11
The Arts of Resistance: Arundhati Roy, Denise Uyehara, and the Ethno-Global Imagination

Ketu H. Katrak

> *To look the other way can also be a type of hate crime.*
> Denise Uyeahara, *Maps of the City and Body*[1]

> *Democracy has become empire's euphemism for neo-liberal capitalism.*
> Arundhati Roy, *An Ordinary Person's Guide to Empire*[2]

In contemporary times, violence is a daily reality entering our lives via personal encounters and television, which viewers take in from the comfort and safety of their homes, even as violent images are replayed at times with troubling relish. Violence includes not only physical fights and drive-by shootings, sometimes of innocent bystanders; it is also embedded in the outrages of hunger and poverty hidden from affluent neighborhoods and gated communities, whether in the United States or among the wealthy in other parts of the world. The gap between rich and poor is increasing in the North and the South, as globalization and liberalization of economic policies bring benefits to a few while the majority struggle for the bare necessities of life. As noted in *Globalization and its Discontents*: "In the last ten years of unbridled corporate globalization, the world's total income has increased by an average of two and a half percent a year. And yet, the numbers of poor in the world has increased by one hundred million. Of the top hundred biggest economies, fifty-one are corporations, not countries."[3]

In this contemporary climate of social injustice and violence,[4] I believe that the creative worker (artist and activist) plays a special role – through the tools of writing and performance, visual and expressive forms of music and dance – in representing political realities and in raising social awareness that can inspire social change. Artists heighten public sensitivity and enable a variety of ways in which ordinary citizens can, to use Edward Said's phrase, "speak truth to power." Words, sound, movement, gesture, props, voice,

244

video footage evocatively expose political realities of racial discrimination and hate crimes, particularly in the aftermath of 9/11.

In this chapter, I demonstrate that artists and activists with an "ethno-global" vision (a term I have used elsewhere) share a socially and politically responsible vision that is grounded, first, in their own ethnic and local communities and simultaneously reaches globally across national boundaries.[5] It takes a kind of ethno-global artist in our contemporary world of social injustice and violence to intervene against state-sanctioned violence as well as against the violence of those representations. I analyze the powerful and politically engaged work of Denise Uyehara, a Japanese American writer and performance artist whose voice resonates with other ethno-global writers such as novelist and political activist Arundhati Roy, whose essays unravel social injustices committed against the poorest tribal communities in India.[6] Uyehara is an award-winning performance artist whose work has been presented nationally and internationally in Tokyo, London, Vancouver, Helsinki, and Beijing. Her artistic investigations, as stated in her promotional brochure, explore "what it means to be a woman, an Asian American, a bisexual, and human being, not necessarily in that order. [She] is interested in what makes us in our migration across borders of identity." Both artists embrace social responsibility as an integral part of being artists. Although separated geographically, Uyehara and Roy are linked politically in their enduring commitment to increasing our common humanity. After a brief discussion of Roy's political essays, I will consider Uyehara's performance art to demonstrate the ways in which an ethno-global artist-activist can illuminate social inequities both within and outside the nation.

The rhetoric of "democracy" and the "small things" inside "big heads"

> All kinds of dissent is being defined as 'terrorism' "
>
> Arundhati Roy, "Come September"[7]

Living under democratic regimes (I use the word "regime" deliberately) in India or in the United States, Arundhati Roy is highly critical of the abuse of the words "democracy", and "freedom." Roy's essay, "Come September" (available via Pacifica Radio and on the web), presented in 2003 as a speech at Riverside Church in New York City in the very belly of the beast, presents a courageous and persuasive argument against US Empire building in the twenty-first century.[8] The reader/listener comes under the power of her prose, as in the magical realism of her Booker prize-winning novel, *The God of Small Things,* a linguistic *tour de force* as well as a searing social critique of age-old prejudices called "traditions" (another misnomer in this context where traditions are used to justify oppression of women and low-caste people) in India's caste- and class-ridden society.[9]

Roy's comment quoted as an epigraph at the beginning of the chapter, namely that "democracy has become Empire's euphemism for neo-liberal capitalism" reveals that such "democracy" is very far from its ideals of freedom and justice. Further, exporting democracy is a misnomer for occupation. Indeed, Roy's searing critique of the abuse of "democracy" and freedom is represented performatively in Uyehara's "Vigil" and other performance pieces in *Maps of City and Body: Shedding Light on the Performances of Denise Uyehara* (hereafter, *Maps*). Roy's personal participation in the anti-Narmada dam project in India testifies to her activism, fuelled by her words and her willingness to put her body in the line of action. Similar to the contempt shown by the Indian government for the poorest of its tribal citizens whose homes and livelihoods were slated for submersion by the Narmada Dam, is the racial profiling and hate crimes that Uyehara represents in her piece "Hate Crime" that occurred in the climate of fear and hatred after 9/11 in the United States. Both artists give voice to the poor and often silenced people on the lowest rungs of society. Silence in the face of injustice is not an option for either artist. They demystify slogans of "freedom," "democracy," and "due process of law" that often remain theoretical for the poor and marginal in India and the United States.

National allegiance as citizens is often betrayed and Uyehara and Roy reach towards a broader than national commitment to "an ethical pledge" as noted by Sara Warner, "as artists and as citizens to fight for social change" (I am grateful to Warner of Cornell University for pointing this out to me via email correspondence.) Allegiance is not to a nation but to "a global citizenship to humanity in general." A coming together of an ethical and political vision enables both artists to work towards challenging social injustices in their local communities that have resonances for progressive peoples across national boundaries.

Roy's direct interventions via essays is as useful a model of creative/activist writing as they are a challenge to the strict academic separation of the literary from the activist. Roy's literary voice permeates her activist writings, and provides a model of how effective a passionate, personal voice can be in conveying historical facts, statistical data, and experiential material from struggles on the ground. Roy's essays are published in *The Cost of Living*; *Power Politics*; and *War Talk* among other texts.[10] Uyehara's performance pieces and her creatively written, poetic, edgy, ironic, and hard-hitting scripts portray with incredible subtlety and profundity the daily "otherness" of immigrants and minorities, and how their very lives and livelihoods are threatened in the racial climate of post-911. The events of 9/11 sadly filled immigrants with a palpable terror of their own dark-skinned bodies, their accented English speech, their clothing, hair, beards, turbans, all of which branded them unjustly as the "enemy." As Arab-American Tamadhur Al-Aqueel remarks in "Vigil", "Will there ever be closure for the September 11th attacks? Probably not. But one thing's for sure: There is no closure for being the target of racism" (*Maps*, 42). Such a comment is itself a striking critique of the failures of democratic ideals for all citizens.

Roy's critique of the lies that are spread like a disease by mainstream media is comparable to the important work of Noam Chomsky and Howard Zinn. Her activist work as a concerned and highly articulate citizen with a creative literary voice has earned her the distinguished Lanaan Foundation award for exceptional public service, an honor also conferred on prominent public intellectuals such as Edward Said. "There is an intricate web of morality, rigor, and responsibility," comments Roy, "that art, that writing itself, imposes on a writer."[11] Roy describes "the Free Press" as "that hollow pillar on which contemporary American democracy rests." She deconstructs the lies fed to the public, such as justifying "the Doctrine of Pre-Emptive Strikes, a.k.a. The United States Can Do Whatever the Hell it Wants, and That's Official." Roy even called for the "faces of United States government officials to be placed on the infamous pack of cards of wanted men and women" ("Instant-Mix Imperial Democracy").[12] This has not occurred, she said, "because when it comes to Empire, facts don't matter [...] The facts can be whatever they want." She renamed the "Coalition of the Willing" who "supported the US invasion of Iraq as "the Coalition of the Bullied and Bought."[13] Roy ended this essay with the following words:

> You have a rich tradition of resistance [...]. Hundreds of thousands of you have survived the relentless propaganda you have been subjected to, and are actively fighting your own government. In the ultra-patriotic climate that prevails in the United States, that's as brave as any Iraqi or Afghan or Palestinian fighting for his or her homeland.[14]

Roy's concerns with the "small things" articulated in her novel extends into the social realities that should concern ordinary citizens, such as impoverished farmers and "the banks of the river that smelled of shit and pesticides bought with World Bank loans."[15] Her detractors on the Indian side believe that she betrays her own class in critiquing mass dam projects that will after all benefit that class. And they ask how she could decry the nationalism behind the nuclear tests. "'These are not just nuclear tests, they are nationalism tests' we were repeatedly told [...] If protesting against having a nuclear bomb implanted in my brain is anti-Hindu and antinational, then I secede. I hereby declare myself an independent, mobile republic. I am a citizen of the earth."[16] For Western critics, her intense scrutiny of the World Bank and globalization are regarded in general as a famous person campaigning for certain issues of the day. But Roy's commitment is more than a passing interest; she is as respected and as marginalized as other important progressive voices, such as Medha Patkar in India and Noam Chomsky in the United States. Roy is as critical of Western power and dominance as she is of India's complacency and cruelty to its poorest.[17] Her commitment is not just a part of a trend, but has become her life's work, as is evident in her political essays.

In a speech at UCLA soon after her "Come September" speech in New York city, Roy reiterated that progressive people need to work together to "dismantle the working parts of Empire; to find the joints that hold them together and to break them." While former colonies fought imperialism through nationalism, today, because of corporate globalization, nationalism is less effective. Roy spoke about the need to create new modes of civil disobedience and about wanting people to focus on the issues that she talks about, and not on her, or her celebrity. She discussed "the psychology of dissent and how to keep the energy going." All her writings are about "the conflict between power and powerlessness and how dissent can keep power on a short leash." The only thing worth globalizing, she remarked, is "dissent." And it is important for progressive people to make dissent into "a kind of political culture." They must ask questions publicly and demand public answers. As she imagines a "god of small things" in her novel, so also she re-imagines another kind of political world where people in power who do not want to be accountable are compelled to do so, and where civil disobedience functions "to dismantle Empire limb by limb." *War Talk* ends with Roy's impassioned plea "not only to confront Empire, but to lay siege to it. To deprive it of oxygen. To shame it. To mock it. With our art, our music, our literature, our stubbornness, our joy, our brilliance, our sheer relentlessness – and our ability to tell our own stories."[18]

Roy as an ethno-global artist is rooted in Indian reality and also reaches out globally. The situation of oppressed tribals in India echoes the discrimination enforced by US "exclusion laws" against Asian groups through the nineteenth and twentieth centuries, not to mention earlier outrages of slavery and segregation. Roy participated boldly in the campaign against the Narmada Dam project that threatened to submerge the meager homes and livelihoods of the poorest tribal groups in India. As a public celebrity, Roy's speaking out against this injustice led to the Indian government trying to silence her through a prolonged legal case against her, resulting in a one-day jail sentence in spring 2002. Roy writes about this ordeal in her essay, "On Citizens' Rights to Express Dissent."[19] A documentary film entitled DAM/AGE, by Aradhana Sen chronicles this historic event.[20]

In a conversation about Roy's political involvement against the forced removal of the poor tribals in India, my 96-year-old neighbor, Eula Guthrie, made an astute parallel with a recent story of forced removal in a poor neighborhood of Los Angeles, where small, barely surviving businesses were forced to move by profit-motivated developers who are not responsible for rehabilitating displaced businesses. Eula's voice and vision connects the realities of poverty that are present not only in remote India but also in nearby Los Angeles, and are visible to those who are aware of social injustices.

As with Eula, a vibrant political individual even at 96, so also with other creative workers and performers who interpret and make connections among stories of disempowerment wherever they take place. However, in today's political climate, there are only stark oppositions – as "Bush the Lesser" (as Roy terms him in *An Ordinary Person's Guide to Empire*)[21] declared to the world after 9/11, "you are either with us or against us" in the war on terror. There is no middle ground, and no room for debate or interpretations of what the "terror" is, or what creates "terrorists." Meanwhile, in the United States., civil rights are threatened, and news reporters are "embedded" within US military so that their reports are "acceptable" for mainstream media. It is a telling fact that in the ongoing war in Iraq, no body bags carrying American soldiers are shown on television, as they were during the Vietnam War, for fear of public outcry and outrage. The government "shields" the public by hiding the deaths of US soldiers and Iraqi civilians.

Roy remarks in "Come September":

> Donald Rumsfeld said that this mission in the War against Terror was to persuade the world that Americans must be allowed to continue their way of life. When the maddened King stamps his foot, the slaves tremble in their quarters. So, standing here today, it's hard for me to say this, but "The American Way of Life" is simply not sustainable. Because it doesn't acknowledge that there is a world beyond America.[22]

Roy most often ends these hard-hitting speeches with hope:

> Perhaps things will get worse and then better. Perhaps there's a small god [as in the title of her novel, *The God of Small Things*] up in heaven readying herself for us. Another world is not only possible, she's on her way. Maybe many of us won't be here to greet her, but on a quiet day, if I listen very carefully, I can hear her breathing.[23]

Roy's "small things" evoke a conjuncture of opposites with Uyehara's performance piece, "Big Head". In the hands of these artists, the big and small come together in a paradoxical realm of human possibilities battling social injustice. Uyehara, at the end of "Big Head" imagines how "our small heads [are] made big" with lessons of the past and the present in order to equip us to intervene more effectively in the future. Above all, remembering past outrages enables us to speak up and to remain vigilant to threats of repetition of past crimes.

Memory and the body: writing history on the body

Roy is an example of what I argue Uyehara does in performance via her body that accesses memory in unique ways and performs both from the personal memory banks as well as from the social testimonies of peoples of color. As a solo actor, she embodies their histories, at times by literally writing their words (projected via video footage) on her body. Her empathetic representation of histories of prejudice and fear draws in the reader/audience who witness a magnificent achievement of ethno-globality. Uyehara's performative/activist work embodies, literally via her body in performance, what Roy analyzes discursively in her essays.

Uyehara, like Roy suggests a new kind of politics, even a reconceptualization of politics not as taking place in some remote location and detached from peoples' daily lives. Politics is personalized as it has an impact on ordinary folk, as in where they live, where they can afford to live, and where they may be targets of racial insults, prejudice, or even physical attacks. Uyehara remains forthright in her condemnation of racism, even as she evocatively expresses and performs her poetic writing as well as in quoting from real peoples' experiences. The latter in her text *Maps* includes the stories of Egyptian American Shady Hakim, or Boston-born Muslim-American Edina Lekovic, who was affronted with the question as to how being a Muslim would effect her job as Editor of UCLA's student newspaper, *The Daily Bruin*. Both Uyehara and Roy's accomplished art does not bombard the reader/viewer with didactic writing; rather both use their literary voices with metaphoric even surreal images to unravel racial and economic injustices.

Uyehara is politically astute and honest enough to recognize that: "I live in a city [Los Angeles] where I can walk down the street and never know we are creating carnage in other countries. And yet I know" (p. 15). With this awareness, she asks, "So, how am I to be accountable as a US citizen? What is my responsibility as an artist?" Her response in three parts beautifully and usefully expresses her own aesthetic-political analysis of the role of art and artists/activists in contemporary times. The writer's role is, firstly:

> To respond. Responding begins the act of live discussion and interaction [...] Demonstrations, rallies, teach-ins, and artist gatherings have taken place across the country and the world [in the aftermath of the US invasion of Iraq...] We (artists, activists, and regular folk) are reminded how important it is to speak up, to ask the necessary, difficult questions. We see how powerful our voices have become, and also the risk involved in speaking up. Response is patriotic.
>
> (*Maps*, 15)

Next, she highlights how such "response" is highly effective via performance which:

> provides a necessary vessel for remembering [...] How we remember is as important as "the facts". Performance, and the larger category of art provide a central site from which to remember what happened to our bodies

and the bodies of our neighbors. What will our bodies tell of these times? Not the official history that goes down in pages of authority, but the history of those who found simple ways to resist and speak their conscience, who invested new ways to record their voices.

(Maps, 15–16)

And thirdly, and "most importantly" for Uyehara, "performance is transformative. It provides new ways of expressing or imagining a culture, situation, or struggle. It challenges us to imagine a new world in which to live. Performance gives us new ways of seeing" (p. 16).

One of the most significant contributions of Uyehara's artistry is her particular performative representations of history, of the Japanese-American internment as that resonates with other unjust discriminations against minorities in the United States and world-wide. Although Uyehara's recording of history through memory, testimony, documentary evidence, facts, and indeed imagined realities when facts have been erased deliberately is not new in drama, I argue that she has a special performative talent in creating a collage of voices including hers and the voice of others in testimony to unravel history in fresh and new ways. This technique, indeed talent, in exploring and bringing new insights into "official" (often limited or untrue) histories is also evident in Roy's essays. Performance art about internment led many artists including Uyehara to express a kind of weariness with this subject. However, Uyehara in *Maps* brings fresh perspectives into this old outrage by contextualizing it and connecting that nightmare of history to another more recent one on 9/11 and hate crimes thereafter. I explore how Uyehara "creates" as she remarks, "from an interdisciplinary, non-linear place, because this is how memory works in me" (p. 125).

Uyehara's performed writing recreates cultural and political memory both as in concrete objects, letters, newscasts, as well as in abstract memory – emotional and that retained in the body, in the very bones, muscles, and cells. Uyehara "has been working with representations of memory," as she remarks, "through the casting of light. Even though we live in a digital age, I tend to favor simple light sources – slide projections through fish tanks, matches, Christmas lights, and so on. I gravitate toward light sources found in everyday life because my stories are from that world. So, for example, I use an overhead projector throughout *Maps* to cast live images over myself and the theater walls" (p. 127).

Uyehara's work profoundly explores "how memory marks the body" (p. 125) from her early collaborative participation with the Sacred Naked Nature Girls (from 1993–99). Further, Uyehara probes "how our memories warp with time and how the use of lighting can represent these memories" (p. 127). She worked with visual and performance artists as well as with dance to actualize these abstract concepts of memory on stage. Uyehara is also interested in "how people pass on memory, and how we teach each other" (p. 130). This is played out in her work from the personal memories of family members and letters from internment, to the interrelated political memories of surviving personal violence as in domestic abuse in "My Best Friend" (pp. 102–8) that enacts "the touch" (a clenched fist about to strike), and "the touch" (a gentle caress of the cheek); to larger national commemoration of wars, of redress and reparation to internment survivors. Such abstract concepts are brought to life on stage via the use of what Akilah Oliver (one of the Sacred Naked Nature Girls) termed "flesh memory" and going deeply into women's "simple acts of resistance to violence and a woman's right to her own body. We explored how voicing this violence empowers a woman" (p. 127). Echoes of this early collaborative work are heard throughout Uyehara's work as in "The Vanishing Point" (pp. 92–7). Here, the theme of sexual intimacy comes up in a serious, longing, and humorous voice as two parallel lines that are never supposed to meet come together and Uyehara is undaunted in confronting female audience members with their ability to have orgasms while simply sitting in front of her in the auditorium!

Performative and published achievement of *Maps of City and Body*

The recent publication of *Maps of City and Body* is a most welcome addition to the available texts of Asian American dramatists though most performance work does not find its way into print. *Maps* is Kaya Press' first in a series of texts by performance artists. I have successfully taught this text to undergraduates who find strong reverberations for their life experiences in Uyehara's moving words and scenarios. *Maps* is captivating on the page thanks to Uyehara's poetic and imaginative language, along with photographs of performances, stage directions, and process notes. Chay Yew in the Foreword notes that Uyehara is:

> never afraid to challenge her own aesthetic [...] she writes with an exquisite and deliberate passion and intellect. Her grasp of the theatrical is uncompromisingly unique and bold; her characters are realized and complex, pulsing with life and urgency; her beautiful, imagistic language floods the stage and fills our senses [...] All of Uyehara's work, from her plays to her performance pieces is defiantly original, dynamic, honest, insightful and deeply moving. To be in the company of Denise Uyehara's words and her art is to be in the eye of the storm. It's a wonderful place to be.
>
> (*Maps*, 5–7)

Maps is divided into three sections that cover the range of Uyehara's solo and collaborative work for over a decade, and each section is thoughtfully arranged, almost like a volume of poems. The first section is entitled, "Big Head," next, "Maps of City and Body," and "Collaborations, Actions, and Public Art Investigations." "Actions" include public art in which Uyehara participates with other progressive artists and community members in the Los Angeles area. For instance, "Shadow Water Project" uses water imaginatively to leave traces marking out shapes of bodies with their shadows as they walk through a park (p. 138). The water imprints also look like police outlines of murder victims' bodies. Here, there is a collective awareness of globalization in terms of shrinking resources in the world and an attempt to be cognizant of the disproportionate use of natural resources by wasteful lifestyles. Other collaborations include "Draw the World," where "artists from different countries gather after an international crisis and collectively draw a map from memory" (p. 140). Another provocative collaboration, called "Kissing: Asian Public Affection," "investigates social norms by getting people to break them in public places" (p. 148). Performed in Little Tokyo, Los Angeles, this work was followed by an art installation with video and visual artists.

This section on "Collaborations" in *Maps* includes a brief, though highly significant mention of Uyehara's participation (1993–99) with the Sacred Naked Nature Girls, who explored the dimensions, limits, and resistances of the female body, and what one of the members called "flesh memory." I believe that this work, although in the tradition of in-your-face kind of performance art with nudity on stage has had a profound and lasting impact on Uyehara's evolving artistry and continues to inform her work. One of her central concerns, namely accessing memories via the body continues in her contemporary work as well as in her workshops and teaching.[24] What began as the nude body to be exposed is now still the exposed (though clothed) body in *Maps* that finds itself on different points of a personal and sociopolitical map (whether in Santa Ana, Orange County where "Hate Crime" took place, or in Helsinki on "The Lost and Found" project). The voice remains the same – intellectually engaged with the most crucial social issues of the day brought to life via Uyehara's performing body.

Maps concludes with a "Discussion" (rather than "interview") with dancer/scholar Yutian Wong "that began on October 15, 2001, and continues to this day" (pp. 153–63). Each section concludes with "Process Notes" that indeed "shed light on the performances of Denise Uyehara" as per the text's subtitle. Useful information is found in "Chronology" of Uyehara's work, 1989–2003 (pp. 165–9); and a "Bibliography of Solo Performance and Authored works" (171–5) In reverting tradition, "Acknowledgments" are placed at the end of the text (pp. 178–9).

Maps of City and Body is a compilation of new and earlier solo and collaborative work. "I knew that my work," comments Uyehara, "was actually a fusion of various actions, texts, images, songs and stories" (p. 125). *Maps* incorporates scenes from her earlier work with provocative titles such as *Headless Turtleneck Relatives* (first performed in 1992). This includes a masterwork entitled "Charcoal," depicting her grandmother's suicide. Uyehara notes that "before most performances, [she] talks to her ancestors: it helps [her] to remember that I am part of a larger cycle of life, death, birth." (p. 134). Another earlier work entitled, *Hello (Sex) Kitty: Mad Asian Bitch on Wheels* (1994) includes pieces like "Best Friend," "Vanishing Point," and "Passed On" that reappear in *Maps*. *Hiro* (1994) is a published drama[25]; and *Big Head*, first performed in 2003, has been extensively performed across the United States.

Newer pieces such as "First Kiss," "7-Eleven Man," "First Political" were developed when *Maps* was first performed at Highways performance space in Santa Monica under Chay Yew as director. Uyehara discusses this as "a workshop process" with Chay Yew's key interventions in finding "an emotional arc for the show" (p. 133). Uyehara notes now instead of her earlier practice of simply paring down her words, she learnt to "simply stay true to the words committed on paper [...] [She] began to trust her writing more, the way an actor ideally trusts the words of a playwright" (p. 133). Hence, her words performed come across in the strong voice and body of a writerly performer and someone performing her own writing.

Uyehara's personal work, even when it relies on family stories and often imperfect memories of childhood, has distinct echoes for a wider humanity. In *Maps*, she uses her personal body, on which histories and testimonies of other peoples of color are written, by holding an incredibly simple prop – a blank sheet of white paper on which words are projected. She moves the paper across her body while video footage of histories of internment, and voices reporting racial incidents appear on her body. Her personal material comes together with the histories of other oppressed peoples as the words float over her skin. The coming together of her personal material with her very bodily personhood as the words float over her body is most moving on stage. The collage effect enables the audience to read the words while viewing Uyehara's body as it invites the words onto her personal space.

Uyehara comments that "as an artist, the act of writing/ performing/ witnessing what is happening now is a first step toward making sure it is remembered" (p. 67). "Vigil" brings different communities of color together to share their "common goal of uniting against injustice [...] The silent footage shows a gathering of Arab Americans, Japanese Americans, and Muslims, all holding candles, faces from the very old to the very young holding tiny, flickering lights. An amplified guitar plays quietly in the background as various voices from the community speak." Uyehara moves a blank piece of white paper all over her body as historical video footage of Arab-Americans, internment survivors, and Palestinians roll over her skin. The words become imprinted on her body, which will retain their memory. This recalls for me the scene in Maxine Hong Kingston's fictional memoir, *The Woman Warrior*, in which the words are literally etched, carved on the back of the legendary woman warrior.[26] Similarly, here, Uyehara's body is the vehicle that remembers history. At the end of that scene, the paper is crumpled gently, then lovingly smoothed out again.

"Vigil" opens with the voice of an Egyptian American who is shocked to learn that Adel Karas, a 48-year-old Coptic Christian grocer from Egypt, was killed in his San Gabriel store, four days after 9/11:

> That's when fear began to be a bigger part of my life. Because suddenly it hit real close to home, where a member of my immigrant community had been killed, and we weren't sure whether it was just a random act of violence or a hate crime [...] And then every time I would drive on the freeway and see a flag or a bumper sticker, I would wonder, what does that mean to that person? Does that flag mean they want me dead?
>
> (*Maps*, 39)

Another striking voice in this piece is that of Palestinian Lu Lu Emery, who lives in Canyon County:

> No matter how little you are, how insignificant you are, you still have a say [...] even if you talk to your next door neighbor and educate them [...] about your country, about your rights, about peace issues [...] you don't have to be this famous person, you don't have to be an artist, you don't have to be an author, you don't have to be a poet, you can be this average person.
>
> (*Maps*, 41)

During these testimonials, Uyehara

> begins to fold the paper on which the video footage is seen until it becomes a small letter. Denise walks into the audience with the folded letter. She speaks to an individual in the audience. "Could you do me a favor? Could you keep this for me until the end of the war?" She gives them the letter. "Thanks."
>
> (*Maps*, 44)

"The end of the war" could refer to any of the many wars of recent history. Then, the stage directions read: "As if on a departing train, she raises her hand to wave goodbye" (p. 44). She leaves empty-handed, with no luggage, evoking the forced evacuation of Japanese Americans told to leave their homes and most of their belongings behind, at times only taking their bodies and whatever they could carry with them on trains transporting them to unknown destinations.

Uyehara notes that for her, "the process of creating 'Big Head' has been like typing a letter to send out into the world. Like the blue type of my great uncle's carbon-copied correspondence, may its resonance remain for posterity" (p. 64). History gets personal as she notes that she has her "own kind of letter in 'Big Head' – a blank piece of paper. By using it to capture images of a vigil, I give it memory." This creates a direct link between the present and history as embodied in her great-uncle's letters from internment, in which he urges his readers to "never forget" that injustice, and to fight for others if and when the time comes. Linking hands from internment to post-9/11 hate crimes.

After 9/11, the Japanese-American community came out in large numbers to hold a vigil to remember the victims and to take a public stand against allowing an outrage such as internment to be imposed on Arab Americans or any other racialized group. Uyehara notes that "since the vigil in fall 2001, the United States government has imprisoned over 2000 people, mostly non-citizens," held them as "enemy combatants" without charge, and some have been "secretly deported" (p. 65). Their names were not released and they were simply "disappeared."

Roy's challenge in "Come September" to the United States' so-called agenda of exporting democracy to the world and of using war as a response to terrorism are given life on stage in Uyehara's performance as in "Hate Crime", located in the "Big Head" section in *Maps of City and Body*. In performance, she uses an incredibly inventive prop – a clay animation of a "big head" – projected onto a video screen (pp. 48–50). The head and face are contorted with pain, as if the figure is screaming, falling, hitting and withstanding blows. The image of the "head" is evocative and ironic. We, with our big and small heads, are often incapable of using our rational brains, and are capable of violence in which the head can become a target of attack.

"Hate Crime" was inspired by a newspaper article that caught Uyehara's attention and imagination. Uyehara enters the stage dressed in white, and her body is caught in the "bright, white light of the video projector over the screen. She slowly raises her hand above her head, bringing it down to caress her cheek. She repeats this action several times." A newscaster's typically flat "recorded voice announces that on October 21, 2001, a South Asian American man and his family were assaulted by several East Asian youths in Orange County. Sundeep, a 27-year-old physical therapist who chose not to give his last name, was leaving a karaoke lounge with his family, where they had just celebrated his birthday. He encountered six to seven youths, approximately 16–22 years of age, who were standing just outside the club. Two of the youths were female, the rest, male." The newscaster's neutral tone of voice shifts to the emotional and passionate plea by Sundeep's wife as the physical assault begins. "Sundeep's wife screamed: 'What did we do to you? We never did anything. Why are you hitting us?" (p. 48). The response from one of the girls watching (who is equally part of the hate crime as Uyehara notes in the epigraph at the beginning of this essay) is "They're going to get you motherfucking Middle Easterners" (p. 48).

Sundeep as a South Asian man suffers the fate of mistaken identity and racial profiling.

As the hitting continues, Sundeep "suffers the worst injuries of the group. His body is bruised, and his jaw so severely broken that it had to be wired shut for eight weeks" (p. 49). The irony of the victim's mouth being "wired shut" is deeply poignant. On stage, as we hear these words, we see Uyehara as she "runs one hand from her cheek down the front of her body, as if dissecting it, then lowers the hand to her side. She turns both hands, palms facing out. She looks at the audience" (p. 49). Without words, her body aspect speaks evocatively through the "dissecting" action cutting her body and the palms outstretched to the audience as if appealing for compassion, as well as outrage at this crime.

In the text, Uyehara includes notes on the process of creating this piece. With the help of videographer Ben Estabrook, she "edited her footage into a ghostly simulacrum of a hate crime. The folks at EZTV helped [her] to figure out how [she] could project the clay man over [her] body" (p. 67). In the performance, "the image of the clay figure is projected onto Uyehara's body as she moves inside and outside the image, sometimes attacking it, sometimes being attacked by it [...] [the image is alternating between] embracing her and being torn apart by her." To represent the attacker and the attacked in one single body is a *tour de force* in itself; it asserts our common humanity. Then on stage, the "chanting in a loud whisper" begins with the repeated lines, haunting and chilling:

What does it take to hate a body? What does it hate to take a body?
What does it take to hate a body? What does it hate to take a body?
What does it take to hate a body? What does it hate to take a body?
What does it take to hate a body? What does it hate to take a body?

(*Maps*, 50)

The repetition and the interchange between "take" and "hate" evokes both the physical assault of "taking on" a body, that is, hitting it, and the psycho-social feeling of "hate," an integral part of the political climate of fear and terror against the "other." "Taking on" a body in a violent encounter is propelled by hatred rooted in prejudice and ignorance. Such "hatred" provides the fuel for attacking the other, who lives among the general population but remains unknown, and perceived as a threat only because of the skin-deep pigmentation of a generic Middle Easterner (read terrorist). In the moment of attack, the superficial skin-deep connection between the South Asian and the Middle Easterner (although there are other deeper and more meaningful linkages via religion, even music, and folk tales) is near fatal in the politically charged climate of hatred and fear. The irony multiplies as one recognizes that this "hate crime," as others in US history, takes place on a misidentification – such as Vincent Chin, a Chinese American mistaken for a Japanese

American auto worker, who was killed via severe beating with a baseball bat in Detroit in 1982. Similarly, Sundeep's brown skin tone, though only skin deep, translates brutally into his being the target of a "hate crime."

Sundeep's family is part of the wave of South Asians who entered the United States under the post-1965 immigration laws. They brought their professional skills and attempted to make a better life than the one left behind. Of course, the wider reach of progressive politics also makes Uyehara's reader/viewer aware of the poverty in the South that is partly responsible for migration to the North. These immigrants may be colonized internally by images of a glorious life in the United States and ignorant of the realities of racial and class inequities that they will face.[27]

Uyehara created the clay animation figure using a digital video camera mounted on a tripod on [her] kitchen table. "Creating a clay man and then tearing him apart was a quietly horrific process," she remarks in her notes in the text. "How does the body respond to blows? What happens to my fist if I hit a person? What does it take to hate a body?" (p. 67). Uyehara literally embodies the hate-blows, and moves jaggedly as the racial slurs and physical blows fall on her body. Reviewer David C. Nichols remarks:

> Uyehara, a virtual conflation of Laurie Anderson and Myoshi Umeki, here addresses imperiled democracy in the present war-shrouded landscape. Contrasting the WWII incarceration of Japanese Americans with interviews culled from ethnic groups now regarded as "the enemy," she uses indirect means to bring her central thesis into bas-relief [. . .] Uyehara's mastery amounts to an extended *coup de théâtre* [. . .] Only at the end does Uyehara's urgent purpose flirt with didacticism, overstressing the already firmly established point. This hardly negates her achievement, though, and fans of trenchant, comprehensive performance art should race to this must-see limited engagement.[28]

Uyehara's art, says Nichols, transforms the bare bones of the horrific act into an artistic work that enters the viewers' consciousness and is deeply affective.

The final image in "Hate Crime" shows Uyehara "facing the screen with both hands almost pressed against it, as if being arrested. The clay figure, now a demolished, amorphous lump, turns to look at her with hollow eyes. She looks at the clay man as he fades to black" (p. 50). I saw "Hate Crime" when it opened at Highways Performance Space in Santa Monica in 2003, and I have seen it again in different venues and have admired its evolution in the hands of this gifted and sensitive artist. In performance, "Hate Crime" is riveting, moving, and poignantly painful.

Naming an assault as a "hate crime" is as powerful and significant in enabling political and legal action to counter it as is the naming of sexual harassment. In the "Hate Crimes" entry in *Keywords of Contemporary America,*

Monisha Das Gupta usefully delineates the parameters of this named social phenomenon as including:

> Speech or action that humiliates, intimidates and attacks individuals on the basis of their race, religion, nationality, gender, sexual orientation, or disability [...] Hurling slurs and physically assaulting racial minorities, or making demeaning remarks about women, gays, lesbians, Jews, or disabled people are examples of such acts and expressions. Such individual acts terrorize entire communities and leave their imprint on public memory. The Federal Bureau of Investigation (FBI) and law enforcement agencies recognize the collective impact of hate-motivated crimes. The FBI also indicates that hate crimes, when compared to other kinds of crimes, tend to be more violent.[29]

Uyehara notes that "Hate Crime" was inspired by the Los Angeles based community of artists and activists who spoke out during the post 9/11 time of fear and hate. She participated with some of the artists/activists, including a group of butoh-style artists who "walked down Third Street Promenade shopping center on which lay a mannequin gagged with the American flag" (p. 68). At Highways, Danielle Brazell invited local artists to perform "The Gathering: An Alternate Response to Recent Events." Japanese Americans and Muslims "came together to break their fast during Ramadan at the Senshin Buddhist Temple. They discussed not only [their] differences in culture and religion, but [their] common struggle to fight for peace and tolerance on a real, grassroots level" (p. 67). T-shirts expressed such sentiments as: "Our Grief is not a Cry for War" and "Not in Our Name." Community organizations such as South Asia Network (SAN), the Indo-American Cultural Center (now renamed "Artwallah"), and Café Intifada came together in this "alternate response" to war.

In order to contextualize "Hate Crime" in the published text, Uyehara places it in the middle of the section entitled, "Big Head," which takes the reader/viewer into Uyehara's personal history and childhood, the Japanese-American internment, and back to contemporary times. Uyehara's style is inimical: poetry and magical realism come together with realistic street stories that unfold in Los Angeles and Orange County neighborhoods. Two tactile words – fire and ice cream – book-end the beginning and end of "Big Head." These words that can also be read as a chilling "Ice scream", and "Ice Cream Bricks" (p. 23), where the burning bricks have to be cooled with water/ice. The surreal image, "ice cream mixed with fire" is paradoxical and striking. The childhood memory in "Big Head," not very reliable, recalls a burning house and a question about whether it was Denise or her sister who burned her hand "by reaching for the wrong end of a sparkler" (p. 22). The comment, "No, I didn't actually see it [...] but I remember it" could bring up different historical memories. The word "it" (in "I didn't actually see it") evokes internment, 9/11, or any of the horrors of history that one reads about and witnesses even without actually seeing them at that time and place. Not seeing "it" but "remembering it clearly" asserts the power of both memory, however imperfect, and the stories we fill our heads with: "We with our small heads made big, filled with voices from the past and the voices of now. It was a small, poor, imperfect thing that I did – to remember. Maybe I even got the facts wrong" (p. 58).

Uyehara's ethno-global range takes the viewer/reader from her local primary school, with its hegemonic training of "One Country! One Language! One Flag!" (p. 29) to a larger world within multiethnic Los Angeles, a microcosm of the world itself. The pieces in "Big Head" are book-ended by the pledge of allegiance – recited in the beginning in "Fourth Grade Book Report" (pp. 26–9) and spoken in the voice of a fourth-grader by Denise. The piece also includes the ironic recitation of the "pledge" by Japanese-American internees, two-thirds of whom were indeed American citizens. And, "Big Head's final piece evokes the national promise again in "Pledge" (pp. 55–8). And now, quite different from the fourth grade innocence, "clay figures stand side by side in an old-style salute to the flag" (p. 55), evoking military formation, and mindless repetition of the pledge as a mantra that has lost meaning. Nationalism is problematic, especially during times of crisis, when patriotism is touted with flags. In "Come September," Roy describes flags as "bits of colored cloth that governments use first to shrink-wrap people's minds and then as ceremonial shrouds to bury the dead."[30] Earlier, in her volume of essays entitled, *The Cost of Living*, Roy had described "nationalism of one kind or another was the cause of most of the genocide of the twentieth century."[31] She rejects being labeled "anti-national" for speaking strongly against both India's and Pakistan's nuclear tests. Rather, she claims the freedom to be "deeply suspicious of all nationalism and to be anti-national[ist]."[32]

Uyehara's skepticism about national allegiance is performed in "Pledge". Allegiance is often misplaced when offered to a nation that betrays its minorities over and over again in recent history. Uyehara's transition to "Knowledge will be my America. My voice will be my pledge" (p. 56) is her personal, thoughtful, politically rooted voice, rather than the mechanical repetition of pledge-words.

Across the geographic distance between India and the United States, the histories of poor people, marginalized by color or caste among other factors, and the realities of colonialism and imperialism are still with us. However, these histories remain largely invisible within US mainstream culture as Indian-American writer, Meena Alexander notes in her essay, "Is There an Asian American Aesthetic?" Alexander evokes the sharp image of "barbed wire" that was part of a colonial state's machinery, and notes that now, for new immigrants, and people of color, "the barbed wire is taken into the heart."[33] Roy's and Uyehara's performed politics, in their moving representations of the forcibly relocated, and the victims of racism and hate crimes echo the feeling of "the barbed wire taken into the heart."

Notes

1. Denise Uyehara, *Maps of City and Body: Shedding Light on the Performances of Denise Uyehara* (New York: Kaya, 2003), 66. All in-text page numbers refer to this publication.
2. Arundhati Roy, *An Ordinary Person's Guide to Empire* (Cambridge, MA: South End, 2004), 56.
3. Joseph E. Stiglitz, *Globalization and its Discontents* (New York: Norton, 2002), 5.
4. News items of "hate crimes", a contemporary phrase reflecting contemporary realities (even as in the past, the naming of "sexual harassment" was powerful in identifying unacceptable words, touches, behaviors) particularly in post-9/11 times are sadly common. For instance, "On the night of 27th September 2003, Nabeel Siddiqui, 24, a recent graduate of NJIT (in New Jersey), and the only son of his parents, was brutally and mercilessly beaten with a baseball bat by three individuals, in the township of Orange, New Jersey. This attack left him fighting for his life due to severe brain damage, and now he has died as a result of the injuries." Newsflash by email, received on 8 October 2003 for signatures to demand "that the assailants be tried as adults and be awarded the maximum sentence to deter such brutality and hate crimes in the future."
5. I create this phrase, "ethno-global writers" in my essay, "Asian American Cultural Expression," in William A. Little et al. (eds) *The Borders in All of Us: New Approaches to Three Ethnic and Global Diasporic Societies* (Northridge, CA: New World African, 2006), 222–47.
6. Other ethno-global writers include the Nigerian Nobel prize-winner Wole Soyinka, who is primarily a playwright though in recent years has turned to political essays written in a passionately personal voice similar to Arundhati Roy's. In texts such as Wole Soyinka's *The Burden of Memory: The Muse of Forgiveness* (New York: Oxford University Press, 1999), and *The Open Sore of a Continent: A Personal Narrative of the Nigerian Crisis* (New York: Oxford University Press, 1996) Soyinka analyzes South Africa's Truth and Reconciliation Commission, as well as world events such as genocides in Rwanda and the former Yugoslavia. He returns always with lessons for his own native Nigeria that has lived under a series of military dictatorships. Brief respites into "democratic" regimes such as Shehu Shagari's in the 1980's were sadly and ironically worse than military dictators' brutalities on ordinary people. Readers undoubtedly will add their own choices of ethno-global artists in contemporary times.
7. Arundhati Roy, "Come September," *War Talk* (Cambridge, MA: South End, 2003), 45–75. (First presented as a lecture in Santa Fe, New Mexico at the Lensic Performing Arts Center, 18 September 2002. Sponsored by the Lanaan Foundation. Roy also delivered "Come September" at Riverside Church in New York City, 13 May 2003).
8. This essay is a landmark contribution to contemporary thought and would be a useful inclusion in General Education courses in academia.

9. Arundhati Roy, *The God of Small Things* (New York: Random House, 1997).
10. Arundhati Roy, *The Cost of Living* (New York: The Modern Library, 1999); *Power Politics* (Cambridge, MA: South End, 2001); *War Talk: The Algebra of Infinite Justice* (New Delhi: Viking; London: Penguin, 2001); *An Ordinary Person's Guide to Empire*.
11. Roy, *Power Politics*, 5.
12. Roy, *Ordinary Person*, 46.
13. Roy, *Ordinary Person*, 25.
14. Roy, *Ordinary Person*, 37.
15. Roy, *God of Small Things*, 219.
16. See "The End of Imagination," in *Cost of Living*, 106, 109.
17. See "The Loneliness of Noam Chomsky," in *War Talk*, 77–101.
18. Roy, *War Talk*, 112.
19. Roy, *Power Politics*, 87–103.
20. In India as in African nations, projects of large dams have had a record of being white elephants. Rather than enhancing "development" as they promise, they benefit the already well-off in these societies. In the case of the Narmada Dam in the state of Gujarat in India, the dam's water would benefit the wealthy farmers in the region, while rendering the poorest tribals homeless and without livelihood since they depend on the land for that. Roy's decision to join hands with the oppressed in this struggle against the Indian state, which collaborated with the World Bank on this project, brought much attention to their plight. The film documents Roy marching with the tribal people and with other major feminist activist, Medha Patkar, in street protests demanding a stop to the submersion of the tens of thousands of tribal homes, and asking for state accountability against the brutal treatment of these downtrodden "citizens" of India. The Supreme Court was involved in the final decision-making that did not go in favor of the tribals. Roy also writes about the plight of these displaced people forced into crowded cities, and without skills, they become part of the urban poor, forced to live in slums.
21. Roy, *Ordinary Person*, 51.
22. Roy, "Come September," 74.
23. Roy, "Come September," 75.
24. I participated in Uyehara's class on "Asian American Performance and Writing" at the University of California, Irvine and learnt a great deal especially on how Uyehara uses various exercises and techniques to teach how to access memory via the body. Recently, she taught this class again and had her students mount a silent performance in front of the Humanities building, with various props and masks exploring issues of Asian-American ethnicities and identities.
25. Denise Uyehara, "Hiro," *Asian American Drama: 9 Plays from the Multiethnic Landscape*, ed. Brian Nelson (New York: Applause, 1997), 385–474.
26. Maxine Hong Kingston, *The Woman Warrior: A Memoir of a Girlhood Among Ghosts* (New York: Random House; Vintage, 1978).
27. See Genny Lim's play, *Paper Angels* in Roberta Uno's *Unbroken Thread: An Anthology of Asian American Women Playwrights* (Amherst: University of Massachusetts Press, 2000). This drama is set in a detention center on Angel Island in the 1920s and it depicts Chinese detainees whose hopes and dreams face disillusionment. "I came on a ship full of dreams," notes one of

the men, Chin Gung, but when he faces a charge of "liverfluke" which mandates deportation, he commits suicide. Ironically, Chin Gung has been in the United States for 40 years, and considers it his home. He had put his lifeblood in building the railroads and done other field labor in his adopted home.

28. David C. Nichols, "Imperiled Democracy Under the Microscope," *Los Angeles Times*, 28 Feb. 2003: E40.
29. Monisha Das Gupta, "Hate Crimes," in Mari Yoshihara and Yujin Yaguchi (eds), *Keywords of Contemporary America/ Gendai Amerika No Kilwaado* (Tokyo: Chuo Koron Shinsha, 2006). See also, Das Gupta's excellent book, *Unruly Immigrants: Rights, Activism, and Transnational South Asian Politics in the United States* (Durham, NC: Duke University Press, 2006).
30. Roy, *Cost of Living*, 47.
31. Roy, *Cost of Living*, 104.
32. Roy, *Cost Of Living*, 109.
33. Meena Alexander, "Is There an Asian American Aesthetic?" in *SAMAR: South Asian Magazine for Action and Reflection*, 1:1 (1992): 26–7.

12

Narrative Representations of Violence and Terrorism: Tragedy and History in Hanoch Levin's Theatre*

Freddie Rokem

> *A storm is blowing from paradise; it has got caught in his wings with such violence that the angel can no longer close them.*
>
> Walter Benjamin, "Theses on the Philosophy of History"[1]

The play *Murder* by the Israeli playwright Hanoch Levin, which premiered at the Cameri Theatre, the Tel Aviv municipal theatre, in the fall of 1997, is unique in confronting the extremely complex conflict between the Israelis and the Palestinians. Levin's text, along with the performance based on it – directed by Omri Nitzan, the artistic director of the Cameri Theatre – is a compelling theatrical representation of the chain of violence and counter-violence, of terror, revenge, and retaliation that is still feeding this now more than century-old Middle Eastern conflict. In examining this play and some details from the performance, as well as situating it within Levin's *oeuvre* as a playwright, I want to raise a number of issues about the theatre's possibilities for reflecting and commenting on violence and terror, and perhaps even for bringing about a change of attitudes towards such phenomena within the Israeli political and ideological contexts. I want to stress from the outset that I will examine an Israeli play and an Israeli playwright from an Israeli perspective, and, at the same time, I am aware that there is another side. The broader questions I want to raise here, however, are how acts of violence, terrorism, and armed conflict shape the narratives performed on the theatrical stages in Israel. Is it at all possible to portray such acts of violence within an aesthetic framework? And how do these kinds of narratives affect our experiences of the grim realities around us; do they have any likelihood of influencing the political realities surrounding us?

The Oslo accords, signed by Yasser Arafat and Yitzchak Rabin in 1993, aimed at putting an end to the Israeli occupation of the West Bank and the Gaza strip, both conquered in the 1967 war. The Accord laid the groundwork for the gradual establishment of an independent Palestinian state in these territories alongside the state of Israel. At that time, the Israelis and the Palestinians seemed to have started down a path that could have led to a peaceful solution to the conflict. But since 1997 – when *Murder* was first produced, two years after the murder of Rabin (in November 1995), during Benjamin Netanyahu's term of office as the Israeli Prime Minister – this conflict has steadily been accelerating towards its present violent impasse. This acceleration was intensified after the talks between Arafat and Ehud Barak at Camp David broke down, igniting what was to become known as the Al Aksa Intifada in 2000 and leading directly to the election of Ariel Sharon for Prime Minister. There have been many dramatic turning points during this last decade, when optimism has been suddenly and unexpectedly reversed into despair and hopelessness. And the events in the larger global arena, in particular 9/11 and the war in Iraq, have made the developments in the Middle East even more complex.

When *Murder* was first performed, the initial hopes for a peaceful solution had been seriously shattered, but the crisis had not reached its present destructive dimensions. Looking back on this performance after almost a decade, when the circles of violence fed by terrorism from both sides have so radically invaded the daily lives of the two peoples and will most probably continue to do so in an unforeseeable future, the performance, and in particular Levin's text, must be viewed as a cynically fulfilled prophesy. Paradoxically, presenting the grim realities of violence and terrorism on the stage in the way Levin has done can be perceived as a form of acceptance of these realities, as a kind of coalition of despair. However, I believe that this was neither Levin's nor Nitzan's intention. Rather, they wanted to show us how detrimental the circle of violence has become by creating a performance that can hopefully change our perceptions. Perhaps even, by supporting a growing awareness of its destruction, such a production can make a significant intervention. But since they did not succeed in this intervention (and I am of course not blaming them for that), and since the realities are still so overwhelmingly violent and at times even hopeless, and the politicians and generals (on both sides) who are leading us deeper into the vicious circles of violence are still in power, the first question that (regretfully) remains to be raised is how the relationships between such realities and a specific play such as *Murder* are constituted.

In addition to being informed by the obvious and immediate contexts of the current conflict, performance narratives about violence and terror and, by implication, our experiences of such events, are constructed on the basis of cultural memories and associations that are nourished by roots that go much deeper than the events themselves. On a daily basis, narrative and experience, like Siamese twins, are inevitably bound together, constantly feeding on each other. On both the Palestinian and Israeli sides, we frequently hear the bomb blasts and the ambulances from our homes before we hear about them in the news, and we constantly worry about family and friends. Terrorism is compellingly real and anguishing. Narratives about violence and terrorism – the ways in which we talk about these events and in which they are reported by the witnesses on the spot – gradually become integral aspects of our experience of such events, creating a fearful symbiosis from which it is extremely difficult to break away. In other words, the "raw" materials of violence have barely been perceived before they become integrated into the more comprehensive narrative of the conflict with its clear-cut ideological and political implications. The theatre – and Levin's work is an example of this – is sometimes able to question this double bind between the experience and the narrative.

The stories we tell and perform about acts of violence also have cultural traditions based on a vaguely defined but constantly reinforced and redefined collective consciousness. It is what Freud, in "Moses and Monotheism," somewhat ceremoniously termed "the inheritance of memory-traces of the experience of our ancestors."[2] Even the news broadcasts about these events, with which we are fed on a daily basis are shaped by these powerful and constantly re-emerging master narratives. These master narratives, in a constant interaction with media coverage and with the stories we tell each other about our everyday experiences, have made their imprints on the narratives of performances like *Murder*. But at the same time as this performance reinforces the already existing cultural and ideological stereotypes created by such cultural traditions (which I will exemplify later by looking at Levin's adaptation of a biblical narrative), it also attempts to break away from them, by exposing them.

The issue I want to raise here can thus be formulated as a constant, sometimes even obsessive quest for the point of rupture where culturally inherited narratives can be reformulated, exposed, and critiqued. In situations of heightened conflict (and the Middle East never tires of inventing new variations on these themes) the artistic representations reflecting the painful realities more or less as they are happening frequently become trite and even sentimental. What Levin attempted to do – and in this he was only partially successful, at least with regard to the audiences who saw the performance in 1997 – was to balance a dangerous tightrope between a somewhat sensational depiction of violence while at the same time subverting its inner core, attempting to show us how meaningless the narratives of violence and terror can become. This is an almost impossible feat, in particular as it obviously did not serve as an antidote to violence.

Taking *Murder* as my point of departure, I also want to examine the structure of narratives of violence in Levin's theatre and reflect on the broader cultural and historical connections in such narratives after World War II. This is an extremely complex topic and I hope that the Israeli perspective, in particular in its links to the ongoing conflict with the Palestinians but also in connection with its deeper cultural roots, can contribute to a more general understanding of the issues connected to the representation of violence in the theatre.

Tragedy and history

There is a constant dialectical interaction between events that gradually crystallize in our collective consciousness as "history" and artistic representations of these events in various media. This dialectical interaction makes it possible to perform history. Such artistic representations usually also aim at assigning some kind of meaning or even coherence to acts of violence, hatred and revenge. Such violent events, most poignantly during World War II, but also before and after this twentieth-century watershed, must first of all be viewed as a fundamental failure of what we consider to be basic human values and ideals. By trying to confront these failures, narratives depicting such events have the potential to become today's tragedies. Due to the moral issues involved, and as Hayden White has so convincingly shown us, because of their common narrative dimensions, history and tragedy are in a constant dialogue with each other. However, there are no fixed rules according to which these dialectical interactions between history and tragedy and the narratives they create can be formulated.

Traditionally, theoreticians, most notably Aristotle who argued in his *Poetics* that tragedy is more philosophical than history, have attempted to keep the notions of history and tragedy at a safe distance from each other, emphasizing the boundaries between them. Shakespeare's *Hamlet*, however, is an interesting example of a complex negotiation between tragedy and history. Historical events surround or frame those parts of the play that we usually perceive as the tragedy. The wars initiated by young Fortinbras that are briefly mentioned in the first scenes of the play remain a potential threat to the stability of the kingdom throughout, although of a very different kind from the threat of old Hamlet's ghost, who influences the events as a supernatural agent. Fortinbras belongs to the world of politics and his wars and conquests serve as the backdrop to the tragedy of Hamlet, though not the tragedy itself.[3]

Only when Fortinbras makes his entry in the last scene of the play do the historical events actually cross the threshold of the stage, invading its core. Horatio wants to "speak to the yet unknowing world / How these things [i.e., the heap of corpses on the stage] came about" (5.2.358–9). Fortinbras nonchalantly responds: "Let us haste to hear it" (5.2.365), but then he does not take the time to listen or to reflect on what we have just seen. He does not even pay attention to the details of the tragedy that has just come to a close, but rather orders the bodies to be taken up, because this sight "Becomes the field, but here shows much amiss" (5.2.381). Immediately afterwards, in the very last line of the play, Fortinbras commands: "Go bid the soldiers shoot" (5.2.382). Horatio, on the other hand, wants to put the bodies "high on a stage" (5.2.357), turning this grim spectacle into some form of theatre after having evoked the flight of angels that will sing Hamlet to his rest. The demands of history, represented by Fortinbras, and the demands of tragedy, represented by Horatio, are kept apart even if they are simultaneously on-stage.

In the post-World War II theatre, it has gradually become more and more impossible to maintain this kind of relative separation. Performance narratives about violence, terrorism, and military conquests are now situated in a scandalous repetitious loop – W. B. Yeats wisely termed it a "gyre"[4] – where it becomes increasingly difficult to distinguish between the Horatios and the Fortinbrases, as the spiral of violence not only repeats itself but constantly seems to become more powerful and threatening. History has become the stuff from which tragedy is formed.

Georg Büchner's *Danton's Death* is an important milestone in the emergence of such a more total fusion between history and tragedy. In the plays of Heiner Müller, it has even become a basic structural principle. *Hamletmachine*, beginning with the words, "I was Hamlet. I stood at the shore and talked with the surf BLABLA, the ruins of Europe in back of me,"[5] is a play that constantly reminds us of the gyre where history and tragedy inevitably converge. This fusion is also the strategy of Levin's *Murder* and of many of his other plays, perhaps with the difference that Levin is not, like Müller, writing with the ruins behind his back. Rather, Levin has situated himself and his characters in the middle of the rubble, where history is continuously taking its tragic course. From this position, it is extremely difficult to enable us to hear the gradually enfeebled voice of the Horatios. This, I would argue, is not an artistic failure, because the challenge to reflect on the destructiveness of history through narrative seems to be almost insurmountable. Rather, it seems that Levin knew very well what the stakes are in trying to delineate the site where the discourses of violence and performance confront each other, or where history as terror, on the one hand, and narrative as tragedy, on the other, can fully blend.

In his well-known meditation on the Paul Klee painting *Angelus Novus*, Walter Benjamin expressed a similar concern about the site where, for a short moment of danger, the beauty of art and the terror of history merge. According to Benjamin, the face of the angel of history "is turned toward the past," which means that the catastrophes of history are not, as for Müller's Hamlet, situated behind his back, because there would then be a risk that we might not notice them. Rather, according to Benjamin, the wreckage of history piles up in front of the angel, while it "would like to stay, awaken the dead, and make whole what has been smashed."[6] Klee's angel staring at us with its wide open eyes is hurled into the future by a wind coming from Paradise, where history supposedly has had its beginning. Klee's angel, just like Levin's play *Murder*, represents a form of aesthetic imagination which makes it possible for the artist to formulate the cruel dynamics of history, fixing our gaze on the debris of history that keeps piling up in front of us.

The performance of *Murder*

Before his death from cancer at the age of 56 in 1999, the Israeli playwright and director Hanoch Levin published 56 plays as well as a number of satirical reviews. Of the plays, approximately 30 were performed during his lifetime, most of them directed by Levin himself. He also published poetry, short stories, and children's poems.[7] Already in the late 1960s, in the satirical review called *Queen of the Tub* (*Malkat Ambatia*), Levin expressed his extreme iconoclasm with regard to the then more firmly established Zionist "myths." *Queen of the Tub* led to unprecedented reactions from right-wing and religious critics and spectators, who frequently even reacted violently during the performances themselves, sometimes physically endangering the actors. The specific showing for which I had bought tickets in 1970 was cancelled because, during the first show that evening, one of the spectators had thrown a brick onto the stage. A few years later, in the mid-1970s, this offending audience member became one of the extreme right-wing settlers in Hebron in the occupied West Bank. After 19 performances, The Tel Aviv Cameri theatre – where *Queen of the Tub*, like *Murder* almost 30 years later, was performed – succumbed to the actors' fears of this kind of violence as well as to public opinion, and closed the production. *Murder* was given a much quieter reception, even if it is in many ways more radical and more provocative than *Queen of the Tub*. The cultural norms and expectations had clearly changed during this period.

   Queen of the Tub criticized the glorification of the 1967 war and its young soldier victims. The section of the performance based on the story from *Genesis* (Chapter 22) of Abraham's sacrifice of his son Isaac was one of the central reasons for the violent protests. According to Levin's radically rewritten version of this archetypal text, which concerns the willingness of the fathers to sacrifice their sons for a higher cause, Abraham tells his son about the sacrifice he is about to make long before reaching the mountain and asks him to forgive him in advance, because, as Abraham says, he is only doing what God has asked him to do. Isaac tells his father not to worry about this because he understands that this is God's will, and therefore his father should not have a bad conscience. Through this scene, Levin criticizes the tacit agreement between fathers and sons in Israel that may lead members of the younger generation, because of their naive patriotism, to pay with their lives. As Abraham and Isaac approach the mountain where the sacrifice will take place, the angel of God calls out to save the young boy, but Abraham is almost deaf and cannot hear this voice. Had it not been for Isaac's ability to convince his father that God really wanted him to save his son, this incident, Abraham concludes in Levin's version of the story, may have ended badly. Abraham asks what will happen if other fathers have to kill their sons. To this Isaac answers that there may always come a voice from heaven. However, as the secular audience very well knows, the chances that this may happen are rather slim.

After Levin's subversive presentation of the biblical story, a young boy who has obviously just died sings to his father:

> Father dear, when you stand over my grave,
> Old and tired and forlorn here,
> And you see how they bury my body in the earth
> And you stand over me, father dear,
>
> Don't stand then so proud,
> And don't lift up your head, father dear,
> We're left flesh facing flesh now,
> And this is the time to weep, father dear.
>
> So let your eyes weep for my eyes,
> And don't be silent for my honor here,
> Something greater than honor
> Now lies at your feet, father dear,
>
> And don't say you've made a sacrifice,
> For the one who sacrificed was me here,
> And don't say other high-flown words
> For I am very low now, father dear.
>
> Father dear, when you stand over my grave
> Old and tired and forlorn here,
> And you see how they bury my body in the earth –
> Then you beg my pardon, father dear.

> (*Queen of the Tub*, xix–xx)

This song was included in the production of *Murder* 27 years later, in a completely different context. In *Murder* a Palestinian boy who has been killed by three Israeli soldiers sings it to his father, transferring the lamenting voice from an Israeli victim to a Palestinian one. In 1970, Levin presented the son as a victim of his own Israeli father, and this raised violent opposition. In 1997, in *Murder*, the young Palestinian boy singing this song had been the victim of the brutal violence of the Israeli soldiers and this hardly raised any reactions at all; it almost seemed "natural."

Murder begins with three Israeli soldiers sadistically beating a Palestinian youth to death. When the boy's father enters, he asks "why?" – why have you done this to my son? The question "why?" will echo throughout the whole performance, assuming that there is an underlying rational order on the basis of which such actions are carried out and that it is possible to construct a logical narrative to depict them. But Levin's play repeatedly illustrates that there is no answer to this question. Paradoxically, it is the impossibility of providing an answer that constitutes the narrative of this play. There is no clear causal or logical narrative that can be extrapolated from the ways in which history "happens," no rationality behind these events.

Not unexpectedly, the Palestinian father who tries to protest against the atrocity he senses has just been committed receives a cynical reply from one of the soldiers:

Flushed Soldier:
There is no why here.
You understand yourself.
We came here to make a search,
he resisted.
We tried to calm him down.
He didn't calm down,
he cursed and hit.
All of a sudden he pulled out a knife.
There was a struggle, he fell.
In the end, he died.

The Father:
This eye of his . . .

Flushed Soldier:
There was a struggle.

The Father:
The eye is outside. It's gouged out.
Somebody stabbed him in the eye.

Flushed Soldier:
Somebody stabbed him in the eye?

The Father:
How can you stab a person in the eye?
Everybody knows what an eye is.
If a grain of dust gets in,
the eye can't stand it,
and here . . .

Flushed Soldier:
In the heat of the struggle, maybe.

(*Murder*, 215–16)

The soldier presents a narrative about what he and his fellow soldiers have just done, but from the point of view of the spectators it is clear that they are lying, because we have just seen them molesting and killing the Palestinian boy, who was begging for his life. An opening scene like this immediately exposes an unsettling gap for Israeli audiences, because, even if we know about this kind of violence, we are not used to seeing a Palestinian in the position of a victim on the Israeli stage. Israelis are supposed to be the victims, not the Palestinians. The opening also determines the way in which the performance presents the initial culpability of the present conflict. This "history" begins with a Palestinian victim who is placed by association in the position of the biblical Isaac.

As the father tries to find out how his son died, and in particular in his desperate efforts to find narrative closure for his son's life, he wants to know what his son's last words were when the messenger of peace appeared. In the Cameri performance of *Murder,* a TV screen came down from above with a news broadcaster or a politician making an announcement. He was played by Yossi Yadin, one of Israel's most famous actors and the brother of the famous archeologist who excavated Massada, the symbol of Jewish resistance during the Roman conquest in the first century AD.

> *Messenger*:
> The time of murder is over.
> The time of murder is over.
> The furious rage has come to an end,
> winds of reconciliation are blowing,
> the time of murder is finished.
> People look
> at the bad times
> and ask: How could we?
> How was that possible?
> Our children will not understand,
> our grandchildren will laugh,
> our great-grandchildren will not know
> what it's about.
> They will study history
> with a shrug.
> With a smile of waking
> from a deep sleep,
> people say
> to one another: peace.

> (*Murder*, 218–19)

Such a meta-theatrical narrator – a kind of politician governing the fictional world, creating its narratives – is in fact a speaking, technological version of Benjamin's interpretation of Klee's painting *Angelus Novus*, the angel of history. This narrating figure, a witness who has a privileged knowledge of the events, appears in many of Levin's plays. This character is always placed in a position of mastery over the fictional world in which he presides and here he repeats the basic *Gestus* of the dialogue between the father and the soldiers. Even "peace" is one of the things that just happen, like the death of the young boy, and for those who will study history in the future there will not be any rational reason for this change. There will not be any answer to the question "why?"

However, for the bereaved Palestinian father, this "peace" does not carry any consolation:

> Peace?
> You say "Peace,"
> and wound my heart.
> If it's peace, why is it an hour late?
> And if it's late and my son is already dead,
> what good is peace to me?
> No, the word "peace"
> pains me.
> I hear cheers,
> and they stab me.
> I hear laughter,
> and I drop my face
> into my hands and weep.
> I lost my son.
> Those aren't just words:
> "my son," "I lost."
> That was my son,
> you killed him, how will I ever have peace?
>
> (*Murder*, 219)

In these passages Levin makes us painfully aware that history has to be written with words. The word "peace" will, according to the messenger, express a sudden change – a kind of waking up from sleep or a dream. For the Palestinian father it has totally different connotations. These words, with which history is written, cannot express the pain of what has happened. The harsh ironies between what has been shown on the stage and the words used to describe it expose a discrepancy between the event and the words that will reappear in many different versions throughout the play.

Before the soldiers go off to have a healthy salad for breakfast, the bereaved father discovers something unusual on his dead son's body:

> His pants are down,
> and there's a cut here
> that somebody started
> around his penis.

> (*Murder*, 220)

The boy has obviously also been sexually molested. But the father is not interested anymore to know "who did it / or what for":

> You won't be able to answer
> "in the heat of the struggle."
> That's a cut that they started
> when he was already lying
> unconscious or dead.
> I won't ask what for.
> I only want to ask
> what were his last words.

> (*Murder*, 221)

The demands of the father are gradually minimized. However, this wish to hear what the young boy's last words actually were will only be fulfilled in the epilogue of the play. The first scene ends with the arrival of an officer who asks if the father has seen the people who killed his son. The father's answer shows that, in order to live in what has now been officially classified as "peace," he too has to tell a lie:

> *The Father*:
> No, sir. I came in
> when the room was empty.
> I saw them going away.

> *The Officer*:
> Take the corpse and bury it
> without any noise or fuss.
> We won't allow demonstrations.

> *The Father*:
> Before, you didn't allow demonstrations
> because of the war.

> *The Officer*:
> Right. And now because of the peace.
> We're generally in favor of quiet.

The Father:
      Yes, sir.
      My son is quiet and I am quiet.
      From now on, quiet will reign here.

          (*Murder*, 222)

At this point the dead boy stands up, addressing his father with the song of the dead son from *The Queen of the Tub*, demanding that his father beg his pardon. But instead of eliciting this gesture of pardon, which would imply that the Palestinian father also carries some kind of guilt for putting his son's life at risk, the boy's cruel death and the father's forced silence serve as the lethal triggers for the violent revenge that will develop from the first scene. This song could also be seen as the last words of the son, for which the father has been asking. It is a lyrical passage connecting the mourning over the dead son and the act of revenge, creating the kind of aesthetic disruption (what Brecht called *Verfremdung*) necessary for the representation of violence.

According to the written play, the second scene takes place three years later, but in performance it felt as if it takes place directly after the death of the Palestinian boy. In this scene, the father performs his act of vengeance on a newly married couple celebrating their first love on a beach, claiming that the groom was one of the soldiers who had killed his son. The bride desperately tries to defend her newly wed husband, but to no avail. After killing the groom he rapes the bride, and before taking her life he asks her:

      Why do your eyes keep
      staring at me?
      Why don't you scream?
      Why don't you get up and run away?
      (*The Bride lies still on her back, groaning softly*)
      You're breaking my heart,
      my girl. My boy.
      My children.
      (*Aims the gun at her face. She whispers, almost swoons*)

Bride:
      Why? ... Why? ...

The Father:
      "Why." We're long past
      the question "why."
      The question "why"
      shouldn't be asked.
      The question "why" belongs
      to other times.

          (*Murder*, 231–2)

The question "why?" has now been transformed into a form of lamentation, as the father has to face his own cruel deeds. Even if the causality that has triggered them is somehow overdetermined, as an act of blind revenge, the question "why?" can apparently not be asked anymore.

The third and last scene of *Murder* takes place an additional five years later and it begins with two Palestinian workers peeping into apartments in the Jewish neighborhood where they are employed as menial laborers. They are not at all connected to the characters in the first two acts. This is a new and even more violent beginning with the same theme: violence and revenge. There is a sudden explosion, for which the workers are evidently not responsible, but in spite of this, one of them is lynched and decapitated by an angry Israeli mob led by three whores. Now the messenger from the first scene returns, announcing in more or less the same words with which he announced that peace has come, that

> The time of calm is over.
> The time of calm is over.
> The winds of reconciliation have flown away,
> war is at the gate.
> People look
> at the illusory calm
> and ask: How could we?
> How was that possible?
> Our children will not understand,
> our grandchildren will laugh,
> our great-grandchildren will not know
> what it's about.
> They will study history
> with a shrug.
> With a smile of waking
> from a deep sleep,
> people say
> to one another: To arms.

> (*Murder*, 242–3)

The officer's comment to one of the whores ends the scene:

> Straighten your dress,
> and lift up your head.
> At this moment you have entered history,
> you are inscribed in the annals of our nation.

> (*Murder*, 243)

For the officer, the violent actions of this woman have been transformed into the stuff from which history is made, the kind of history that our grandchildren will study in school, but without being able to understand how these things were possible. There is no rational principle according to which these historical events have taken place, and yet they will be "inscribed in the annals of our nation." Levin has, as in most of his plays, created an uncanny amalgamation of the seemingly trivial – the whore straightening her dress and lifting her head – with what must obviously be considered something sublime – the history of the nation. Levin combines the low and the high, showing that the mistake of the whores even influences the illogical fluctuations between peace and war. The killing of the young couple on the beach was also the result of such a mistaken identity. The father believed that he had found the guilty soldier, while in fact the narrative constantly stresses its own disjointed characteristics. The big time gaps between the scenes, at least in reading the play, reinforce this disjointed quality. And accordingly, Levin shows, this is how history comes about. The narrative consists of disconnected events which, when joined together, are assembled into the irrational debris of history.

The performance ends with a short epilogue, which, according to the script, took place two years later, but which for the spectators is an undefined, perhaps even idyllic future with a group of children playing happily on the sand. The sand has covered the stage throughout the performance, and is gradually perceived as a kind of universal signifier for the Middle East. This epilogue is also based on an incorrect identification, which actually leads to an illusory closure. An old man appears on the stage, mumbling something that resembles the word "father." He stumbles on a blind man who turns out to be one of the soldiers from the first scene and who thinks he is able to identify the old man:

> I don't see you, I've gone blind
> in the war that came after that one,
> but I owe you an answer:
> Your son didn't curse, he wept,
> and he couldn't stop shaking.
> His last words, and they
> will peck at my brain as long as I live,
> were: "Have pity on me,
> I want papa."

> (*Murder*, 243–4)

Only when it is too late will the soldier reveal the last words of the Palestinian youth, but the youth's father is no longer there to hear them.

The kind of sincerity needed to acknowledge the unjust suffering of the victim will only be possible, Levin seems to imply, when those for whom this recognition can also become a form of consolation have already passed away. This is a (theatrical) closure showing that (historical) closure is actually impossible.

Hanoch Levin's narrative strategies

The history books that will be written on the basis of Levin's version of the present conflict between the Israelis and the Palestinians will present a series of loosely connected violent events that have been triggered by irrational fear and by the blind lust for revenge. But these events basically cannot be contained by a causal narrative structure. From the point of view of the victims on both sides, these forms of violence are completely uncalled for and have no rational explanation or cause. They defy the classical narrative formula of our view not only of tragedy, but also of most events that occur around us, a view based on the notion that there must be a cause for everything that happens. A narrative structure can be schematically formulated as a conditional statement: "*If* so-and-so is the case, *then* something like this or that will happen." The most important aspect of this formula is the idea of conditionality and even predictability, of being able to say "If – then." However, what Levin has done in most of his plays is to create a narrative sequence in which there is only a series of "thens," presenting a series of meaningless and even unconnected catastrophes that are, to paraphrase Benjamin, endlessly piling up as debris from the failures of history: "Where we perceive a chain of events, he sees one single catastrophe which keeps piling wreckage upon wreckage and hurls it in front of his feet."[8] This pile is constituted by wreckages that are not necessarily causally connected. Levin invites us to view the events from the point of view of the angel. Furthermore, Levin's narrative formula can be seen as an ironic reversal of the classical device of the *Deus ex Machina*, through which the humans, for no apparent reason, are salvaged from an impossible fate by an unexpected, supernatural intervention. For Levin, the unexpected is realized by an endless series of catastrophes in which no one is saved.

The classical paradigmatic example of a narrative presenting this kind of total arbitrariness of human fate is the book of *Job* in the Bible. Job experiences a series of misfortunes, which – were it not for the bet between God and Satan with which the biblical narrator begins this story – are totally meaningless, at least from Job's point of view. Levin directed his own play based on the book of *Job*, called *The Torments of Job* (to my mind one of his most interesting plays), at the Cameri theatre in 1981. It begins with Job seated together with his friends after having had a festive meal, praising the riches of God. Levin has discarded the bet between God and Satan in order to test Job, but in spite of this, he has created a disjointed narrative supposedly proving the existence of God, though in his own subversive fashion. At the same time, Levin has placed these events at a specific juncture in history, showing – in a way that is quite different from *Murder* – how the tragedy of Job's misfortunes fuses with the historical.

According to Levin's Job, who has just finished a festive meal, there is enough food on the table to satisfy the Beggars, who eat from the leftovers as well as the Beggars of the Beggars, who eat from what is left of these leftovers. Only the Beggarly Beggar, the poorest of them all, gets nothing to eat at this time. But, he says:

> Be patient, my friend,
> And someone will surely puke in your hand.
> Well, somehow we manage to live.
> There's a God in the sky.

> (*Torments of Job*, 56)

This is no doubt an unorthodox way of introducing the theological arguments at the heart of this biblical text. At this point in Levin's play, the Messengers of Poverty, who announce the loss of Job's wealth, begin to arrive. But here also, as with the story of Abraham and Isaac, Levin iconoclastically elaborates freely on the classic original. Job's belongings are confiscated because a new Emperor has seized power in Rome, which is of course not the case in the emphatically a-historical rendering of the story of Job in the Bible. Levin, on the other hand, has situated this narrative within a concrete historical time frame, around the time of the life of Jesus, that "other" Jewish victim of the whims of history. In *The Torments of Job* Levin examines the points of contact between the irrationality of the events through which Job gradually becomes totally victimized and the comprehensive and well-known narrative framework of the suffering of Jesus, which will supposedly bring the Messianic era and the end of history. In *Murder*, on the other hand, the realities of the Israeli–Palestinian conflict are molded into a similar irrational narrative frame, showing that in the future the present chaotic situation will be studied as history.

After the Bailiffs in *The Torments of Job* have emptied the banquet hall and stripped Job to his underwear, he cynically remarks:

> You forgot my gold teeth.
> I've got some gold teeth in my mouth.

> (*Torments of Job*, 62)

To this the leader of the Bailiffs answers, in the style typical of Levin, mixing the sublime mythical story with seemingly trivial everyday realities:

> Don't be ridiculous.
> Don't try to make us into monsters.
> We're all just human, part of the group,
> We all go home to our wives at night,
> To our slippers and a hot bowl of soup.

> (*Torments of Job*, 62)

This scene, at least for an Israeli audience, triggers associations with the Nazi period, when all the belongings of the Jews were confiscated, and finally their gold teeth were taken out as well. However, the Bailiffs claim that, since they enjoy the pleasures of everyday life, they will not commit such atrocities. Job's response to his situation shows how Levin has re-written the despair of the biblical Job. His words express a modern, existential, and nihilistic, Beckettian view of life, while the echoes from the biblical text can still be clearly heard:

> Naked came I from my mother's womb and naked came
>   my mother
> From her mother too.
> Shuddering, we emerge, one from another,
> A long line, naked and new.
> "What shall I wear?" asked my mother in the morning
> But when the day was done,
> Naked was she borne to the pit.
> Now I too stand naked, her son.

> (*Torments of Job*, 62–3)

At this point the Bailiffs, who had already left the scene, suddenly return and seize Job by the throat, violently taking out his gold teeth with a pair of pliers. Their initial compassion has gradually been transformed into a spectacle of extreme cruelty. Job – like the Palestinian youth, the newly wed couple and the workers in *Murder* – becomes a victim without knowing the reason for his victimization. Catastrophes just happen, and when, as in the case of the gold teeth, the possibility of something terrible is announced, the catastrophe is usually alleviated for a short moment, only to become realized with a vengeance when the victim naively believes that he is safe. When the catastrophe takes place there is never, Levin implies, an answer to the question "why?"

Levin's use of this narrative structure in his play about Job, just as in *Murder*, is not just based on introducing a single misfortune, but rather on a pattern of violence that is constantly repeated, bringing new and unexpected misfortunes. However, at least from the perspective of the victim, there is no presentation of any form of obvious causality or reason. This completely erratic narrative structure, I would argue, reflects something central to the Jewish experience of history – to the extent that it is possible to talk about such a category without essentializing it – in particular during the twentieth century, which has been seminal for the creation of a collective Israeli historical and cultural consciousness. This culturally conditioned pattern, which has become so deeply ingrained in narratives of violence and terror, gives additional force to the narrative structures developed by Levin. Jews have, one could even claim, internalized the experience of irrational misfortunes initiated by others, for which there has been no apparent cause and for which there has been no prior warning. In *The Torments of Job*, Levin transformed this experience into an effective theatrical narrative, and in *Murder*, he extended it to include the Palestinians as well. Levin implies that the collective trauma of the Jews has gradually brought the Palestinians into the same scandalous gyre.

After Job has also lost his children, a Roman Officer sent by the new Emperor announces that

> The god of the Jews is null and void, wiped out.
> All who believe in him are heretics and rebels.
> To reinforce the new belief and make it crystal clear:
> All those who believe in the god of the Jews will have
> A spit stuck up their rear.

> (*Torments of Job*, 77)

Job's friends, who according to the biblical text are completely firm in their belief in God, are now facing a terrible dilemma. After considering the fact that their fields have to be harvested and that their children are still young, they somewhat unwillingly compromise their convictions with regard to God's existence. Only Job, who has nothing left to lose, is willing to acknowledge such a possibility, and he is immediately punished with the pole stuck into his behind on which he hangs for the rest of the performance as if he has been crucified. Besides situating the biblical narrative in an historic context, *The Torments of Job* asks a question, to which Levin will finally give quite a cynical answer: under which circumstances is it possible to prove the existence of God?

Just after Job has been punished for still believing in God, a Ringmaster (the owner of a circus) appears, saying that it is too bad

> For such a performance as this to go to waste.
> All those potential tickets mutely crying out

Like the souls of unborn children dying out.
Not to mention the educational worth
For those who still think god exists on earth.
                    I've run musical circuses in all the most
Important capitals of Europe.
I can even say that I've run Europe.
[...]

Five hundred dinars to the royal treasury
For the right to put this man
In my circus.

(Torments of Job, 84–5)

After tough bargaining between the Officer and the Ringmaster, the "torments" of Job, hanging on the pole become the main attraction of the circus. The theatre, and art in general, Levin implies, cynically exploit human suffering. The Ringmaster, just like the Messenger in *Murder*, serves as the metatheatrical narrator and witness, cynically commenting on how the present is connected to a historical chain of events. But there is one problem the Ringmaster cannot control: Job is going to die before the circus has been able to make any real profit from this spectacle. The moment Job dies, the Ringmaster even curses him, and while the circus abruptly disperses, Job vomits. Now the Beggarly Beggar quickly returns and hungrily licks the vomit:

Just like I said: a little patience
And somebody finally pukes. Yes,
Somehow we manage to live.
There's a god in the sky.

(Torments of Job, 91)

This form of divine "benevolence" is very different from the biblical narrative in which Job's possessions are returned and he even gets new children. In Levin's version of this narrative only the dead can now be heard, just as the Palestinian youth in *Murder* sings to his father from the grave. In Levin's play about Job, the dead appear as angels from the afterworld, paraphrasing the words of Sonya in Chekhov's *Uncle Vanya*: "But there is mercy in the world / And we are laid to rest."

In Chekhov's own play Sonya speaks from a position that to her is like death, but is not death itself:

We shall rest! We shall hear the angels, and see the heavens all sparkling like jewels; we shall see all earthly evil, all our sufferings, drowned in a mercy that will fill the whole world, and our life will grow peaceful, gentle, sweet as a caress.[9]

This gentle sweetness is, however, not at all the *locus* towards which Levin directs his narratives, either in *Murder*, or in *The Torments of Job*, or in most of his other plays. Instead, Levin has radically rewritten the sentimentalizing gesture at the end of Chekhov's play, letting us see the angel rather than just hearing it sing, like Sonya or Horatio in *Hamlet*. In the world of Benjamin, Müller and Levin, as well as that of many other contemporary artists, the angels are not only heard but are seen as well, and while we look at them, they are at the same time looking at the tragedies of history. The angel has become a central figure – a poetic *figura* as well as a creature – that modern culture has valorized to epitomize the fusion between a personal, individual tragedy and the collective consciousness expressed through a historical process. For Chekhov the angel was primarily a figure (in this double sense) for religious redemption. For Benjamin and Levin, the angel actually serves as the locus where tragedy and history merge, and through which meaningless loss by a scandalous default becomes the foundation for a collective narrative and a sense of history. The *oeuvre* of Levin (though I have only been able to examine a few of his texts here) is characterized by a sense that this narrative is reflecting the repeated and erratic failures of the individual to identify a viable alternative to such forms of terror, violence, and revenge.[10]

In closing I want to add another story, the story of Paul Klee's painting, *Angelus Novus*. Walter Benjamin purchased this painting for approximately 14 dollars in 1921 and it became one of his most valued possessions. After Benjamin fled the Nazis from Berlin to Paris it was brought to him there, and before he fled to the Spanish border, where he committed suicide, he deposited the Klee painting, together with the manuscripts that have become known as the "Arcades Project," in one of the two suitcases that George Bataille hid from the Nazis in the Bibliothèque Nationale. Following the war, the painting made its way to Theodor Adorno in New York, who brought it back to Frankfurt before passing it over to the Kabbala scholar Gershom Scholem, one of Benjamin's oldest and closest friends in Jerusalem. Since the death of Scholem's widow, Fania Scholem, the Klee painting has become part of the collections of the Israel Museum, also in Jerusalem.

As we watch this painting (and Benjamin no longer had the painting in front of him when he wrote his text on *Angelus Novus*), the angel, whose back, according to Benjamin, is turned towards the future, is curiously looking back at us. And since his face, Benjamin says, "is turned towards the past," we are actually looking into the empty space of that future while our backs are turned to the past. In the space between the viewer and the painting, just in front of the angel, but also in front of us while we are looking at the painting is the constantly growing debris of history, the failures that keep piling wreckage upon wreckage. The theatre of Hanoch Levin not only forces us to contemplate the destructiveness of this wreckage, but also induces us, sometimes against our own will, to acknowledge that we are positioned right in the middle of it. *Murder,* with its insistence on the totally erratic nature of the narrative of violence does exactly that, while simultaneously, buses are exploded by suicide bombers and targeted assassinations and bombings frequently kill innocent victims. Levin has shown us that if these patterns are not broken and deconstructed we will all live in constant fear of the next act of violence, a basic attitude or state of mind that obviously only generates more violence.

Notes

* This chapter is dedicated to my friend Muhammad, his wife and his two daughters, living in Ramallah, Palestine. The chapter was originally presented at the PSi8 conference at NYU in April 2002. I want to thank Peggy Phelan for inviting me to this conference and for encouraging me to write this study. It has previously appeared in *Theatre Journal*, 54:4 (2002): 555–73, and I want to to thank Harry Elam, who was editor then.

1. Walter Benjamin, "Theses on the Philosophy of History," in Hannah Arendt (ed.), *Illuminations: Essays and Reflections*, trans. Harry Zohn (New York: Schocken, 1968), 257–8.
2. Sigmund Freud, "Moses and Monotheism," in Carl Jung (ed.), *The Origin of Religion*, The Pelican Freud Library vol. 13 (Harmondsworth: Penguin, 1985), 345.
3. All Hamlet citations are from *The New Cambridge Shakespeare* edition (Cambridge: Cambridge University Press, 1985). Line numbers from the play will be noted in the text.
4. William Butler Yeats, "The Second Coming," *Selected Poetry* (1967; Basingstoke: Macmillan – now Palgrave Macmillan, 1999). The first verse of this poem reads: "Turning and turning in the widening gyre / The falcon cannot hear the falconer: / Things fall apart; the center cannot hold; / Mere anarchy is loosed upon the world, / The blood-trimmed tide is loosed, and everywhere / The ceremony of innocence is drowned; / The best lack all conviction, while the worst / Are full of passionate intensity."
5. Heiner Müller, *Hamletmachine and Other Texts for the Stage* (New York: Performing Arts Journal, 1984), 53.
6. Benjamin, 257–8.
7. All quotations from Hanoch Levin, *The Labour of Life: Selected Plays*, trans. Barbara Harshav and with an introduction by Freddie Rokem (Stanford, CA: Stanford University Press, 2003). Page-numbers will be noted in the text.
8. Benjamin, 257.
9. Anton Chekhov, "Uncle Vanya," *Plays*, trans. Ann Dunnigan (New York: Signet, 1964), 231.
10. In my *Performing History: Theatrical Representations of the Past in Contemporary Theatre* (Iowa City: University of Iowa Press, 2000) I have analyzed the angel as a witness of the past and as a meta-theatrical figure in more depth.

13

Not So *Innocent Landscapes*: Remembrance, Representation, and the Disappeared

Mark Phelan

> *A photograph is not only an image (as a painting is an image), an interpretation of the real; it is also a trace, something directly stencilled off the real, like a footprint or a death mask.*
>
> Susan Sontag, *On Photography*[1]

> *Hope relates to a future that is always yet to come but may never arrive. To remain hope, hope must desire the possible, yet it wants what escapes the domain of possibility, the plenitude of a presence that cannot be appropriated.*
>
> Edith Wyschogrod, *An Ethics of Remembering*[2]

> *Of the 30 years of the North's conflict and atrocity a small group of people stood apart: they were the "missing", the "disappeared" – absent and yet somehow still present.*
>
> David Farrell, *Innocent Landscapes*[3]

> *each hooded victim / slashed and dumped.*
>
> Seamus Heaney, "The Grauballe Man"[4]

It is a classic trope within Irish cultural representation that the repressed past returns with a vengeance. In Seamus Heaney's poetry, peat bogs are repositories of the past; mudbuttery metaphors for memory banks, not only in the way that they preserve and petrify art(i)facts from the past, but also in the way that the past's interrelationship with the present is reflected and re-enacted in the bog's cycles of crusting and cracking under the sun: a process that periodically enables it to yield up its memories, its secrets, its silences – the lacunae of history – in the fossilized forms of the "Tolland Man,"[5] the murdered girl of "Strange Fruit,"[6] or the ancient "salty and white" butter of "Bogland."[7]

Terry Eagleton has commented on the way in which objects preserved within bogs "reveal the past as still present," describing how such artifacts are "caught in a kind of living death":[8] a paradoxical plight that resonated with Heaney's childhood memories of ancient Irish elk skeletons and fresh butter being dug out of bogland; retrievals that encouraged him to read bogs as "the memory of landscape"; "a landscape that remembered everything that happened in and to it."[9] The publication in 1969 – a vintage year – of *The Bog People* also profoundly affected Heaney. Its arresting images of immolated Iron Age bodies; petrified men and women sacrificed in fertility rites to a Mother Goddess who required fresh victims every winter to warm her bogbed was redolent of killing fields closer to home:

> Taken in relation to the tradition of Irish political martyrdom for that cause whose icon is Kathleen Ni Houlihan, this is more than an archaic barbarous rite: it is an archetypal pattern. And the unforgettable photographs of these victims blended in my mind with the photographs of the atrocities, past and present, in the long rites of Irish political and religious struggles.[10]

It seemed these terrible numina were twinned as "The Troubles" burst onto the streets of Derry and Belfast and the subterranean savagery of distant Jutland bogs became a murderous metaphor for bloodletting loosed in the local "man-killing parishes"[11] of the North of Ireland, though this particular cycle of sacrifice became more routine than ritual as each decade bled into the next. Imagery of excavation, memory and landscape later became a leitmotif haunting Heaney's later work; his most recent collection, *District and Circle*, is "ghosted by other 'underground' presences from Heaney's past"[12]: presences self-reflexively exhumed from his earlier work, notably in "The Tolland Man in Springtime."

Heaney is not alone, however, as the imaginative *topos* of Irish cultural history has been profoundly influenced by landscape.[13] This tradition can be traced back to Ireland's ancient oral literature of *dinnseanchas*[14] which, indeed, continues to contour the rough fields of contemporary Irish poetry: "The whole landscape a manuscript / We had lost the skill to read."[15] It also shapes the imaginative geography of Irish literature[16] and semiotically saturates the visual fields of tourism, advertising and heritage industries as well as virtually the entire *mise-en-scène* of modern Irish drama so that collectively it has become "an unseverable aspect of self."[17] This condition has been further compounded and intensified by Ireland's colonial experience, "after all," Edward Said notes, imperialism "is an act of geographical violence"[18] and consequently space, place and landscape are transformed into symbolic sites of ideological struggle between colonial power and decolonizing nationalism:

> National identities are co-ordinated, often largely defined, by "legends and landscapes"; by stories of golden ages, enduring traditions, heroic deeds, and dramatic destinies located in ancient or promised home-lands with hallowed sites and scenery [...] Landscapes [...] provide visible shape; they picture the nation.[19]

In all postcolonial societies, landscapes are highly politicized products of the imperial encounter as space and geography are organized according to the ideological imperatives of settlement before being resisted and reorganized by the colonized.[20] They are never merely scenic devices nor picturesque backdrops, but discursive, dialectical sites of contestation; ideologically inscribed texts of conquest and resistance. The extent to which landscape has been interpellated by these forces is evident in the sheer ubiquity with which territorializing tropes are used in Irish cultural and political discourse, as "these competing projects of mapping [...] could take the shape of the Ordnance Survey or the idealization of the West in the Revival."[21] The role and representation of Ireland's landscape, therefore, has long been symbolically central to the construction and deconstruction of its national(ist) identity. "You and your landscapes! Tell me about worms!"[22] excoriates Estragon from his rural wasteland; it is a laconic line laced with Beckett's own anti-essentialist leanings and a self-reflexive reference resonant with Irish audiences whose own horizon of expectations have long been conditioned by this enduring visual aesthetic.

Within Irish culture, then, the *topography* of landscape has an extraordinarily heteroglossic *textuality*: it is a palimpsest, a memory bank, a manuscript – a mnemonic terrain that registers and remembers repressed traumas. The earthy connotations of this imagery in turn invoke the contemporary theoretical, political and historiographical projects of revisionism and their Benjaminian brushing against the grain of Irish history, to unearth and expose the occluded histories and silenced voices of the subaltern: those written out of history. It is within this field of revisionist retrieval; of an archaeological (re)examination of the past, that David Farrell's remarkable photographs of the excavation for the secret graves of the Disappeared intervene.

The Disappeared – those secretly abducted from their homes in the North to be murdered and buried in unmarked graves in their "homeland" in the South – represent one of the darkest stains of the past four decades of conflict and, in themselves, pose a profound problem to the myriad processes of remembrance and commemoration to have emerged as part of an evolving process of post-conflict resolution. In fact the fates of the Disappeared pose ethical and epistemological questions for the very act of representation itself, for how can you express what is ineffable? How can you represent something when the referent has been absolutely annihilated? This philosophical aporia has been familiar since Adorno's famous cry about art after Auschwitz and, in an Irish context, has immanently haunted historical, political, and cultural attempts to engage with the enormity of the Irish Famine, as Christopher Morash and others have pointed out.[23]

Apart from Robert Nairac,[24] all 15 Disappeared were Catholics: victims of IRA internal "policing." However, in contrast to the explicitly public spectacle made of the bodies of "ordinary informers" executed by the IRA, whereby their bound, bagged, and beaten bodies were left lifeless on remote rural roadsides as gruesome warnings to the rest of the community, the fates of the Disappeared were steeped in secrecy and silence. It has been suggested this was done for reasons of political expediency, as admission of guilt in these particular murders would have been deeply embarrassing to the Republican movement.[25] Accordingly, such was the "sensitive" nature of these "stiffings" that the IRA adopted a policy of denial: they claimed to have no knowledge of the victims' fates and continually frustrated attempts by the families to discover the truth behind their relatives' disappearance and to secure the return of their bodies for burial.[26] Paradoxically, this cold, Creon-like denial of the burial rites demanded by the Disappeareds' families was a desecration of one of the most sacrosanct traditions within Republicanism: the burial of the dead, a ritual symbolically central to the IRA's commemoration of their own "patriot dead." Moreover, the paramilitary pageantry associated with these occasions cruelly contrasts with the lost, liminal state of limbo to which the Disappeareds' families were consigned as they were denied the same such ceremony and closure.

In the context of the ongoing peace process and the new struggle for post-conflict resolution in the North, it is worth recalling Wole Soyinka's passionate rejoinder on the South African experience of coming to terms with the burden of history, when he declared that "the territory of culpability in the South African instance was not limited to the state."[27] Encouraging signs that this was being recognized in Ireland by the IRA came with their disclosure, under the terms of the Good Friday Agreement (GFA) and following the passing of the *Northern Ireland Location of Victims' Remains Bill*, of the whereabouts of six secret burial sites of eight people they had murdered in the 1970s and early 1980s. Although extensive excavations were carried out at these locations, many of which were scattered throughout the remote, liminal hinterland of the border counties, only three victims were discovered; the rest remained missing.[28] Farrell's photography bears witness to the fates of the remaining Disappeared, as he imaginatively maps out "the territory of culpability" in his extraordinary images.

Innocent Landscapes emerged out of Farrell's visceral reaction to the "violation" of the landscape effected by the enormous excavations made in the search for the Disappeared, which scarred an otherwise picturesque pastoral scene.[29] Numerous images vividly portray this scarification of landscape: drained bogs, dug-up fields, chewed-up clay, uprooted trees, splintered branches, silted-up streams, broken boulders, mangled vegetation, and churned earth, all of which appear almost as wounds on the landscape. The invasion of this pastoral world by modern machinery has a further connotative charge, in that the violence of its excavations mnemonically re-enacts the violence perpetrated on innocent victims by their executioners, as the diggers' defiling of the eponymous "innocent landscape" embodies the IRA's desecration of the "Disappeared" bodies and the denial of their burial rites.

Throughout *Innocent Landscapes*, Farrell self-reflexively exploits the cultural trope of an Arcadian, prelapsarian landscape ubiquitous in Irish cinema, painting, photography, tourism, and advertising.[30] However, the power of his images is generated by his deconstructive representation of space, place, and landscape, which have traditionally been idealized in the visual arts, ever since the "Wild West" of Ireland became a stock, sensational *mise en scène* for nineteenth-century melodrama and a sublime setting for a whole genre of Anglo-Irish romantic literature. The elemental, exotic appeal of the Irish landscape has proved as irresistible to imperialist and revivalist self-fashioning as it has to tourism. The "picturesque primitivism" that informed orientalist and nativist representations alike, now manipulated to market "Ireland": a global brand; an island idyll removed from the "filthy modern tide." It is a paradox presciently critiqued by George Bernard Shaw in *John Bull's Other Island* (1904), which exposes the convergences between colonialism, nationalism, and tourism. Farrell self-consciously plays upon the visual cultural legacy of the Irish pastoral.[31] Many photographs deliberately resemble *Bord Fáilte*[32] images of Ireland's bucolic beauty and are replete with stunning panoramic vistas of undulating hills, country lanes, ancient graveyards, misty mountain streams, primordial bogs, windswept beaches, and the like. Farrell also appropriates the conventions of nineteenth-century landscape painting and his appropriation of this clichéd visual culture is all the more interesting considering that its original international site of consumption was the opening of the new Irish cultural centre in Paris in 2002.[33] In such a setting, Farrell's ironic foregrounding of familiar signifiers of "Ireland" meant that they were jarringly juxtaposed with the traumatic nature of his subject matter, a conflict that aesthetically and ideologically charges these cultural signs with a disturbing valence, one which defamiliarizes and deconstructs their comfortable connotations. In Barthesian terms, the *studium* of these stereotypical images is disrupted by their devastating *punctum*:[34] that these photographs are of the Disappeared. They portray not a pastoral Eden, but a landscape of the dead; each image enframing the archaeology and pathology of violence as the "holy ground" of nationalist rhetoric is profaned by the reality of republican "war crimes."[35]

Various commentators have noted that images of landscape and violence are interconnected themes found throughout Irish literature and the visual arts.[36] However, whilst these tropes and themes are also present in *Innocent Landscapes*, they differ significantly in their semiotic and ideological connotations, for the foundational images and iconography of Irish cultural nationalism and identity are contested visually (in the exhibition) and vertically (in the book) through their archaeological excavation of the past, as well as the politics and pathology of republican violence. Farrell self-consciously draws upon the romantic cult of the landscape cultivated by both imperialist and nationalist discourse, which fashioned images of Irish landscape into potent signifiers of nationhood. In his own words, Farrell provides us with "luscious, beautiful images"[37] which actively solicit romantic responses from their spectators, but only to deconstruct these nationalist tropes by confronting the viewer with the violent, hidden histories of these places, which are sites of some of the worst excesses of Republican violence. The performative affect of this disjuncture is to disrupt the viewer's touristic gaze and deny any simple spectatorial pleasure. Traditionally, in landscape painting and photography:

> The role of the spectator is all-important as the act of seeing corrects and completes the landscape. With the discovery of the vantage point that provides this balance of foreground and background, a "sublime synthesis" occurs: the authenticity of effect takes place in the epiphanic moment in which the unified aesthetic essence of the place shines forth.[38]

However, no "sublime synthesis" is allowed; there is no epiphanic moment of unity: vanishing points lead to voids; evacuated beaches and countrysides appear desolate; cloud patterns foreboding, whilst the initial tranquility of lonely leafy lanes, stilled lakes, waterlogged bogs and overgrown graveyards slowly subsides into silences that suggest unspeakable loss; immanent absence. Experientially, it is harrowing. Each place is haunted by history; by the guilt that lies beneath the innocent landscape; where men and women once knelt in the earth before secret graves before being shot in the back of the head and secretly buried. Nonetheless, the role of the spectator *remains* all important, as the "act of seeing" is performatively transformed into an ethical act of witnessing. "Learning to see is training in careful blindness";[39] an insidious conditioning effect that Farrell exploits and exposes. By altering the aesthetic contract and attacking this condition, he politicizes the viewer's gaze, transforming spectators into witnesses; passive consumers into moral agents.

Farrell's work powerfully offsets the *temenos* of place against the trauma of the events that transpired in these remote, seemingly peaceful locations. As such, his photographs resemble the "literature of testimony" generated after the Holocaust: a body of work whose form and function are eloquently explained by Shoshana Felman: "texts that testify do not simply *report facts* but, in different ways, encounter – and make us encounter – *strangeness.*"[40] Farrell's photography operates in much the same way. His images "make strange" familiar scenes, though more in an ethical than an epic mode, as image after image defamiliarizes the Irish landscape, deconstructing its comfortable connotations and stripping away its romantic accretions to reveal an entirely different "hidden Ireland." This brutal process reflects, and is re-enacted, in the violence of the excavations that are the subject and setting of *Innocent Landscapes*.

It is productive to examine a number of images in a little more detail. Figure 13.1, depicting the site of the excavations for Kevin McKee and Seamus Wright, has the appearance of a vast expanse of land that has been systematically cleared for a specific reason, perhaps for agricultural or archaeological purposes. The former possibility has a particularly haunting association given the ironic assumption that this disturbance of the earth has been caused by tillage, for the purpose of sowing the seeds for next year's crop. Such a scenario radiates a reassuring sense of the turn of seasons, the natural cycle, man's harmonious interaction with the natural world. Such a sylvan sense of harmony and tranquility is spatially enacted in the symmetrical structure of the photograph's composition: in the balance between the field in the foreground and the backdrop of the sky (though its darkness hints of something else). The photograph's sense of poise and balance

Figure 13.1 Wilkinstown, 2001
Source: Photograph by David Farrell. Courtesy of the artist.

Figure 13.2   Ballynultagh, 2001
Source: Photograph by David Farrell. Courtesy of the artist.

generated by its compositional and spatial equilibrium, underscores these idyllic associations and engenders a comfortable sense of *studium* that is shared by a further sequence of similar images which play upon pastoral idealizations of the Irish countryside. However, this sense is violently ruptured by the following photograph's *punctum* (Figure 13.2), the element that breaks or punctuates the *studium* present in the foreground of the image: the seemingly innocuous strip of tickertape that the viewer belatedly realizes is a police cordon marking the possible location of a grave. This *punctum* is not registered immediately; it appears at first that the tape penning off the area could perhaps be part of an archaeological dig, another familiar imagistic trope of Irish landscapes that at first earths the voltage of *punctum* with its comfortable connotations of antiquity.[41] These archaeological resonances are reinforced by the image's proximity to another photograph depicting the patterned, overlapping footprints in clay at the excavation site, an image resembling the whorled forms of ammonite fossils, ancient sea-shells preserved millennia ago by rapid burial with prehistoric sediment. However, the *punctum* eventually registers and shatters the *studium* when it becomes clear that the ticker tape delimits a crime scene and a possible unmarked grave.

Susan Sontag's claim that "photographs furnish evidence [...] In one version of its utility, the camera record incriminates"[42] is worth recalling here, for, as in a number of other images, Farrell deliberately deploys the evidential style of police crime scene photography (Figure 13.3). In this image, there is an almost palpable sense of belatedness, that we have arrived too late at the scene of a crime. The yellow cord of the police cordon warns us to back off, to keep out, to maintain our distance, yet ineluctably, we feel propelled by our voyeuristic curiosity to wonder what lies beyond the cordon: what secrets does it conceal, what crime was committed there, what remains of the victim? However, the knowledge that this was one of the huge number of excavations that failed to find anything serves to highlight the futility of the searches and foreground the forensic uselessness of these photographs as documentary evidence.

Figure 13.3   Bragan, 2001
Source: Photograph by David Farrell. Courtesy of the artist.

Figure 13.4 Templetown Beach, 2001
Source: Photograph by David Farrell. Courtesy of the artist.

Another image of Templetown beach (Figure 13.4), the site of the fruitless search for the body of Jean McConville, also recalls the above image of a crime scene. There is an extraordinary openness to this photograph's depiction of the beach, its shoreline and the sea as it eventually merges with the sky at the point of the horizon. The thin, white ribbon of breaking waves separating shore from sea visually recalls the cordon in the previous image, as this too is a (postcard-picturesque) crime scene; however, the vastness beyond this cordon is the sea rather than a forest. The clear stratification of the sky, sea, and sand in the image expands and explodes its sense of spatial depth and further foregrounds the futility of excavations in so huge an expanse of space. This is further intensified by the feeling of impossibility of finding a body in such a liminal location that is neither aqueous nor earthy, but a terraqueous non-space. Moreover, this specific site had changed significantly over the 31 years since Jean McConville had supposedly been buried there. A car park and tourist facilities had been built, whilst geological experts advised that "an extreme storm event" in 2002 significantly changed the topography of the site after washing "away part of the car park and roadway [so that] the high embankment west of the car park was eroded and undermined."[43] The car park itself was removed by the excavators but was later rebuilt after the search was abandoned, even though Jean's body had not been found. This also vestigially haunts these images, with the sense that her missing body was somehow being re-entombed in Templetown; a placename derived from the Irish *"Baile an Teampaill"* – townland of the church.

A similar sense of vastness radiates from images of Ballynultagh; one of the locations furthest removed from the North, though hauntingly, its Irish derivation *"Baile na nUltach"* means "place of the Ulstermen."[44] One image depicts an open field that extends to the horizon and resists camera efforts to contain or constrain its proportions (Figure 13.5). The black sky overhead is palpably heavy, reinforcing the ominous sense of space. The field itself retreats to the fringe of the forest, which rolls over the hills to the sky beyond, sky creating an unfolding sense of depth and distance that cannot be contained by the camera's gaze and that seems to have burst the cordon in the foreground of the picture through the land's resistance against any attempts to bound or measure it.

In these and other images, Farrell consistently relies on the conventions of landscape art. There are no obvious markers identifying these sites as the resting places of the Disappeared or the scenes of their murders, save for some images of simple shrines improvised by family members who attended the excavations. This lack of recognizable landmarks foregrounds the futility of searches in these remote locations which, in turn, reminds the viewer why they were selected in the first place. The IRA subsequently blamed its difficulty in pinpointing the exact location of their victims on the lapse

Figure 13.5   Ballynultagh, 2001
Source: Photograph by David Farrell. Courtesy of the artist.

of time and the deaths of its volunteers, but the obvious difficulties in remembering and retracing these remote locations were even acknowledged by some of their victims' families "You could see how they would get the exact site wrong in a place like this, where there are no landmarks."[45] Accordingly, Farrell gives no guidance to his viewer beyond a simple list of the victims accompanied by images of annotated ordnance survey maps used to indicate their approximate resting places. In decoupling each image from any clear references to the specific location or the victim supposedly buried there, Farrell capitalizes on the connotative power of photography which irresistibly entices us to interpret and to decode each image's significance. Photographs elicit our deepest instincts to narrate; to imagine what happened before and after an image was taken. Art critic John Berger observes that, "Before a photograph, you search for *what was there*."[46] This generic dynamic of object/viewer exchange freights these images with an altogether more haunting valency given their context and content.

This formalist aesthetic contract between viewer and image forces us to visually excavate the picture; to seek its meaning, a performative process that reproduces and re-enacts the excavators' search for the missing bodies of the Disappeared. However, what Farrell forces us to see is that there is, indeed, *nothing* to see: the emptiness of the expanses of bogland and beach, forests and fields, foregrounds the futility of the excavations, the absence of physical remains, and the absolute annihilation of each individual. For Roland Barthes, photographs are "never anything but an antiphon of 'Look,' 'See,' 'Here it is,' "[47] but Farrell disrupts this aesthetic, documentary, and evidential function: his images instead point to absence: "Look, see; here it isn't." They are haunted and hollowed by the absence of their referents, as Farrell uses photography's inherent "dialectic of presence and absence"[48] and inverts it. It is from this paradox that his photographs draw their extraordinary power; they undo the epistemological essence of photography, whereby "inherent within the photograph is the particular requirement for the physical presence of the referent."[49] Within Farrell's images, the referent is the *absence* of physical presence: a dynamic that reflects the uncanny experience of visiting the sites of these excavations when Farrell recalled how the "contradictory feelings of presence and loss were intense, overwhelming."[50]

For Barthes photographs authenticate existence; they are "certificates of presence,"[51] although this materiality is paradoxically permeated by the absence of the object/moment they record. It is this epistemological irony that Farrell deploys brilliantly, as the significance – and signification – of *Innocent Landscapes* resides primarily in its explicit dialectic between the material presence of the photographic artifact on one hand, and the profound sense of a doubly absent photographic subject – the Disappeared – on the other. In this way, the innate ambiguity of photography, given its paradoxical sense of materiality and emptiness, and the simultaneity of presence and absence, provides a profoundly apposite, ethical medium for representing the Disappeared. This is perhaps most palpably experienced by the viewer in several images of impromptu shrines constructed *in situ* by the families of the Disappeared who kept vigil at the excavation sites. These simple shrines of flowers, candles, cards, religious pictures and icons, pebbles from the sites, and personal memorabilia also featured framed pictures of the dead. "All portrait photography is performative"[52] and in these instances their affective force is arresting, even devastating. Framed against a violated landscape – and in the absence of any physical remains – these pictorial and psychical traces remind the viewer of each extinguished human life in what are otherwise evacuated scenes. They accentuate the incorporeal presence of the dead whilst at the same time, intensifying their aching sense of absence. It is an excruciating double bind. The desecration of the Disappeareds' bodies and their families' denial of a Christian burial is aesthetically reproduced in these images as Farrell's *photographs of photographs of the dead* doubles the semiotic sense of deferral and the denial of closure, which mnemonically mirrors (in an infinitesimal way) the experience of the victims' families.

For Barthes, all photographs are haunted by their "spectrum": "that rather terrible thing which is there in every photograph: *the return of the dead*."[53] Photography, like a mechanical medusa, mortifies the flesh, movement, and time itself with a single glance. It embalms its subject with a spectral presence of death and denotes the limitations of representation. Sontag, too, is fascinated with the way photographs possess, and are possessed by, "ghostly traces."[54] She describes all photographs as "memento mori,"[55] elaborating that they are as much an *aide memoire* as a testament to loss: their sense of presence, through the preservation of the past, serving to accentuate the actual disappearance of the original referent. This immanent, elegiac essence is what moves Eduardo Cadava to classify photography as "a mode of bereavement"; for every image conjures "the return of the departed."[56]

It is the return of the dead in photography that troubles Cadava, Barthes, and Sontag and which, in a different context, traumatizes the families of the Disappeared. The return of the dead is one of the most powerful human drives as Robert Pogue Harrison's magisterial *The Dominion of the Dead*[57] eloquently argues. Harrison hails the human compulsion to bury our dead as "*the* generative institution of human nature,"[58] a cultural practice that paradoxically connects humanity whilst irrevocably separating us from each other. As "one of the most primordial of human institutions,"[59] the corpse serves to separate the dead from their remains, but in doing so it also blurs the ontological gulf between the living and the dead. For the living, liberation from the dead can only come in the form of funeral rites which physically and psychically separate us from the dead:

> Funeral rites serve to effect a ritual separation between the living and the dead, to be sure, yet first and foremost they serve to separate the *image* of the deceased from the corpse to which it remains bound up at the

moment of demise. Before the living can detach themselves from them the dead must be detached from their remains so that their images may find place in the afterlife of the imagination.[60]

However, in the absence of a body this process cannot take place, leaving bereaved families with "a catastrophic loss, their grief could not find its proper object; hence the work of mourning by which the dead are made to die was destined to fail."[61] Consider the testimony of Margaret McKinney (75) whose Disappeared son Brian's remains were eventually found, exhumed and re-buried with his kith and kin 21 years after he had been disappeared:

> I can't begin to explain the difference getting his body back made [...] I've a grave to visit where I can plant wee flowers. I feel Brian is with me now. Before that, I had no life. I'd sleep in Brian's bed, with his coat wrapped around me. I changed completely as a person. I lost interest in my husband, I neglected my other children, and I lost my faith. I was so angry with God. I'd go to the chapel and inside me was screaming at the tabernacle, 'How can you be there when you're doing this to me?' I took the picture of the Sacred Heart off the wall and smashed it. I've found peace since I buried Brian.[62]

The burial of the dead performatively shapes the communion of the living; without it, the deceased dwell in an infernal nether world whilst their families, in the words of Margaret McKinney, are banished "to Hell; the darkest, darkest corner of Hell."[63] It is only with the burial of her son that Margaret secures "peace" after two painful decades of purgatorial suffering which lasted as long as her son remained in the realm of the undead: "to become truly dead they must first be made to *disappear*. It is only because their bodies have a place to go that their souls or images or words may attain an afterlife of sorts among the living."[64] It is this plight that makes the Disappeared, in Farrell's words, "stand apart" from all the other victims of the Troubles; the IRA's deliberate withholding of their victims' bodies a crueler action than their actual murder. "Where's the glory, killing the dead twice over?"[65] Sophocles's seer Tiresias declares in *Antigone*. Hegel hailed *Antigone* as the "perfect example"[66] of tragedy because its conflict between the individual and the state, divine and human law, creates a perfect dialectic of concordant ethical positions; neither the right of the family nor that of the *polis* is wrong; but "absoluteness of the claim of each."[67] This celebrated reading of Sophocles's play is central to Hegelian dialectical philosophy about kinship, ethics, nature, and gender, but it is reductive: it ignores the fact, highlighted by Harrison, that "to be human means above all to bury."[68] This is an elemental and essentia*list* truth about human beings. Post-structuralist or postmodern disavowals of such totalizing humanistic claims collapse in the face of this fact; of this force. (Even) Peggy Phelan concedes: "we are, despite our best intentions, stuck with essences and essentialism. And perhaps never more fully than when the body of the beloved has vanished."[69] This is the fundamental human compulsion that drives Antigone to her destruction, and one that is no less compelling today as the phenomenal success of Alice Sebold's *The Lovely Bones* "testifies to our cultural fascination with the dead and to our collective need to recuperate the lost body even in the process of giving voice to loss."[70] It is the denial of this fundamental human right that makes the Disappeared an example of what Derrida defines as the absolute crime. "The absolute crime does not only occur in the form of murder," but of an "unforgiveable" action, one which involves "depriving the victim of this right to speech, of speech itself, of the possibility of all manifestation, of all testimony."[71] In less secular terms, Father Gordon McKendry condemned these crimes as acts "of satanic proportions."[72] Indeed, condemnation was ecumenical and Monsignor Denis Faul's – a tireless campaigner on behalf of the Disappeared – deathbed wish was for Ian Paisley's assistance in appealing for the return of the dead, with which he duly obliged. If nothing else, both men shared the Christian *telos* of the resurrection of the body so that the desecration of the Disappeareds' bodies was absolutely sacrilegious; a violation of divine law and that of human decency. At the funeral of Jean McConville, 31 years after her murder, Monsignor Thomas Toner's funeral oration expressed this spiritual abhorrence in no uncertain terms:

> In the history of our Troubles, there can be no more despicable act than the abduction, murder and casual disposal of the body of Jean McConville and the subsequent plight of her ten children. It is our most shameful example of the moral corruption and degradation that violence generates in the human spirit.[73]

The Catholic celebrant connects the material "corruption and degradation" of Jean's mortal body with a moral and spiritual corruption of the human spirit. Others viewed this corruption, this contamination, in more explicitly political terms:

> Creon, you were too tough;
> the state is dead.
> That girl's brother,
> he'd no right burial
> and now his body
> stinks on our altars,
> stinks in our homes too.[74]

Monsignor Faul associated the degradation and desecration of the Disappeareds' bodies with the political corruption of the *polis*; insisting that the return of their bodies was required for the healthy restoration of the larger body politic: "it is absolutely necessary for the whole of Northern Ireland that these bodies be given back."[75] For Faul, the personal and the political; private suffering and the public sphere were ethically and urgently connected. There could have been no more powerful embodiment (with all that term connotes) of this than the events of 5 April 2004. In Dundalk's County Court on this day, and only a few miles from where her body was discovered, the inquest into Jean McConville's death took place three decades after she had been disappeared. The same day in Dublin, Seamus Heaney's *The Burial at Thebes* – a reworking of *Antigone* – opened at the Abbey Theatre in Dublin as part of the national theatre's centenary celebrations. Though all theatre is haunted, or ghosted by traces of past performances[76] I cannot think of a more compelling choreography of two very different public performances: one theatrical, the other legal; in two civic buildings with a shared historical role in interrogating issues of law and justice. After all, "The law *is* theatre," says Sartre.[77] Moreover, the Abbey's origins are imbricated in the revolutionary nationalism that led to (partial) independence; it was the first ideological apparatus of the infant state-in-waiting where in 1904, Kathleen Ni Houlihan embodied in the form of Maud Gonne[78] exhorted the young men of Ireland to die (and kill) for her. A century later, the consequences of her call to arms were now being examined in a courthouse on the Irish border. The County Coroner's description of the "cold, calculating and totally brutal" murder of Jean McConville and his condemnation of the killers, whose actions "lacked any basis of humanity,"[79] could not have given graver import to Antigone's opening speech that night in the Abbey:

> What's to become of us? ...
> There's a general order issued
> And again it hit us hardest.
> The ones we love, it says,
> Are enemies of the state.
> To be considered traitors –
> [...] Polynieces is denied
> Any burial at all.
>
> Word has come down from Creon
> There's to be no laying to rest,
> No mourning, and the corpse
> Is to be publically dishonoured.
> His body's to be dumped,
> Disposed of like a carcass. [80]

That night, the Abbey Theatre's stage was once more haunted by history as crepuscular shadows of gunmen ghosted Heaney's play with politics, performances, and presences that not even the "perfect exemplar" of tragedy could cathartically purge. That night, Creon's claims to be "acting ever in the interests of all citizens"[81] as justifications for his sacrilege rang as hollow as republican rhetoric about their roles as defenders of a beleaguered Catholic community in the North; whom they murdered in greater numbers than the security forces or loyalist death squads. That night, against the inhuman inviolability of Creon's edicts, Antigone's elemental appeal that human "daylight" laws must not prevail over those that were divine and her defense of "a reverence that was right"[82] echoed the McConvilles' dignified appeals to be allowed "to bury their loved ones with Christian dignity."[83] Three days later the IRA's annual Easter statement buoyantly suggested that "this year's Easter Commemoration Parades across the country will see bigger numbers than ever marching to remember our patriotic dead."[84] There was no mention of Jean McConville or the rest of the remaining Disappeared, as Creon's cold decree that Polyneices be "forbidden any ceremonial whatsoever. No keening, no interment, no observance of any of the rites"[85] still held sway for the families of Columba McVeigh, Brendan Megraw, Kevin McKee, Seamus Wright, Danny McElhone, and Charlie Armstrong.[86]

To this list of the Disappeared could be added another name – Patrick Duffy, who was abducted, shot dead, and "buried in a bog outside Buncrana" over the Donegal border by the Derry brigade of the IRA in 1973. However, this triggered widespread anger and revulsion and demands for his body to be returned, which the local leadership resisted. Indeed, at a public meeting, Duffy's murder was defended in Creon-like terms by that "colossus of the struggle in Ireland,"[87] Barney McFadden, who "lambasted local priests who had condemned the killing as heartless,"[88] for which he was loudly applauded by hundreds of supporters, including IRA commander Martin McGuinness,[89] the present-day Deputy First Minister of the Northern Ireland Assembly. However, they were eventually forced to capitulate and Duffy's body was exhumed and returned to his family for burial. Extraordinarily, the most vociferous opposition to his "disappearance" was voiced by local IRA "internees in Long Kesh," outraged because:

> the IRA had committed a gross human rights abuse by denying the Duffy family's right to *habeas corpus*. The internees were disgusted by this because the Special Powers Act allowed the prison authorities to keep and bury the bodies of any prisoners who died or were killed inside the jails.[90]

Habeus corpus: "thou (shalt) have the body." A law designed to protect citizens from being unlawfully detained; to prevent "the innocent in gaols";[91] the suspension of which allowed the British government to introduce internment and other repressive legislation regularly commemorated by the republican movement. "Thou (shalt) have the body": a moral, legal, political, and ethical edict the IRA violated, thereby replicating the same oppressive measures it accused its enemy of: an irony not lost on some volunteers who considered these practices as inimical with "liberational" struggle: "I changed my mind and came to the conclusion that the secret grave was an abhorrent practice that ran freedom struggle dangerously close to being on a par with those intent on maintaining the status quo.[92]

Republican commentator, Deaglán Ó Donghaile, argues that the IRA had "hoped that, in the absence of a body, total blame will be attached to the victim" and moreover, "that by making a person vanish they [could] erase all memory of the crime."[93] It was this "abhorrent practice"[94] that compounded their crime, making them murderers of the memory of their mortal victims[95] and causing unspeakable grief for their families: "I don't think there will ever be an end to the story until we get the remains, until we can actually have a funeral and be able to grieve."[96] These sentiments were shared by all the other Disappeareds' families: "Maybe this hell is coming to an end [...] but its hard to hope,"[97] admitted Vera McVeigh. Her 17-year old son Columba, whose name means "dove," had been abducted in 1975 on Halloween night when, according to old folk beliefs, evil spirits were abroad. The fact that the Irish festival of *Samhain* "was pre-eminently a commemoration of the dead"; an occasion of "family reunion, [and] also a reunion with the ancestral spirits of the family"[98] must have tormented the McVeighs the rest of their lives, especially after three separate excavations failed to find their beloved son in the soggy peat of Bragan bog, barely 20 miles from his birthplace. The family of Eamon Molloy were luckier.

As the sun rose on 28 May 1999 on the ancient graveyard of Faughart, located high on the hills of County Louth along the Irish border, an early morning mist, which had settled on the cemetery shrouding its stone crosses and statues, slowly began to lift to reveal, before the astonished gaze of a Catholic priest and the Irish Gardai, a brand new coffin beneath the gnarled boughs of the holy laurel tree beside St Brigid's well. Only hours earlier under the cover of darkness an IRA unit had carried the coffin into the graveyard where they deposited their incongruously light burden before disappearing back into the darkness. The casket they left behind was "covered in rosary beads, red holy ribbons and assorted medals"[99] and contained the earthly remains of Eamon Molloy, their former comrade. He had presumably been exhumed hours earlier from some other secret location where he had lain undisturbed for 24 years until this morning, when a new political body – the Independent Commission for the Location of Victims' Remains – came into being and resurrected his own that bright May morning. The members of Eamon's family who later travelled to Faughart found a Gothic scene that one bereaved relative of the Disappeared likened to "something out of a Bram Stoker novel."[100] It was an extraordinarily evocative and apposite description. After all, *Dracula* "shows that the past is never completely done with";[101] the undead state of its eponymous protagonist mirroring

Molloy's, for in the absence of a corpse, the "missing body of the deceased person [the Disappeared...] remain[s], in effect, *undead*."[102] The "lesson" of Stoker's most celebrated novel, according to Declan Kiberd, is "obvious: what is repressed will always return"[103] which returns me to my opening sentence and reinforces Farrell's representation of these innocent landscapes as Ireland's political unconscious. Unspeakable traumas lurk beneath the surfaces of these "luscious" images; inconvenient truths lie buried below beautiful landscapes so that they can be forgotten, absented, excluded; left to Estragon's worms.

Is this why the IRA broke with tradition by leaving their victim in consecrated ground: to assuage their guilt, to reconcile the past, to return the undead to the communion of the dead? But why Faughart? Because of its propinquity to the border or to where Molloy had previously been buried? Situated on the ancient entry point into Ulster, it lies along the hallowed "Gap of the North," which in the *Táin Bo Cuailgne* was heroically defended by Cuchulain who slew several men at Faughart in a feat of arms: "*Focerd*" or "the Good Art" of killing.[104] Perhaps the men who executed and exhumed Molloy to bring him here reflected on this? Cuchulain's heroic actions had helped model modern Irish republicanism and Patrick Pearse's vision of sacrificial martyrdom. These very ideals are embodied in the statue of the dying Cuchulain which stands in Dublin's GPO to commemorate the "blood sacrifice" of the Easter Rising. However, Cuchulain and Pearse's heroic stands against the "ungovernable sea" of superior power seems far removed from actions of armed masked men, a dozen strong, abducting the diminutive 4' 9" Jean McConville from her hysterical children before beating, shooting, and burying her in an unmarked grave; orphaning her ten children and "dismembering"[105] her family in the process. From another perspective, Faughart feels appropriate, as the gallous myth of the "Good Art" of killing clashes with modern dirty deeds and continued blood sacrifices that beget more bog bodies, that some day, will return. And hopefully, to a new political landscape where the cycle of killing has ceased. Until then in Faughart, one imagines that a vampirish Kathleen Ni Houlihan will be perfectly at home in her Gothic surroundings.

Faughart is also an important historical and religious site: where Ireland's last High King, Edward Bruce (the brother of Robert), was killed fighting English armies in 1318. It is also the birthplace of St Brigid whose shrine and holy well still attract pilgrims to attend its ancient penitential stations situated alongside her sacred stream, after which many of her devotees tie votive rags and ribbons to the holy laurel tree where the IRA left their more macabre offering. As local men in all likelihood, they were probably well aware of all this. Perhaps they also knew of the laurel's significance which dates back to pre-Christian times and that it later became a subversive nationalist symbol[106] used to decorate "streets, houses, parade floats and even people," reaching its apogee in the Monster Meetings of Daniel O'Connell's Repeal movement which sought to abolish the 1801 Act of Union between Britain and Ireland. Presumably they did *not* know that laurels represented "amity, peace, and regeneration"[107] unless, that is, they were resurrectionists rather than republicans.

> And then we washed him,
> Took olive branches
> and green laurel leaves
> to crown and lap him in – we burnt him then.
> Next we dug his native earth
> and raised a big mound
> for all to see.[108]

There has been no such closure or ceremony for the rest of the Disappeared, in light of which Farrell's images are poignant, moving examples of what Barthes described as the "evidential power"[109] of photography. However, *Innocent Landscapes* also presents a profound challenge to the myriad projects of remembrance and commemoration that are part of the present process of post-conflict resolution. In recent years, immense effort and energy has been invested into addressing the enormity of what happened in the North over nearly four decades of conflict. This has generated various projects, notably the publication of *Lost Lives*, the Linen Hall Library's *Troubled Images* database, the Arts Council's nascent *Troubles Archive*, and an extraordinary efflorescence of books, exhibitions, documentaries, docudramas, public testimonies, legal enquiries, "cold case" police investigations, civic and community commemorations, which are usually underpinned by extensive archival research and the accumulation of huge amounts of historical data and detail. *Innocent Landscapes* presents a poignant challenge to such materialist approaches to the past as Farrell's evocative, poetic register contrasts with these predominantly empirical, positivist attempts to record, represent, and redress the past.

As mentioned earlier, Farrell deliberately deploys the evidential style of police crime scene photography to highlight these photographs' worthlessness as evidence. In doing so, he exposes the limitations of documentary, forensic, and empirical efforts at representing or redressing the traumatic aftermath of the North's political conflict. Indeed, punctuating *Innocent Landscapes* are reproductions of the ordnance survey maps signaling the approximate locations of the burial sites which were passed onto the state and used by the authorities to conduct the digs. But the empiricism of cartography and the scientific precision of the ordnance survey are redundant for the excavations failed to find most of the bodies. Maps are no match for the vagaries of memory nor the verdure of nature. Farrell has regularly revisited these excavation scenes over the years to record how they have been reclaimed by nature and returned to their original anonymity. On his first visit back, almost a year after the initial digs, he recalls that weeds three or four feet in height had sprung up; part of natural changes that once more make it impossible to orientate oneself in the shifting grounds of memory and landscape. Set in similar border country, Brian Friel's masterpiece *Translations* also excavates historical trauma, as landscape and language, history and memory are again violently unmoored from each other, leaving "a civilization [...] imprisoned in a linguistic contour that no longer matches the landscape [...] of fact."[110] It seems apposite to recall Friel's gentle admonition which is almost premonitory of Farrell's *Innocent Landscapes*. As ordnance maps and the verisimilitude of photography[111] are exposed as positivist technologies that can neither locate nor represent the Disappeared: a failure that reveals the need for new forms of representation, new forms of *witnessing*.

In *Regarding the Pain of Others*, her recent sequel to *On Photography*, Sontag explores the ethics and politics of photography in a world where atrocity is an aestheticized, everyday event. "Being a spectator of calamities taking place in another country is a quintessentially modern experience"[112] she argues, so that images of violence have become depoliticized as we have become inured by their ubiquity; insulated against their power to shock. Sontag's humanist critique of modernity resonates with philosophers Paul Virilio and Jean Baudrillard's more dystopian diagnosis of the postmodern condition. Both men maintain that modern warfare – the ultimate exhibition of violent images – is hyper-mediated like never before; watched by millions of "attentive telespectators"[113] in a world where speed and technology have collapsed time and space. We live in a world supersaturated with images, where an omni-present media, videoscopic technology, satellite telecommunications, digital culture, tele-presence, and real-time streaming have flattened the world into a "squared horizon."[114] The technological advent of real-time streaming is perhaps the most compelling example of this global spatio-temporal collapse, but if we consider the idea of "real time streaming," with its emphasis on speed, simultaneity, instantaneity, and re-contextualize it in relation to Heraclitus's concept of time-as-river, we are left with an extraordinarily potent metaphor for the power of photography: for a photograph freeze-frames time, stills this stream, fixes the Heraclitean historical flux in silent, two-dimensional form.

Farrell's unspectacular and elegiacally understated images memorialize that which was largely unseen and not remembered through media representations of the Troubles, and which was actively repressed by the republican movement. Though a much maligned word, there is an indisputable *morality* to Farrell's images, a morality does more than merely move an empathetic viewer for they have the poetic power to politically challenge spectators who have been depoliticized by exposure to more explicit representations of violence and death.

Innocent Landscapes is much more that a photographic necrology of the Disappeared: it negotiates the "shifting nature of landscape and memory, presence and loss"[115] and excavates ethical and epistemological issues about history, remembrance, forgiveness. It contests our very modalities of knowledge and representation, and in the process reveals the limitations of documentary and materialist forms of remembrance and their capacity to deal with traumatic historical events. Farrell uses the evidential force of photography to emphasize the ethical imperative for it to bear witness, to testify, to memorialize. His images cannot (re)present the Disappeared or (re)produce their bodies, and can never recover or reverse the absolute annihilation of their existence but they remind us all of our ethical responsibility to remember them.[116] If the silence of victims, according to Lyotard, can be attributed to the inequitable power relations of the speakers (*différend*), then *Innocent Landscapes* represents an important ethical intervention by an Irish artist to redress this. Farrell's efforts to commemorate and represent the fates of the Disappeared is inversely proportionate to the IRA's executions, which sought to disappear their victims, disperse all evidence, and destroy their memory. It is also ethically inverse with the way photography is used to violate, even in death, the "excluded" of Barbara Lewis's chapter in which she describes how the bodies, memories, and identities of black lynch victims – the "excluded" – are further "violated and effaced" by trophy photographs and postcards taken by their oppressors. Farrell fulfills the ethical duty outlined by Edith Wyschogrod in *An Ethics of Remembering*, whereby the "professional historian, the journalist, the filmmaker or keeper of archives has the responsibility to coerce the dead other into speech yet to uncover in language the antidote to violence."[117]

Finally, I would like to explore how this moral mode of witnessing, referred to above, serves as a new form of representation and to examine the metonymic function of these photographs which serve to "document and mark loss, instead of simply substituting or sublimating it through representation."[118] Moreover, the publication of *Innocent Landscapes* as a highly acclaimed book allowed Farrell to experiment with the visual syntax of his images and to forge a compelling narrative from his exhibition. All galleries effectively place their exhibits under house arrest due to the precise, particular positioning of each individual artwork, although the ambulatory viewer has a certain amount of agency to negotiate any exhibition. However, the particular visual grammar and precise narrative structure of the published version of *Innocent Landscapes* completely altered this aesthetic contract by forcing the reader to view the images in a precise sequence and which made the ethics of its aesthetic all the more effective.[119] In its book form, *Innocent Landscapes* provides a compelling answer to the question posed by John Berger, "[C]an we think in terms of a truly photographic narrative form?"[120] for its narrative mnemonically recalls and physically re-enacts the process of excavation it portrays. This is achieved by the grammatical arrangement of text and image, as the book follows a precise sequence: a photograph of a landscape, a placename, an image of an annotated map, another landscape photograph, the name of the victim, and a final series of images portraying the excavations carried out at the aforementioned location for the aforementioned victim(s). The whole process is repeated for each victim, in each location. Consequently, all of the images are vertically compressed on top of each other, creating a sense of depth that is incrementally excavated with every leaf turned in the book, so that the process of reading re-enacts the physical act of excavation depicted in each image. This sense of reading as an archaeological act is heightened by the palimpsest-like appearance of the text as the faintly translucent quality of the paper allows multiple images to coexist on the same page. Accordingly, in excavating the text, the reader also seeks to exhume the missing remains of the Disappeared and, in doing so, shares (in a small way) the experience of the excavators and the families, in that they are all ultimately denied closure by the failure to unearth any of the missing bodies.

As physical bodies are metonymic of presence,[121] a profound sense of absence suffuses our entire experience of reading the book. In this way, *Innocent Landscapes* is a remarkable example of what Pierre Macherey psychoanalytically likens to the unconscious of a text. Arguing that all texts are marked by contradictions, silences, absences, that they are haunted by the presence of what they cannot say, he contends that the task of the critical reader should be to locate these lacunae within texts and to make their silences speak: "We should question the work as to what it does not and cannot say, in those silences for which it has been made."[122] This sense of *presence in absence* is a key determining influence in the shape and structure of any text, and Macherey's paradoxical formulation that what a work says is, in fact, what it does *not* say can be appositely applied to *Innocent Landscapes,* in that what these photographs reveal is what they do *not* reveal.

Innocent Landscapes is remarkable for its recognition of the relatives' need for closure and the wider struggle of the North's communities to deal with the legacies of violence, as well as its rejection of a documentary photographic aesthetic in favor of a more poetic, synecdochic style. Farrell's work seeks not to reproduce what was lost, but instead "helps us to restage and restate the effort to remember what is lost," an ethical act of recovery that "rehearses and repeats the disappearance of the subject who longs always to be remembered."[123] Whilst photographs serve to "authenticate" presence for Barthes and to "certify" experience for Sontag, Farrell's visual poetics of witnessing is predicated on the authentication and the certification of *absence.* There is no closure brought about by the vertical, archaeological excavation of the book, a physical, performative, personal experience that reinforces

and mnemonically re-enacts the way in which closure, burial, and the return
of the dead are denied the families of the Disappeared.

* * *

To conclude, it is only fitting to recall Margaret McKinney's testimony that
only with the burial of her son Brian was she able to "find peace" and with
it, forgiveness. In his study of the ethics and politics of forgiveness, Derrida
critiques the (recent) tendency of states to instrumentalize forgiveness (as
demonstrated recently in relation to Ireland with the British Prime Minister Tony Blair's apology for the Irish Famine and more farcically, Danish
apologies for the Vikings[124]). Formal apologies and pleas like these demean
the very idea of forgiveness and preclude meaningful political and historical
understanding. According to Derrida, "forgiveness forgives only the unfor-
givable":[125] it is absolute and unconditional; it makes no demands in its
enunciation and is beyond the gift of states, courts, or judicial bodies: "if
anyone had the right to forgive, it is only the victim and not a tertiary
institution"[126] Forgiveness "is not, it *should not be* normative, normalizing.
It *should* remain exceptional and extraordinary, in the face of the impossible."[127] This honorable Abrahamic moral tradition is embodied in the
attitude of the mother of Brian McKinney to her son's murderers:

> It might sound crazy that I am grateful just for the right to bury my
> younger son after 21 years, but the anger at the IRA is gone now [...]
> I often think about his killer coming to the door. I would really like that
> to happen. I would forgive him completely, because I have my peace now
> and maybe it is time he had his too.[128]

Released from her traumatized state, Margaret eventually "found peace" and
relief from decades of grief. It is tempting in the broader context of the peace
process to draw a poignant parallel between the personal and the political; with the return of the Disappeareds' bodies and the restoration of the
body politic in the form of the Northern Ireland Assembly. Derrida counte-
nances against reducing forgiveness in this way to a facile "political therapy
of reconciliation";[129] and Farrell equally eschews such sentiment. It should
be remembered that in the same "miraculous" week in which the political bodies of the Good Friday Agreement were resurrected and those iconic
images of Ian Paisley and Martin McGuiness laughing together as First and
Deputy Leaders at the 2007 (re)opening of Stormont were beamed round the
world, another mother of the Disappeared, Vera McVeigh, quietly expired.
Born the year the Boundary Commission finally agreed the border of partition, she had spent the past three decades with the vain wish "to bury my
son before I die myself": she even had his name engraved on a headstone
"it's just waiting for him."[130]

The Disappeared are minor figures; marginalized and forgotten; buried figuratively and literally. The most celebrated theorist of absence Paul Ricoeur, in his examination of memory, history, forgetfulness and forgiveness, declares:

> I remain concerned by the troublesome spectacle that too much memory gives here and too much forgetfulness gives there, to say nothing of the influence of the commemoration and the abuses of memory – and of forgetfulness. The idea of a politics of a just memory remains, in this regard, one of my declared civic themes.[131]

It is this idea of the "politics of a just memory" that is captured and crystallized beautifully in *Innocent Landscapes* in its elegiac ethics, its understated morality, and its performative poetics of loss, longing, and hope, as using photography's inherent sense of presence and absence, Farrell transforms spectators into civic witnesses. Out of the these desecrated landscapes, from the framed emptiness of each image, the disembodied, ghostly aura of the dead – the Disappeared – are summoned to remind us to remember *all* the victims of violence.

Notes

1. Susan Sontag, *On Photography* (London: Penguin, 1979), 50.
2. Edith Wyschogrod, *An Ethics of Remembering: History, Heterology, and the Nameless Others* (Chicago: University of Chicago Press, 1998), 242.
3. David Farrell, *Innocent Landscapes* (Stockport, University of Kansas Press: Dewi Lewis, 2001), XX.
4. Seamus Heaney, "The Grauballe Man," *Opened Ground: Poems 1966–1996* (London: Faber & Faber, 1998), 116.
5. Buried in a Danish bog for over 2000 years after being ritually sacrificed, the Tolland Man lay undisturbed until 1950 when two turfcutters discovered his perfectly preserved corpse. His body was later exhumed and examined by the celebrated archaeologist P. V. Glob, who later produced a book *The Bog People* (1969). This classic study of Iron Age man featured dozens of images of mummified bog bodies, which inspired Heaney's poem (*Opened Ground*, 64–5).
6. 'Here is the girl's head like an exhumed gourd [...] Murdered, forgotten, nameless, terrible / Beheaded girl, outstaring axe / And beatification.' Heaney, "Strange Fruit," *Opened Ground*, 119.
7. 'Butter sunk under / More than a hundred years / Was recovered salty and white. / The ground itself is kind, black butter.' Heaney, "Bogland," *Opened Ground*, 41–2. Drawn from three of Heaney's earlier poetry collections (*Door into the Dark* [1969]; *Wintering Out* [1972]; *North* [1975]) these three poems belong to a larger body of "bog poems" which excavate imagery of landscape, history, and memory, often in relation to the political violence of the Troubles.
8. Terry Eagleton, *The Truth about the Irish* (Dublin: New Island, 1999), 32.

310   *Violence Performed*

9. Seamus Heaney, "Feeling into Words," *Preoccupations: Selected Prose 1968–1978* (London: Faber & Faber, 1980), 54. See "Bogland": "Our pioneers keep striking/ Inwards and downwards,/ Every layer they strip/ Seems camped on before." *Opened Ground*, 42.
10. Heaney, "Feeling into Words," 57–8.
11. Heaney, "The Tolland Man," *Opened Ground*, 65.
12. David Wheatley, "Orpheus Risen from the London Underground," *Contemporary Poetry Review*: http://www.cprw.com/Wheatley/heaney.htm (Accessed 12 Sept. 2007.)
13. See Gerry Smyth, *Space and the Irish Cultural Imagination* (Basingstoke: Palgrave, 2001).
14. A Gaelic term referring to toponymic placelore, aetiological poetry, and onomastic legends explaining the origins of places.
15. John Montague, "A Lost Tradition," *Selected Poems* (Oxford: Oxford University Press, 1982), 108.
16. The motif of "blighted land"; of an indigenous landscape violated by its experience of industrialization in the novels of Northern Catholic writers, for instance, is synecdochic for colonial despoliation. See Edna Longley, *The Living Stream: Literature & Revisionism in Ireland* (Newcastle-upon-Tyne: Bloodaxe Books, 1994), 90.
17. John Wilson Foster, *Colonial Consequences: Essays in Irish Literature and Culture* (Dublin: Lilliput Press, 1991), 30.
18. Edward Said, *Culture and Imperialism* (London: Vintage, 1994), 271.
19. S. Daniels, *Fields of Vision: Landscape Imagery and National Identity in England and the United States* (Cambridge: Polity Press, 1993), 5.
20. See Helen Gilbert and Joanne Tompkins, *Post-Colonial Drama: Theory, Practice, Politics* (London: Routledge, 1996), 145; Brian Graham, "Ireland and Irishness: Place, Culture, Identity," in Brian Graham (ed.), *In Search of Ireland: A Cultural Geography* (London and New York : Routledge, 1997).
21. Scott Brewster, *Ireland in Proximity: History, Gender, Space* (London: Routledge, 1999), 125.
22. Samuel Beckett, *Waiting for Godot* (London: Faber & Faber, 1965), 61.
23. Christopher Morash, *Writing the Irish Famine* (Oxford: Oxford University Press, 1995).
24. Even this number is the best estimate and the IRA have only admitted to ten, including Robert Nairac, a British officer abducted from a bar in South Armagh and later executed. To the Commission for the Location of Victims' Remains, the IRA gave information on six sites for eight victims.
25. For example, Jean McConville was a single mother of ten young children, whilst other victims came from venerable republican families whom the IRA leadership wished to avoid embarrassing.
26. Ed Moloney's recent book on the Troubles, *The Secret History of the IRA* (London: Penguin, 2003), explores this in considerable detail.
27. Wole Soyinka, *The Burden of Memory: The Muse of Forgiveness* (Oxford: Oxford University Press, 2000), 27.
28. Jean McConville's remains were later discovered in August 2003 by a man walking his dog at Shelling Beach, County Louth, which was near Templetown.
29. "The first thing I noticed was how the field seemed to be almost violated; rough tyre tracks; groups of rocks piled here and there; a solitary silver birch abandoned on its side, its roots holding on to that circle of earth which had once held

it firmly in place." David Farrell, "Introduction," *Innocent Landscapes* (London: Dewi Lewis, 2001). Over 85,000 square metres of land was excavated by An Garda Síochána in the course of this process

30. See Liam Kelly, *Thinking Long: Contemporary Art in the North of Ireland* (Kinsale: Gandon Editions, 1996); Kevin Rockett, Luke Gibbons and John Hill (eds), *Cinema and Ireland* (London: Routledge, 1988); Joep Leerssen, *Remembrance and Imagination: Patterns in the Historical and Literary Representation of Ireland in the Nineteenth Century* (Cork: Cork University Press, 1996); Luke Gibbons, "Synge, Country and Western: The Myth of the West in Irish and American Culture," in *Transformations in Irish Culture* (Cork: Cork University Press, 1996), 23–35; Bernadette Quinn, "Images of Ireland in Europe," in Ullrich Kockel (ed.), *Culture, Tourism and Development: The Case of Ireland* (Liverpool: Liverpool University Press, 1994), 61–76.

31. Although this genre which dominated Romantic and Victorian periods is attacked for its idealism and escapism, it should be remembered that in medieval times the pastoral was often concerned with "social, moral [and] religious matters." See Helen Cooper, *Pastoral: Mediaeval into Renaissance* (Ipswich: Rowman & Littlefield, 1977).

32. The Irish Tourist Board.

33. The College des Irlandais, founded in 1578 and later part of the University of Paris, was refurbished and reopened in October 2002 as an Irish cultural centre. *Innocent Landscapes* was the first exhibition to be displayed there and was featured in a festival designed to inaugurate the opening of the building.

34. Both terms derive from Barthes' seminal study on photography, *Camera Lucida* (1981; London: Vintage, 2000). *Studium* is a term he gives for a "kind of general interest" in a photograph but "without special acuity" (p. 26). The *punctum*, (prick, sting, wound), is the element that punctuates the *studium* and which "shoots out of [the photograph] like an arrow and pierces me" (p. 26).

35. In an article on the "Disappeared" former IRA Volunteer Anthony McIntyre recants his youthful support for the secretive murder and burial of informers as "legitimate" actions in time of war to advocate that the IRA now acknowledge these deeds as "war crimes." "Disappeared and Disapproved," *The Blanket*, 6 Nov. 2003: http://www.phoblacht.net/disappeareddisapam.html. Accessed 1 October 2007.

36. See Luke Gibbons, *Transformations in Irish Culture* (Cork: Cork University Press, 1996), 117; essays by John Hill and Luke Gibbons in *Cinema and Ireland*, 147–93, 194-257; David Lloyd, "Ruination: Partition and the Expectation of Violence in Allan deSouza's Irish Photography," *Social Identities*, 9:4 (2003): 475–509.

37. *The View*, RTE 1, 16 Oct. 2001: http://www.rte.ie/tv/theview/archive/20011016.html

38. James Buzard, *The Beaten Tracks: European Tourism, Literature and the Ways of Culture, 1800-1918* (Oxford: Clarendon Press, 1993), 188.

39. Peggy Phelan, *Unmarked: The Politics of Performance* (London: Routledge, 1993), 13.

40. Shoshana Felman and Dori Laub, *Testimony: Crisis of Witnessing in Literature, Psychoanalysis and History* (London: Routledge, 1992), 7.

41. See Luke Gibbons for the ways in which such images of antiquity served to bring about a convergence between cultural nationalism and political separatism, *Cinema and Ireland*, 233

42. Sontag, *On Photography*, 5.

43. Indeed, it was this process of erosion that eventually facilitated the discovery of McConville's body much later: at the specific location where she was "eventually found 'the erosion has lowered the beach by 10" to 12" in recent times.'" Police Ombudsman for Northern Ireland, "Report into the Complaint by James and Michael McConville Regarding the Police Investigation into the Abduction and Murder of their Mother Mrs Jean McConville," 8. See: http://www.policeombudsman.org/publicationsuploads/JEAN%20McCONVILLE%20Final%20report%20August%202006.doc

44. I am grateful to Dr Patrick McKay of the Northern Ireland Place-Name Project based at Queen's University for his assistance in translating and decoding these placenames.

45. "Never Lose Heart, 'Disappeared' Families Told," *BBC News Online*, 4 June 1999: http://news.bbc.co.uk/1/hi/uk/358523.stm (Accessed 2 Sept. 2007.)

46. John Berger and John Mohr, *Another Way of Telling* (New York: Vintage, 1995), 279.

47. Barthes, 5.

48. Don Slater, "Photography and the Modern Vision: The Spectacle of 'Natural Magic,'" in Chris Jenks (ed.), *Visual Culture* (London: Routledge, 1993), 219–37.

49. Derrick Price and Liz Wells, *Photography: A Critical Introduction*, ed. Wells (London: Routledge, 1997), 27.

50. Farrell, Introduction, unpaginated.

51. Barthes, 87.

52. Phelan, 35.

53. Barthes, 9 (my italics).

54. Sontag, *On Photography*, 9.

55. Sontag, *On Photography*, 15.

56. Eduardo Cadava, *Words of Light: Theses on the Photography of History* (Princeton, NJ: Princeton University Press, 1997), 11.

57. Robert Pogue Harrison, *The Dominion of the Dead* (Chicago: University of Chicago Press, 2002).

58. Harrison, xi (my italics).

59. Harrison, 92.

60. Harrison, 147–8.

61. Harrison, 144.

62. Susan Breen, "If I'm on Holiday, I'll Stop the Car if I See Someone like Him," *Sunday Tribune*, 9 March 2007. The same sense of closure is also articulated by Jean McConville's daughter after her mother's body was discovered: "it is such a relief that she is finally home, and has received the Christian burial she deserved [. . .] I hope this brings closure for me and my children, because they have also had to live with this nightmare for many years [. . .] Maybe now that my mother has been returned home, I can start to be a better mother to my own kids, who have suffered, because of what the IRA did all those years ago." "Our Mum is Now at Peace," *Sunday Life*, 2 Nov. 2003.

63. "Hope for the Families of the Disappeared," *BBC News Online*, 9 May 1999: http://news.bbc.co.uk/1/hi/uk/338183.stm (Accessed 2 Oct. 2007.) The fate of the McVeigh family was described in the Irish Parliament by TD Austin Currie in the following terms: "On Hallowe'en night 1975 a young man, Columba McVeigh [. . .] left his flat in Dolphins Barn [. . .] took nothing with him, apart from the clothes he was wearing. He never came back; he disappeared. He left behind in Donaghmore, County Tyrone, a father, mother,

two brothers and a sister. Their long purgatory began that Hallowe'en night." "Criminal Justice (Location of Victims' Remains) Bill, 1999: Second Stage," *Parliamentary Debates (Dáil and Seanad)*, 5 May 1999: http://www.gov.ie/debates-99/5may99/sect8.htm (Accessed 5 Oct. 2007.) The priest who conducted the funeral service for Eamon Molloy remarked of the families of the remaining Disappeared: "They're left really in a no man's land, in suspended animation. We're hoping the digs will be resumed and, hopefully, this will be soon to put this tragic episode finally to rest and enable the families to begin to put their lives back together." "Victims' Sons Lose Faith in Search," *BBC News Online*, 15 April 2004: http://news.bbc.co.uk/1/hi/northern_ireland/725358.stm (Accessed 2 Sept. 2007.)

64. Harrison, 1, my italics. Consider Anna Armstrong's description of her purgatorial state of loss which was intensified as the IRA still had not confirmed their role in disappearing her father, Charlie: "But it's worse not having a body. I know my father is dead but I'm still searching. If I'm on holiday or crossing O'Connell Bridge in Dublin, I'll stop the car if I see somebody who looks like him." *The Sunday Tribune*, 19 March 2007.

65. Sophocles, Antigone, *The Three Theban Plays: Antigone, Oedipus the King, Oedipus at Colonus*, trans. Robert Fagles (London: Penguin, 1984), 112.

66. Miriam Leonard, "The Uses of Reception: Derrida and the Historical Imperative," in Charles Martindale and Richard F. Thomas (eds), *Classics and the Uses of Reception* (Oxford: Blackwell Publishing, 2006), 120.

67. Tom Paulin, *Ireland and the English Crisis* (Newcastle-Upon-Tyne: Bloodaxe Books), 28.

68. Harrison, xi.

69. Phelan, 35.

70. Laura E. Tanner, *Lost Bodies: Inhabiting the Borders of Life and Death* (Ithaca, NY: Cornell University Press, 2006), 215–16.

71. Jacques Derrida, *On Cosmopolitanism and Forgiveness*, trans. Mark Dooley and Michael Hughes (London: Routledge, 2001), 62. All references are from Microsoft Reader version of the text.

72. "Funeral of Disappeared IRA Victim Held," *BBC News Online*, 3 Sept. 1999: http://news.bbc.co.uk/1/hi/northern_ireland/436993 (Accessed 3 Sept. 2007.)

73. Dona Carton, "IRA Victim Laid to Rest After 31 Years," *Sunday Mirror*, 2 Nov. 2003.

74. Tom Paulin, *The Riot Act: A Version of Sophocles* Antigone (London: Faber & Faber, 1985), 51.

75. "IRA Urged to Locate Bodies," *BBC News Online*, 28 Aug. 2003: http://news.bbc.co.uk/1/hi/northern_ireland/3189437.stm (Accessed 4 Sept. 2007.)

76. Marvin Carlson, *The Haunted Stage: The Theatre as Memory Machine* (Ann Arbor: University of Michigan Press, 2003).

77. Jean Paul Sartre, cited in Herb Blau, "The Detestable Screen, and More of the Same," *Ciné-tracts: A Journal of Film and Cultural Studies*, 3:4 (1981): 59. "From the earliest Greek tragedies [...] the exposition and resolution of juridical proceedings have always had latent if not blatantly manifest dramatic value." Jody Enders, *Rhetoric and the Origins of Medieval Drama* (Ithaca, NY: Cornell University Press, 1992), 2.

78. The mythical figure of Kathleen Ni Houlihan appears throughout the literature, poetry, and song of the Irish Revival as a political symbol of a suppressed nation. In Yeats and Gregory's play of the same name (1902) she ratifies the republican

belief that blood sacrifice is needed if Ireland is to gain her freedom from British rule. Maud Gonne (1886–1953) was the beautiful firebrand, feminist, and political revolutionary for whom Yeats's wrote the title role. Gonne's performance in what proved to be the authors' most nationalist play had an incandescent impact on Dublin audiences, famously prompting one spectator to speculate, 'I went home asking myself if such plays should be produced unless one was prepared to go out to shoot and be shot.' Stephen Gwynn, cited, A. N. Jeffares, A Commentary on the Collected Poems of W. B. Yeats (London: Macmillan, 1968), 512.

79. "Family Seeks IRA Admission," *BBC News Online*, 5 April 2004.
80. Seamus Heaney, *The Burial at Thebes* (London: Faber & Faber, 2004), 2.
81. Heaney, *Burial*, 10.
82. Heaney, *Burial*, 41.
83. "Search for IRA Secret Graveyard Under Way: Families Plead for Information": http://scathan.journalexchange.net.1pt/archive/1995/07_july/07.1995.ira.html (Acessed 7 Oct. 2007.)
84. "Support your Easter Parade," *An Phoblacht*, 8 April 2004: http://www.anphoblacht.com/news/detail/4163 (Accessed 27 Oct. 2007.)
85. Heaney, *Burial*, 11.
86. Moreover, Republicans had actively campaigned for almost half a century to have the remains of executed IRA man, Tom Williams, who had been buried in the grounds of Crumlin Road jail in Belfast in 1942. Their campaign successfully culminated in 1995 with the exhumation of his body and its burial with full republican honours in Milltown Cemetery on the Falls Road. These events were described by Danny Morrison, Sinn Féin's former Director of Publicity as, "represent[ing] a kind of closure, just as the release of all the political prisoners will allow us to draw a line through the past and move forward. Move permanently away from repression, discrimination and armed struggle, and towards justice and equality." The Disappeared, yet again, conspicuous by their absence. Danny Morrison, "Williams was Hanged for Trying to Establish Right to Commemorate One's Dead," *Sunday Tribune*:http://www.nuzhound.com/articles/morr1-30.htm (Accessed 15 Aug. 2007.)
87. This is Martin McGuinness's description of McFadden, whom McGuinness has regularly and warmly paid tribute over the years. "Barney McFadden: Father of Derry Republicanism in Derry Dies," *An Phoblacht*, 10 Jan. 2002.
88. Liam Clarke, "IRA Demands Disclosure, Yet it Offers Too Little, Far Too Late," *The Sunday Times*, 26 June 2005. For further details, see David McKittrick, Seamus Kelters, Brian Feeney, Chris Thornton and David McVea (eds), *Lost Lives: The Stories of The Men, Women and Children who Died as a Result of the Northern Ireland Troubles* (Edinburgh: Mainstream, 2001), 383.
89. Martin McGuinness made it a matter of public record that he was an IRA commander in Derry in this period during the recent Bloody Sunday hearings.
90. Deaglán Ó Donghaile, "Derry's Disappeared," *Derry News*,11 Sept. 2003.
91. Seamus Heaney, *The Cure at Troy: A Version of Sophocles's* Philoctetes (London: Faber & Faber, 1990), 77.
92. Anthony McIntyre, "Disappeared and Disapproved," *The Blanket*, 6 Nov. 2003: http://www.phoblacht.net/disappeareddisapam.html (Accessed 2 Sept. 2007.)
93. Ó Donghaile, 11 Sept. 2003.
94. Anthony McIntyre, "Disappeared and Disapproved."

95. See "Against the Murderers of Memory: Pierre Vidal-Naquet," *Questioning Judaism: Interviews by Elisabeth Weber*, trans. Rachel Bowlby (Stanford, CA: Stanford University Press July 2004), 26–38.

96. "New Victims Search Welcomed," *BBC News Online*, 1 May 2000: http://news.bbc.co.uk/1/hi/northern_ireland/730737.stm (Accessed 8 Sept. 2007.)

97. *Sunday Tribune*, 9 March 2007.

98. E. Estyn Evans, *Irish Folk Ways* (London: Routledge & Kegan Paul, 1957), 277.

99. "Unshrouding the Truth" *The Guardian*, 29 May 1999.

100. *The Guardian*, 29 May 1999.

101. Declan Kiberd, *Irish Classics* (London: Granta Books, 2000), 397. Joseph Valente suggests an even more direct connection between Stoker's novel and the spectre of nationalist violence, of "Dracula as a complexly apposite metaphor for the patriotic summons delivered by Cathleen Ni Houlihan: each figure reproduces itself by dealing death." *Dracula's Crypt: Bram Stoker, Irishness, and the Question of Blood* (Urbana: University of Illinois 2002), 58.

102. Harrison explores this in relation to the plight of Argentina's Disappeared, *Dominion*, 144.

103. Kiberd, 396.

104. Toby Harnden, "Molloy's Body Left in Ancient Graveyard," *The Telegraph*, 29 May 1999.

105. "Search for the Disappeared Restarts," *BBC News Online*, 2 May 2000: http://news.bbc.co.uk/1/hi/northern_ireland/732948.stm (Accessed 8 Oct. 2007.)

106. This political encoding of the laurel was enhanced by the government's long standing prohibition of more explicit nationalist symbols in the forms of flags, emblems, insignia, etc.

107. Gary Owens, "Nationalism Without Words: Symbolism and Ritual Behaviour in the Repeal 'Monster Meetings' of 1843–5," in J. S. Donnelly and Kirby Miller (eds), *Irish Popular Culture 1650–1850* (Dublin: Irish Academic Press, 1998), 252.

108. Paulin, *Riot Act*, 58.

109. Barthes, 106.

110. Brian Friel, *Translations*, in *Plays One*, ed. Seamus Deane (London: Faber & Faber, 1996), 419.

111. Aerial photographs and x-ray equipment were also used in these excavations, but to no avail.

112. Susan Sontag, *Regarding the Pain of Others* (New York: Farrar, 2003), 18.

113. Paul Virilio, *Desert Screen: War at the Speed of Light* (London: Continuum, 2002), 1. See also Jean Baudrillard, *The Gulf War Did Not Happen*, trans. Paul Patton (Indiana: Indiana University Press, 1995).

114. Virilio, 20.

115. David Farrell, "Innocent Landscapes": http://www.galleryofphotography.ie/exhibitions/davidfarrell/sections/exhibition.html (Accessed 2 Sept. 2007.)

116. Indeed, the importance of this is eponymously emphasized in the 1998 report compiled by the Northern Ireland Victims Commissioner, Sir Kenneth Bloomfield, *We Will Remember Them*; see: http://www.nio.gov.uk/bloomfield_report.pdf

117. Wyschogrod, 243.

118. See Vivian M. Patraka, "Spectacular Suffering: Performing Presence, Absence and Witness at the Holocaust Memorial Museum," in Erin Striff (ed.), *Performance Studies* (Basingstoke: Palgrave Macmillan, 2003), 82–96.

119. In this context, it is interesting to note the symbolism of these digs for the Disappeared which were (initially) heralded as a landmark signifying the possible end

of 30 years of the 'Troubles'. Moreover, these events were often described with pronounced textual metaphors belying the strong sense that history was being re-written and redressed. Consider the following *Irish Times* editorial: 'Many [...] will see in this yielding up, a declaration that a chapter has been closed; that this is a balancing of the books or a tidying up of unfinished business, that the dreadful deeds which are represented in these unmarked graves are a thing of the past and will be no more." Quoted in Conor Cruise O 'Brien, *Memoir: My Life and Times* (London: Profile, 1999), 451–2. It seems all the more appropriate then, that Farrell's photographs are also published in book form which helps commemorate this particular historiographical moment.

120. Berger, 279.
121. See Peggy Phelan, *Unmarked*, 148.
122. Pierre Macherey, *A Theory of Literary Production*, trans. Geoffrey Wall (London: Routledge & Kegan Paul, 1978), 115.
123. Phelan, 147.
124. "Danes Say Sorry for Viking Raids on Ireland," *The Guardian*, 16 Aug. 2007.
125. Derrida, 39.
126. Derrida, 49.
127. Derrida, 38.
128. John Mullin, "Grateful to Bury and Grieve," *The Guardian*, 1 May 2000. Michael McConville, who was only 11 years old when his mother Jean was taken by the IRA, also said he forgave her killers though he would "never forget the hurt and anger we have suffered and I will never forget them taking my mother out that night, the distress that was on my mum's face," "Son's Appeal to Find Mother's Body," *BBC News Online*, 8 Dec. 2002: http://news.bbc.co.uk/1/hi/northern_ireland/2555855.stm (Accessed 5 Sept. 2007.)
129. Derrida, 50.
130. *Sunday Tribune*, 9 March 2007.
131. Paul Ricoeur, *Memory, Hitory, Forgetting*, trans. Kathleen Blamey and David Pellauer (Chicago: University of Chicago Press, 2004), xv.

14
Directing Tourists and Escapees: North Korea's Two Conflicting National Performances

Suk-Young Kim

1 Tourism and human rights as related performances

Since the North Korean state opened the Geumgang Mountain tourist zone to outside tourists in 1998, more than 860,000 South Koreans and other foreign nationals have toured the famous North Korean scenic site.[1] Even though there were occasional obstacles which brought the tourist operation to a halt, both Koreas hailed the tourist project as a successful effort to increase mutual understanding via human exchange.[2]

However, while South Korean visitors have already begun setting their feet back onto what was once forbidden North Korean territory, North Korea's human rights violations are still a long way from center stage in official inter-Korean political dialogues. Despite grass roots-level protests against North Korea's prison camps and persecution of political dissidents, the South Korean state has not confronted the North Korean leadership about these politically volatile issues in order to maintain a smooth, inter-Korean relationship.[3]

North Korea, on the other hand, made it its mission to improve its belligerent national image by presenting itself as a respectable nation to the outside world. This meant that North Korea carefully developed criteria for selecting presentable images to be shown to outside spectators and had its people rehearse according to those criteria. For North Korea, fabricating a desirable national image is a process contingent on Foucauldian bodily discipline requiring a "meticulous control of the operations of the body, which assured the constant subjection of its forces."[4] Truthful to Foucault's statement that "the body was in the grip of very strict powers, which imposed on it constraints, prohibitions or obligations," the North Korean state deemed bodies of its people as ancillary props to the ideal self-presentation while it systematically eliminated unnecessary bodies from its territory.[5] At the same time, the state also ensured that foreigners – tourists, journalists, and aid workers – did not interfere with its plans to stage a highly choreographed

317

self-idolization by prohibiting outsiders from accessing unrehearsed daily lives of the North Korean people. This process of staging an ideal national image according to strict state control necessitated a high degree of regulation intended to differentiate what should be visible and invisible. I argue that North Korea's systematic process of disciplining certain bodies to be invisible while promoting others as the self-reflexive representation of an ideal nationhood is supported by state-sponsored violence, which forcefully determines the mobility of both domestic and foreign bodies within North Korea. For a country which chronically suffers from famine, the state-controlled distribution of necessary resources for survival, most notably food, became a tangible form of state violence to reinforce and perpetuate its power and discipline its subjects.

Disciplining North Korean people and keeping outsiders at bay are mutually dependent processes, but no matter how rigorous the state control may be, the carefully planned tourism at times resulted in unexpected interactions between North Koreans and foreigners. Despite the North Korean state's efforts to reduce the chances of accidental encounter between its people and outsiders, at times the two engaged in spontaneous interactions which the North Korean state could not control.

In the master plan of the North Korean tourist industry, its natural and cultural landscape becomes a stage on which the North Korean state directs the movement of its people and foreign tourists. In this totalizing scheme, tourists are intended to be silent and compliant spectators. The North Korean state perceives foreign aid workers and tourism service staff, operating on the shifting boundary between tightly controlled North Korea and the world beyond, to be undesired but necessary intermediaries. From the North Korean state's point of view, foreign aid workers are undesired parties because they insist on stepping out of the roles of controllable spectators, but they are granted some degree of autonomy so they can distribute much needed resources. North Korean employees providing services to foreign tourists are also in a peculiar situation. They are constantly exposed to visitors, and thereby become unwanted but necessary agents who must mediate between North Korea and the outside world.

But how does North Korea handle the significant contradictions of having a permeable boundary, which allows only a selected few to enter its territory? And how should we understand the various performative modes through which North Korea addresses foreign tourists, aid workers, service providers, and escapees? I attempt to answer these questions by exploring specific performance strategies that demonstrate the North Korean state's desire and motivation to both show and also hide certain elements of its national landscape.

The two issues that I have chosen to explore – tourism and the human rights question – may at first appear to be diametrically opposed realms of North Korean performance, since they explicitly disclose the contrapuntal principles of staging and hiding. At first glance, euphoric tourist sights and the brutal hunting of escapees may appear to be unrelated. But once the performance mechanism of producing what is visible and invisible becomes the focal point of analysis, tourism and human rights issues form a coherent dialogue about North Korean propaganda as a process of making a national performance. Performance, as I see it in this case, derives its energy from the paradoxical stance of the state, which intends to render visibility only to limited areas while making the rest of the nation invisible to the outside world. However, to accept the dichotomized production of visible and invisible as the strategy of the North Korean state alone is to reduce the complex nature of the two aforementioned performances, as such a stance overlooks other agencies involved in matters of tourism and human rights. North Korean escapees, foreign tourists, aid workers, and human rights activists, all bring in competing performance strategies in an effort to create their version of the "real" representation of North Korea.

In exploring how visibility shapes the notion of the real, I appropriate Peggy Phelan's methodological stance which centers on "the way in which the visible itself is woven into each of these discourses as an unmarked conspirator in the maintenance of each discursive real."[6] Phelan's ideas are particularly pertinent to the analysis of North Korean national performance in that she points to the notion of "visible" as the misleading indicator of empowerment: while acknowledging how visibility exerts power over our perception of the real, Phelan points out that "the binary between the power of visibility and the impotency of invisibility is falsifying."[7] Invisibility is not the usual marker of disempowerment if we take the following fact into account: human rights activists often have to become invisible in the eyes of the North Korean state in order to gain access to the realities of North Korean human rights abuses. But once they complete their mission to document the abuses, they will turn the documentation into a cantankerous anti-North Korean performance on the international stage. Human rights violations, for the North Korean state, is not a legitimate undertaking in its canon of national performance, but for activists, who temporarily assume the role of invisible actors in accessing the hidden realities of North Korea, invisibility becomes an empowering performance strategy. If visibility summons up the state's surveillance, invisibility provides ways to subvert them. In this sense, invisibility is just as powerful a performance mode as visibility in regards to documenting North Korean human rights violations.

On the other hand, state-sponsored tourism exhibits North Korea's wishful self-portrait; it visualizes everything the North Korean state deems as its ideal self-image. Similar to Dean MacCannell's notion of tourist attractions as "analogous to the religious symbolism of primitive people," tourist attractions designated by the North Korean state are unmistakable signs of the ideological hierarchy of that society.[8]

This chapter neither claims to investigate the entire modus operandi of the North Korean tourist industry, nor does it appropriate an activist stance, accusing North Korea of violating human rights.[9] Rather, I attempt to determine the overlapping fields of two evidently disjunctive realms of performance – tourism and human rights issues – by looking at North Korea's highly choreographed principles of showing and hiding. Moreover, the contrasting directing principles of showing and hiding, as an image-producing strategy becomes more important than what actually happens to the actors involved in the process. Despite the impression of a solipsistic nation that North Korea may produce to outsiders, the state in fact works faithfully according to the principles of Anthony Giddens' notion of modern "reflexivity" and prioritizes the production of an idealized self-image over the actual experience of people beyond that theatrical facade.[10] This chapter attempts to fill that precise void stemming from the absence of the voices of the actors – South Korean tourists, foreign aid workers, human rights activists, and North Korean escapees – involved in various capacities in creating a national performance out of tourism and human rights issues. In order to meet that objective, I interviewed North Korean escapees, former trainers and managers of service workers, and, finally, participated in a tour of the Geumgang Mountain resort.

2   Inventing silent tourists

After the 1994 death of the North Korean leader Kim Il-sung, North Korea underwent a severe economic crisis, which necessitated that the state compromise by at least partially joining the world market economy.[11] Consequently, gaps and fissures began to appear in the tightly controlled state, leading to a gradual transformation of North Korean society. While hunger was sweeping North Korea, the country opened up limited special tourist zones to visitors. The emergence of a free market economy has changed the way North Korea approaches its resources for tourism. The nation used to take great pride in assigning mythological significance to scenic sights as inspiring, and inspired by, the national leaders' revolutionary deeds.[12] This practice is still alive at the present moment, but, with the deprivation of the financial luxury to prioritize ideological purity over economic needs, North Korea opened up part of its territory to foreign visitors in order to bring new financial resources into its cash-strapped economy.

Many advertisements evidence the fact that North Korea has been making consistent efforts to attract tourists from various parts of the globe: the North Korean National Tourism Administration in 2003 invited the Bangkok-based Pacific Asia Travel Association (PATA) to send a task force to advise them on tourism development, which resulted in a PATA report; in April 2004, after two years of planning, a German travel agency started to sell a tourist package for railway travel in North Korea; beginning in July 2004, North Korean travel agencies started advertising on the internet; at the same time package tours to North Korea via South Korea became available in Canada for the first time.[13]

However, putting economic motivations aside, North Korea seems to have another compelling motivation to encourage tourism: that of constructing a radically different notion of North Korea for foreign spectators. By partially opening up what it regards as the most desirable locations in the county, North Korea is able to reinvent its national image so often associated with military threats and destitution of apocalyptic proportions. The lure of attracting a global flow of finance, together with positive media coverage, certainly propels the clandestine country's desire to stage Potemkin-village-like images for tourists.

In analyzing the motivations for fabricating tourist cities, Dennis Judd and Susan Fainstein wrote:

> The designers of tourist space understandably avoid the troubling aspects of life [...] the main spatial effect of urban tourism is to produce spaces that are prettified, that do not feature people involved in manual labor, that exclude visible evidence of poverty, and that give people opportunities for entertainment and officially sanctioned fun.[14]

Even though North Korean tourist zones seem to reflect the spirit of the above-mentioned fun and entertainment, visitors to North Korea search for a different paradigm of what Judd and Fainstein describe as an attractive tourist space. To many potential visitors, the selling points of this clandestine country are not the conventional commodities of relaxation and sightseeing that many world famous tourist resorts have to offer, but rather the pleasure of satisfying their curiosity about being in a forbidden territory. To some tourists, the attraction of visiting North Korea is to have an experience of glimpsing times gone by. "The Stalinist theme park" – as a BBC reporter aptly referred to Pyongyang – offers a unique experience of traveling backwards in time, especially for those who have lived under socialism elsewhere in the world.[15] Thus, visitors expect to find relics of time, the frozen historical memory ubiquitously present on the city's streets, such as the gargantuan monuments of the North Korean rulers. However, this is an experience that has only been shared by a limited number of South Korean tourists, who still have few opportunities to visit urban areas.[16]

Most South Korean tourists who visit North Korea are limited to seeing vastly rural areas far away from the center of daily activities and real life, which raises the question of how nature is consumed by South Korean visitors. For example, hiking trail maps prepared by the Hyundai-Asan Cooperation, the South Korean business investor in the Geumgang Mountain resort, show a colorful illustration of the Mountain area – an aestheticized version of already innocuous nature – as if to minimize South Korean tourists' perception of North Korea as enemy territory. Urban landscape, with political slogans hanging practically on every building, may provoke negative feelings in South Korean visitors, whereas nature provides the visitors with the harmless pleasure of being in a quixotic land without being exposed to blunt political propaganda.

However, such an assumption immediately encounters challenges once nature stops being natural and becomes political. Tourism is an ideological reconfiguration of nature, thus there is hardly any room for natural elements to remain innocent in the ideological remapping of tourist space. The emergence of nature as a political agency invites sets of related questions: What is the framework that is being used to politicize nature? How does the political framing of nature impact the performances of various participants such as tour guides, tourists, and North Korean inhabitants? I point out two related modes in which the North Korean state frames tourist space, which is to provoke nostalgic feelings in tourists while suppressing multidirectional discussions on political issues in that space. Both are concerned with disciplining South Korean tourists' bodies so that they may become tame actors the state can easily direct.

An attempt to find a happy balance between ideological purity and capitalist gain seems to have led the North Korea state to deploy two plans – making a commodity out of nature and provoking nostalgic feelings for the lost motherland. Significantly, both items do not invite bitter political contention between the two Koreas. Nostalgia, in particular, has an intuitive appeal for South Koreans whose families originally come from North Korea. For example, the South Korean tourist agency selling tour packages to visit Pyongyang identified nostalgia as the central selling point for potential tourists; posters clearly state that the Pyongyang tour was "a rare chance to realize the wishes of separated family members."

In fact, nostalgia is not only the sentimental mode in which to conduct tourist business, it is also a best-selling product in itself for South Korean and Russian tourists. Russian tourists, in particular, find it amusing to encounter reminiscences of the Soviet Union in contemporary North Korea.[17] Leonid Petrov, a Russian tourist writing to his compatriots in an online travelogue, described his visit to Pyongyang as a nostalgia-evoking experience ("on a train with the Kims' portraits you may find yourself like an extra in a Stalinist film"), even though nostalgic feelings are often accompanied by ironic relief that the good old Soviet days are part of the past.

Many South Koreans who go on the Geumgang Mountain tour are senior citizens whose families were originally from North Korea.[18] At one of the summits of Geumgang Mountain, from which one can get a good glimpse of nearby towns on a clear day, I overheard conversations where the tourists expressed: "somewhere there should be our ancestral burial ground." Other elderly tourists told me: "we are originally from this township; you can see the edge of our neighborhood from here." Catering to the nostalgic sentiment, a speaker incessantly blasts music with lyrics such as "welcome brothers," or "we will meet again" when tourists arrive at the checkpoint to enter or depart from North Korea. The lachrymose songs invent and groom forced intimacy for those, like myself, who have no relationship to North Korea.

Likewise, the South Korean tour guides, who are assigned to the group of 20 to 25 tourists to accompany them throughout the tour, constantly emphasize the connection between North and South Korea. As the bus I was on entered the North Korean territory, they urged tourists not to look for differences, but to find similarities between North and South. As the tourist bus passed by dilapidated farmhouses along the unpaved roads, the tour guide urged curious spectators not to point fingers at the scenery and people outside: "Do not look for how backwards North Korea is, but just think about how life was 30 years ago in the South and you will find unmistakable links between the South and the North." While these guides were making good efforts to eliminate any prejudices against North Korea, it was not the condescending attitude towards economically backwards North Korea that marked the South Korean tour guides' direction. Quite the opposite, their direction reflected the North Korean state's desire to regulate the way South Koreans view their country. As directors and dramaturges often do to actors, the guides urged tourists to process visual images in a prescribed way and thereby made the viewing experience a directed one.

This process of editing tourists' vision is coterminous with producing docile bodies out of them; both discipline South Korean visitors to be exposed to sets of limited activities sanctioned by the North Korean state. As Foucault noted, "discipline increases the forces of the body (in economic terms of utility) and diminishes these same forces (in political terms of obedience)."[19] The North Korean state sees the economic utility of the South Korean tourists primarily in their power to consume, while disciplining them to become politically obedient. Such a dual stance toward South Korean tourists can be best seen in the shops catering to tourist shoppers. The only souvenir shop in the tour zone is a replica of what could be any typical shopping center in any urban area around the world: bright lighting fixtures, impeccably made up clerks, pleasant music to stimulate consumerist desire, and a conveniently located foreign currency exchange booth (Figure 14.1).

Moving through this seductive space almost erases the ideological boundary South Koreans have been made aware of for the past 50 years. The consumerist desires provoked by the shopping center in Geumgang Mountain create a sense of misplacement for tourists because the firm boundary between the North and South Koreas becomes malleable and penetrable – to the degree that the two Koreas are more divided by the "relic" boundary than by a functional boundary.[20] Tourists become the liveliest actors in this situation as they are encouraged to ask about products, compare prices, and even document their activities by taking photos – an activity which is forbidden in other spaces in the tour zone.

Figure 14.1   Shopping center in Geumgang Mountain tour zone
Source: Photograph by Suk-Young Kim.

This formation of theatrical space, based on the acculturation of the consumer-driven economy intended by the North Korean state, leaves little room for ideological contention between North Korean employees and South Korean tourists; the only books on display are photo collections of scenic sites, cook books, and collections of children's fairy tales. No ideologically charged products are on display for consumption. In the same spirit, none of the North Korean guides spoke of the benevolent North Korean leaders Kim Il-sung and Kim Jong-Il in an obvious attempt not to incite the anti-North Korean sentiment that is so deeply ingrained in many South Koreans. The tour guides also spoke only of fairy tales and legends related to each site. South Korean tourists are greeted by North Korean tour guides' eloquent storytelling when they reach a summit named Cheonsundae ('the altar of the heavenly fairies' in Korean). The love story between a human being and a fairy is a familiar tale for all Koreans, providing common ground for understanding between otherwise culturally alienated people.

The coordinators of the Geumgang Mountain tour, such as Hyundae-Asan Corporation, South Korean tour guides, and North Korean clerks, make it clear that the conventional verbal propaganda between South and North accusing each other, which has been the dominant mode of communication between these countries since the end of the Korean War, is non-existent in this pleasant fairyland. In lieu of a verbal medium, visual images spoke flamboyantly, sometimes even louder than actual words. The most evident examples of this are Kim Il-sung and Kim Jong-il's instructions

Figure 14.2 Geumgang Mountain features plenty of monumental rocks on which Kim Il-sung and Kim Jong-il's instructions are engraved
Source: Photograph by Suk-Young Kim.

engraved on monumental rocks (Figure 14.2). The sight of nature deployed as unnatural political propaganda generates a question: what happens when natural scenery speaks for the North Korean leaders in bloody red letters, silently, but loudly?

The engraved red letters on the mountain cliffs and stone monuments are ideological scars of the Cold War era that immediately catch the attention of tourists who are not accustomed to seeing communist propaganda. South Koreans have been continuously educated to identify such visual images as alien objects, while North Koreans have been systematically trained from a very early age to see these monuments as sacrosanct objects. For South Koreans, these silent yet politicized monuments evoke strange feelings and memories of war and confrontation, but the tour guides advise them not to make any politically sensitive comments on these monuments displaying the holy scriptures of the Kim rulers.

For South Korean tourists, such advice carries firm authority because of an unhappy precedent involving a politically careless comment. In June 1999, a South Korean housewife told a North Korean tour guide that North Korean defectors live a good life in South Korea, for which she was detained for a week in North Korea before being released back to the South. After this incident, no South Korean tourist has ventured the inconvenience of being detained for exercising freedom of speech. Thus, silence becomes the best prescription for tourists visiting North Korea.

Despite its tranquil surface, silence, when coupled with invisibility, can be the most abrasive mode of directing human bodies. While the North Korean state makes seamless efforts to display content but silent tourists in limited tourist zones, it also makes brutal but futile efforts to hide epidemic hunger by hunting down economic migrants, political escapees, and regulating the mobility of foreign aid workers. Silence is the link that brings together two seemingly opposite types of social actors – tourists and tourist guides, foreign aid workers and escapees – in making North Korean national performance. While tourists are treated with a hospitality that forces them to be silent and compliant spectators, escapees are treated with a hostility that literally mutilates their bodies to become immobile.

In coordinating hospitality towards tourists and hostility towards its own people, violence becomes the state's major mode of directing the bodies of the various parties involved. As the North Korean state silently eliminates undesired bodies when coping with food crises, it simultaneously spins the wheel of lucrative tourism while it forces silence upon tourists, an invisible way of exercising a minor degree of violence.

3   Erasure of undesired bodies: escapees and foreign aid workers

The North Korean state and visitors to North Korea have very different expectations regarding what they want to show and see, which certainly creates conflicts of interest. In order to resolve such conflicts, the state uses its unlimited power to erase undesirable elements from its national landscape, the process of which constantly requires the violation of the North Korean people's human rights. The degree of erasure varies tremendously, ranging from preventing foreign visitors from taking snapshots of unrehearsed local scenes to expelling disfigured people from the capital city, Pyongyang. The former was the experience of Jill Dougherty, the former Managing Editor of CNN International Asia Pacific, when she visited Pyongyang in August 2005:

> One morning, as we were driven out of Pyongyang to a mountain resort ensuring we would be kept far from any interaction with ordinary people,

we asked our guides to let us stop by the road and shoot some pictures of the countryside. Grudgingly, they agreed. Suddenly, the young one, 29-year-old Mr. Jang, sporting a sleek black pompadour and a smirk, told us to stop. "There is an old woman down there," he explained. Presumably, her bent back was not what he wanted on tape.[21]

Although the Western world's prevalent notion of North Korea as an isolationist country might have contributed to Dougherty's impression, defectors I interviewed indeed confirmed more drastic measures taken by the North Korean state to expel invalid people from Pyongyang. Two former residents of the North Korean capital city – one a former military officer and the other a high-ranking official at the foreign trade bureau – told me that the central party systematically transferred invalid people to other job sites outside of Pyongyang.[22]

While the North Korean state makes relentless efforts to veil undesired bodies to make them invisible to outsiders, many impoverished North Korean people voluntarily choose to escape from North Korea before the state takes any measures towards them. Economic hardship, combined with political oppression, triggered a massive exodus of North Koreans leaving the country in the 1990s. As soon as North Koreans, formerly invisible to outside visitors, reach neighboring countries, they become visible objects of the international media and foreign aid workers, creating bad publicity for North Korea. This is why the state mercilessly hunts down the escapees and severely punishes them once they are brought back to North Korea.

The food crisis in North Korea began to surface in the mid-1990s when the centralized food rationing system came to a halt and people were left on their own to make ends meet.[23] Up until this point, the North Korean people relied on the state rationing system as their sole guaranteed source of food. But the lack of consumable food made the state prioritize the military over commoners in its food distribution, thereby leaving many people with no recourse but to face certain starvation. As the crisis was prolonged into the late 1990s, gruesome reports came out of North Korea via escapees about how North Korean people coped with hunger: their testimonies covered a wide spectrum of horrifying stories, ranging from robbery to cannibalism.[24] Children and elderly people died first; then young people disappeared. Some managed to escape to search for food while others perished on the way.

The North Korean state only recently resumed its food-rationing system, and the impact of the food crisis during the nine to ten years when the system came to a halt is hardly fathomable.[25] There have been only speculations about exactly how many North Koreans disappeared during the 1990s famine. As the human rights activist Norbert Vollertsen put it, this was not a natural disaster but a man-made one that could be referred to as "genocide."[26] It was literally a form of state-engineered violence against its own people, the majority of whom depended on the government's food supply to survive. Instead of openly seeking the international community's help, the North Korean state handled the disaster by keeping the situation under cover and making it invisible to the outside world. On the other hand, foreign aid workers wanted to claim some degree of control in aiding the troubled country,[27] only to face an impenetrable barrier shutting them off from the actual lives of the people.[28] For the most part, foreign aid workers were immediately deprived of their mobility once they brought humanitarian aid to the country,[29] and when the North Korean state took control of the distribution process, they did so in such a way that only escalated the crisis.[30]

As the North Korean economy has recently improved, so the state has reclaimed full control over the food distribution system.[31] The state's first step was to refuse foreign donation and expel aid workers. Radio Free Asia reported on 24 September 2005:

> Officials working for non-governmental organizations (NGO) operating in North Korea are bewildered by the nation's official request for them to leave by the end of this year. Senior officials at NGOs, such as Ireland's Concern, France's Triangle Generation and Germany's German Agro Action, are frustrated over North Korea's decision to end their on-going aid activities. Ann Omahony, head of Concern's Asian affairs, said North Korean authorities have asked the NGO to hand over its work to them or to leave the country by Dec. 31. But she says that "it is impossible to do so because it defeats the purpose of the NGO's existence."

North Korea's strategy of keeping NGOs at bay seems to have triggered the intended result. In line with North Korea's request, the United Nation's World Food Program reported that it would halt its decade-old emergency food shipment program by January 2006 and focus on development strategy.

Likewise, on 2 October 2005, *The Guardian* reported in an article entitled "North Korea Turns Away Western Aid":

> North Korea has begun to reverse market reforms by kicking out international relief workers and choking off supplies of food and medical aid in a crackdown that puts millions of the country's children and elderly at risk. In what one resident described as the biggest change in the humanitarian situation in 10 years, the government in Pyongyang is attempting to regain control over the distribution of essential supplies that have increasingly been provided by the market and outside donors. As of yesterday, stall-holders have been ordered to stop trading in cereals, including rice. From now on they will only be sold at controlled prices through the state's public distribution system.[32]

As the report predicted, North Korea resumed the centralized rationing system soon after.

From a performance studies point of view, the hunger crisis in North Korea may be seen as a troubled rehearsal process during which North Korean state and foreign aid workers strive to claim control over their actors – the North Korean people. While the North Korean state intends to retain power over its people by monopolizing food distribution, many foreign aid workers also see their project as a chance to disseminate positive images of the outside world to North Koreans and bring small, but gradual incremental changes in North Korea. Even though there is a perceived difference between the intentions of the North Korean state and foreign aid workers, both share a common desire to use food to mobilize people and train them to act in certain ways. For the most part, the North Korean people have no choice but to become compliant actors, displaying their bodies as silent props on the stage of a tranquil national landscape. For example, when foreign NGOs arrive in rural areas to distribute food, local party representatives require people to wear their best clothes to create presentable images (Figure 14.3).[33]

In this case, clothing choice for North Koreans is not left to individuals' discretion, but rather to the decisions of the state, just as stage actors have to wear costumes specifically tailored for the show. The staged nature of the North Korean people's act, nonetheless, is unstable: ironically the visible actors in a carefully planned performance often accentuate what is invisible,

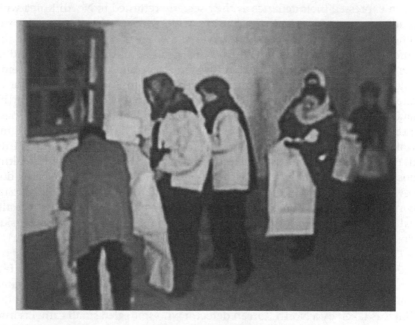

Figure 14.3 North Koreans line up for food rations
Source: *White Paper on Korean Unification* (Seoul: Ministry of Unification, 2005).

that is, the non-compliant bodies who resist institutionalized power – be it the North Korean state or foreign aid workers. As a result, they make what is visible highly susceptible.

For example, some North Koreans assume partial agency by refusing to accept passive roles, and functioning as independent actors when they refuse to comply with directorships by their state or from foreign aid workers. Some Christian NGOs saw missionary work as their primary objective in rescuing North Korean refugees in China. Even though many North Korean escapees gladly accepted help from these missionary workers, most of them did not have exposure to religion in North Korea and found it difficult to accept new religion. As a result, many resist conversion – a resistance against foreign aid workers' desire to direct them.[34] North Korean people fluctuated in between the foreign aid workers and their state in a variety of roles – as compliant actors propelled by the fear of fatal punishment, as economic immigrants in search of chances for survival, and finally, as rebellious escapees motivated by the desire for freedom to denounce the state and express their violated rights.

Despite the North Korean state's astringent strategies to display and conceal its national subjects, the state's failure to provide its people with the minimum calories to survive created a massive exodus of starving North Koreans. These economic migrants often bore a second identity as political dissidents, as they searched for food as well as political freedom. Some of them expressed their defiance as they secretly returned to North Korea with hidden cameras to provide the outside world with rare footage of what the official North Korean media will never show. These returnees became the chroniclers/documenters of the gruesome realities of the food crisis and of the violations of human rights that the state desperately wants to hide. Being able to play various roles is, for some North Koreans – from docile actor to director/producer of a piece with a controversial point of view – one of the fatal flaws in the North Korean state's master plan for directing people. These escapees create a subversive performance out of North Korea's human rights violation by transforming the invisible (starved and persecuted bodies) into the visible (the main focus of their documentary). The danger these stealthy documenters voluntarily expose themselves to is also grave, since being discovered while reporting the "unpresentable" will result in torture and capital punishment. These home-coming escapees are uninvited guests to the ritual that is both public (for North Koreans) and private (for the outside world).

The documentary film *Seoul Train* features the first known attempt to show, close-up, the reality of a provincial North Korean market place. It features shocking images of starved children begging for food and emaciated adults with bare feet scraping for any food residue. The footage used in the film was taken by a North Korean defector who smuggled small cameras into North Korea.[35] As noted by Jim Butterworth, the producer of this film, "the actual footage [...] inside North Korea came from a defector who risked his life and went back in, and if he had been caught he would have undoubtedly been summarily executed. The underground railroad was mostly shot by the activists."[36]

Similarly, the public execution scene, which took place in the Sino-Korean border town of Hoeryeong in March 2005, was taped secretively by an escapee who went back to North Korea with a small portable camera hidden in his pocket.[37] This image was posted on the web for world-wide circulation, providing an undeniable piece of evidence, testifying to the North Korean state's public display of capital punishment. The footage captures a blindfolded man with his mouth fully stuffed with what appear to be pebbles in order to prevent him from pronouncing any criticism, which would work against the executioners' planned scenario. In these public executions, only carefully chosen local North Koreans are witnesses. In this respect, public executions are communal rituals in which inhabitants are forcefully drafted to be spectators as well as ritual participants. Once the executioners carry out the capital punishment, spectators are forced to act out certain gestures denouncing the executed, such as finger-pointing at the corpse or throwing rocks as they abuse the dead verbally. Capital punishment becomes an invitation-based performance in which forced subscribers are also requested to participate in making the ritual complete.

As Dwight Conquergood noted: "executions are awesome rituals of human sacrifice through which the state dramatizes its absolute power and monopoly on violence."[38] It is true that the public executions in North Korea are rigidly controlled state rituals which are visible only to insiders. But the fact that dissidents turn these invisible rituals into an international spectacle to be posted on various internet sites, including CNN, signals a fissure in the coherency of the North Korean state's performance. Dissident performance captures and brings back the executed bodies, which are about to vanish into oblivion, to the world's attention – an accident completely unintended by the state's directorial intentions.

These stealthy documenters, however, have to become invisible during the process of creating a documentary in North Korea in order to gain access to the otherwise unapproachable materials. Even though these documenters assume temporary invisibility and thereby may appear to be abiding by the rules of the state, their invisibility is empowering precisely because of its subversive potential to culminate in dissident performance. Dissident performance is made possible by temporarily complying with the rules the documenters intend to resist – a paradox which shatters the notion that the visible is powerful.

Moreover, public executions illustrate the fact that there is hardly any distinction between the concentration camps and the rest of North Korea. According to the testimonies of refugees who survived North Korean concentration camps, public executions are an integral part of taming the prisoners. In numerous Korean *kwan-li-so* (political penal-labor colonies) and *kyo-wha-so* (prison-labor camps),[39] public executions occur as a part of routine everyday life for the edification of the inmates, who are expected to swallow their defiance and integrity at the moment of execution as a visible marker of their loyalty to the state that imprisoned them.[40]

When compared to the haunting images of the public executions of captured escapees who ran to China in search of food, the choreographed images of well-dressed food recipients shown in Figure 14.3 create a grotesque protrusion into the coherent landscape of official North Korean propaganda. The gap between authentic and fabricated images primarily accentuates the North Korean state's strategy to visualize only limited objects. But more evidently, it points to the state's persistent investment and belief in staging an ideal self-image even in the most extreme of situations, such as the massive death of its people from starvation.

What makes the food crisis a dialogic performance directly related to tourism is that some human rights activists see tourism as an easily available opportunity to enter North Korea in order to raise human rights issues, adding yet another paradigm to the already multifaceted relationship between the two performances. Just as some North Korean escapees with hidden cameras are silent-actors-turned-noisy-producers, some foreign aid workers subvert the North Korean state's intention to use tourism as an ideal showcase. By posing as tourists to gain access to the sites of human rights violations, these foreign workers eventually disclose the North Korean state's atrocious treatment of its people.

For instance, the producers of the documentary film *Seoul Train*, Jim Butterworth and Lisa Sleeth, posed as tourists in the Sino-Korean border region in order to interview and shoot footage of North Korean escapees on the run. By actively seeking images of what the North Korean state wants to hide, these fake tourists step out of the passive roles of spectators who are constantly relegated to viewing choreographed images of North Koreans. Thus, they blur the boundary between actors, spectators, and directors, challenging the North Korean state's exhaustive efforts to monopolize directorial intentions.

As a more recent example, Norbert Vollertsen has attempted to enter North Korea as a tourist with the aim of staging an anti-North Korean protest.[41] Even though the attempt was bound to be aborted because of his cantankerous revelation of the plan in advance, the provocative idea and the frenzied response from the media created enough performative effect for him to view it as successful engagement. The South Korean government declared that it would prohibit Vollertsen from re-entering South Korea in case he adheres to such sensational strategies to incite the South Korean public. Norbert Vollertsen's strategies are based on agitation more than anything else, making a spectacle of resisting silence and invisibility:

> Try to get the attention of the world – nobody knows – so nobody can care. When there is no normal way to get this attention try the unusual: hunger sit-in in front of the Chinese Embassy (Olympics!); set up refugee camps at the Chinese-North Korean border; take provocative acts at the border, jump over the fence, get arrested.[42]

Vollertsen's blunt agitation makes one wonder: Does inviting attention have a value of its own, or does it merely demonstrate an empty propaganda strategy with no tangible reference to the actual issue? Human rights activists and theater scholars will have different answers. From a performance studies point of view, such an action of creating the extreme version of what a conscious person should do, as demonstrated by Vollertsen, creates a simulacrum of angry tourists who have the freedom to resist the silent roles they are assigned to play within North Korea. However, this is impossible in reality, because staging an actual demonstration in North Korea would certainly result in the tourist's detention, or in lucky cases, expulsion. Thus, Vollertsen is virtually rehearsing a show meant to culminate in the actual protest. However, in this case, the rehearsal process is much more significant than the actual, though paradoxically non-existent, performance within North Korea.

Such a predominance of rehearsal over culminating performance strikingly parallels the North Korean state's emphasis on creating theatrical performances rather than actual outcomes. In the coordinated and disciplined rehearsals for mass games and parades, North Koreans learn to subordinate their individual voices to the commands of leadership. Thus, rehearsals are not mere preparations for culminating performances, but rather, are crucial educational processes defining the fulcrum of North Korean society's seeming absence of individual opinions.

The concern for appearance always stands as a superior issue to people's well-being in North Korea. Related to this consistent hierarchy of values is the notion of theatricality – the excess of illusion so as to eclipse reality – which looms large in disguising socialist dystopia. According to Anne Applebaum, in 1929 when Stalin's Politburo first began to discuss the expansion of the concentration camp system that eventually became known as the gulag, there were almost no objections. In the end, only one problem was taken seriously: that the system might look bad abroad.[43]

Unfortunate historical parallels are presented by the North Korean state's obsession with outside spectators, that even though outsiders are not an integral part of either society, they shape the presentation of the society. This chapter selected two varying kinds of performances concerning North Korea, namely tourism and human rights issues, in an attempt to illustrate how various parties involved in both performances create a complex matrix of illusion and reality, visible and invisible, discipline and rebellion.

Notes

1. *White Paper on Korean Unification* (Seoul: Ministry of Unification, 2005), 109.
2. North Korea's bureaucratic measures often interfere with the tourism business. For example, in the summer of 2005, the North Korean state abruptly postponed sales of the much-awaited Gaeseong tourist package for reasons related to the resignation of personnel from Hyundai-Asan Corporation, the South Korean investor and partner in the Geumgang Mountain project. Even though North Korean officials seem to abide by the rules of the market economy, some of the fundamental decisions they make seem to be driven by ideological motivations, which constantly interfere with conducting business in a competitive way.
3. The most controversial move the South Korean state has made regarding the North Korean human rights violations was to abstain from voting for the UN resolution to condemn North Korea's violation of human rights in 2005.
4. Michel Foucault, *Discipline and Punish: The Birth of the Prison*, trans. Alan Sheridan (New York: Vintage, 1991), 137.
5. Foucault, 136.
6. Peggy Phelan, *Unmarked: The Politics of Performance* (London: Routledge, 1996), 3.
7. Phelan, 6.
8. Dean MacCannell, *The Tourist: A New Theory of the Leisure Class* (New York: Schocken, 1976), 2.
9. Ever since George W. Bush made the famous "Axis of Evil" speech in 2001, North Korea has become a hotbed of arguments and debates waged by academics and policy-makers alike. Many have projected their given agenda on North Korea without taking a very close look at the country, either to uphold the moral lessons of how and why communism does not work, or to critique the appallingly dogmatic attitude of the United States and its wartime allies. As a scholar whose primary goal is to bring out new knowledge about North Korea to the world, I find it increasingly difficult to claim intellectual objectivity. As I see it, North Korea is a complex country like anywhere else in the world: on the one hand, it has demonstrated both remarkable achievements in modernizing the postcolonial state; on the other hand, it has committed unthinkable atrocities and violence. The purpose of my research is to bring out these complexities so that we become more sophisticated in our approach, moving beyond extreme positions of idolization or accusation, which have so dominated discourse on North Korea.
10. Reflexivity, according to Giddens, is the quality of modern culture that results from peoples' ability to know what is happening elsewhere in the world and to change their behavior because of expectations of particular consequences. See *The Consequences of Modernity* (Stanford, CA: Stanford University Press, 1990).
11. Even though economic hardships in the 1990s forced North Korea to seek opportunities to cooperate with the outside world, the country's efforts to join the world market economy pre-date the 1990s. Dallen Timothy gives a comprehensive account on how North Korea invited foreign partners to revitalize its domestic economy. See *Tourism and Political Boundaries* (London: Routledge, 2001), 122.

12. There are far too many examples of revolutionary scenery to describe, and it is not an exaggeration to say that revolutionary landscaping has created and determined the spatial hierarchy in North Korea. The holiest site of all is Mansudae, the mausoleum of the late North Korean leader Kim Il-sung in Pyongyang, where tourists visiting the capital city are required to pay their homage to the deceased founding father of North Korea.
13. Report by Korea Trade-Investment Promotion Agency (KOTRA), Stockholm branch.
14. Dennis Judd and Susan Fainstein, "City as Places to Play," in Judd and Fainstein (eds), *The Tourist City* (New Haven: Yale University Press, 1999), 269.
15. Kate McGeown, "On Holiday in North Korea," *BBC Online*, 17 Sept. 2003, 22 June 2007: http://www.news.bbc.co.uk/1/hi/world/asia-pacific
16. The tourist spaces that North Korea opens up for South Koreans are different from those operated for non-South Koreans in terms of location and rationale. Tourists who visit the Geumgang Mountain area are mostly South Koreans, because South Koreans are not granted access to other urban tourist destinations such as Pyongyang or the North Korean side of Baekdu Mountain, located at the far northern side of the Sino-Korean border. Thus, for South Koreans, North Korean urban space is a desired destination because of its inaccessibility. When it was announced that the city of Gaesung would be opened to South Korean tourists, the phone lines at the Hyundae-Asan office jammed due to the overwhelming number of calls from potential South Korean tourists inquiring about the package. Similarly, the pilot tour package to Pyongyang, which was specially designed to enable South Koreans to attend the North Korean "Arirang" mass game (the quintessential North Korean performance with approximately 100,000 performers coordinating their actions to stage synchronized card sections or seminal events in North Korean history) also attracted more applicants than the tour package could accommodate. The pilot tour of Pyongyang admitted ten delegations, each consisting of 150 tourists who stayed in a four star hotel in Pyongyang for one night during the fortnight between 4 and 15 October 2005. Tourists of other nationalities, however, are granted limited access to Pyongyang and other tourist destinations. Their travelogues are valuable accounts for learning about conditions in North Korean cities, even though these non-South Korean travelers are constantly placed under surveillance.
17. The Vladivostok branch of the KOTRA (21 Sept. 2004) report cites some Russian tourists from Khavarovsk who did not necessarily want to visit the late Kim Il-sung's mausoleum to pay homage to the deceased North Korean leader.
18. South Korean students who tour North Korea on fieldtrips organized and sponsored by their schools are exceptions to this trend. These students do not have much choice in selecting the destination of their fieldtrip, and thus are distinguished in their motivation from senior citizens who visit North Korea voluntarily.
19. Foucault, 138.
20. According to Timothy, "relic boundaries" refer to those that no longer function as borders but are still visible in cultural landscape (e.g., parts of the Berlin Wall and the Great Wall of China), 5.
21. Jill Dougherty, "North Korea: A Prism to Soviet Era," *CNN.Com*, 14 Sept. 2005, 22 June 2007: http://www.cnn/com/1005/world/asiapc
22. I conducted interviews with them in August and September 2005 in Seoul and Los Angeles respectively.

23. According to Andrei Lankov, the North Korean food distribution system came to a halt in 1994–95 in the countryside, and in 1996 in urban areas. "North Korea Hungry for Control," *Asia Times*, 10 Sept. 2005, 22 June 2007: http: //www.atimes.com/atimes/Korea
24. For detailed testimonials on cannibalism in the late 1990s in North Korea, see Joeun Butdeul (ed.), *Saramdapge Salgosipso (We Wish to Live like Human Beings)* (Seoul: Jeongto, 1999), 65–7, 132.
25. According to the Associated Press (2 Oct. 2005), North Korea announced its plans to resume full-scale food rationing across the impoverished communist country after ending grain sales. On 1 October 2005, the World Food Program reported that their cereal sales in the markets would cease and that public distribution centers would take over country-wide distribution.
26. Norbert Vollertsen is a German doctor who spent time in North Korea providing medical services from July 1999 to December 2000. He was expelled from North Korea for expressing his opinion against the state's violation of human rights. He currently works as a human rights activist and is widely known for his provocative strategies for demonstrating against the North Korean state.
27. For example, Tim Peters, the director of Helping Hands Korea, a Seoul-based NGO, operates a bread factory in the Sino-Korean border region to alleviate the food crisis in North Korea. He uses a cell phone to call his North Korean contacts to ensure that the bread is appropriately distributed among starving people (Tim Peters, personal interview, 20 Aug. 2005).
28. The World Food Program, which has fed an average of 6.5 million North Koreans in recent years, withdrew its emergency aid from North Korea at the end of 2005, causing concern that monitoring access would be lost. Hankookilbo (9 Oct. 2005).
29. Norbert Vollertsen, in an interview with Chosunjournal.com (30 June 2001), declared that when he joined the NGO Cap Anamur – German Emergency Doctors, "I was not allowed to see behind the curtain of silence in this country. Like all the other foreigners, we were never allowed to look into any prison camp or the like. We were fooled like idiots in regards to the cruel reality of this place [. . .] Before Cap Anamur came to North Korea, other agencies such as Oxfam and CARE pulled out because they weren't allowed to distribute aid directly to the people. They had to turn it over to the authorities, who took complete charge of distribution. Monitoring is impossible. Nobody really knows where the aid is going, except that it is not going to the starving citizens."
30. Yong Kim, a North Korean defector, told me that he had a friend who worked as a warehouse manager in the city of Hoeryong. According to Kim, the warehouse managed by his friend, which he had visited, had piles of aid materials from foreign donors – medical kits, food, clothing – not for distribution to the people who needed them, but mostly for army provision (interview with Yong Kim, Santa Barbara, 13 Sept. 2005). As another example, a retired official in Pyongyang heard of a donation from the South Korean state including 50,000 pairs of shoes and 5000 items of clothing, but he has never seen them distributed (Butdeul, 121).
31. Andrei Lankov states that the North Korean state over the years has made efforts to achieve domestic economic stability by relying on free foreign donations: "the aid-maximizing strategy allowed them to extract some resources from outside donors through diplomatic efforts. Such aid made possible the survival of an economic structure that otherwise would be unviable. For a long time, the role of the overseas donor belonged to the USSR and China, then it was China and South Korea (and some foreign aid agencies), and nowadays it seems that this

role has been enthusiastically assumed by Seoul. The overseas aid is probably not sufficient to kick-start economic development, but it is sufficient to keep the economy afloat, prevent a major famine and also allow for a reasonably luxurious life for the country's few elite – the few dozen families around Kim Jong-il." "North Korea Hungry for Control," *Asia Times*, 10 Sept. 2005, 8 June 2007: http://www.atimes.com/atimes.Korea/GI10Dg01/html

32. Jonathan Watts, "North Korea Turns Away Western Aid," *The Guardian*, 2 Oct. 2005, 8 June 2007: http://www. observer.guardian.co.uk/international/story

33. Yong Kim (personal interview, 13 Sept. 2005).

34. Yong Kim, for example, recollects that soon after he crossed the Sino-Korean border, he was fortunate enough to have met South Korean missionaries who helped him escape to South Korea. But soon after his rescue, the missionaries brought him the Bible and other reading materials on Christianity, which he initially resisted (interview with Yong Kim, Santa Barbara, 13 Sept. 2005.

35. Jim Butterworth, the producer of the film *Seoul Train*, revealed this fact in his interview with the UCLA Asia Institute on 15 January 2005.

36. Butterworth interview.

37. For a detailed background of how the footage became public, see: http://www. atimes.com/atimes/Korea/GD13Dg01.html   http://www. northkoreanrefugees. com/dvd/statement.htm

38. Dwight Conquergood, "Lethal Theatre: Performance, Punishment and the Death Penalty," *Theatre Journal*, 54 (2002): 342.

39. I follow David Hawk's description of these detention facilities. See *The Hidden Guilag: Exposing North Korea's Prison Camps* (Washington, DC: US Committee for Human Rights in North Korea, 2001), 56.

40. For a detailed account of how public execution is used in kyo-hwa-so, see Soon-ok Lee, *Jeung-eon (Testimonial)* (Chicago: Christian Journal, 2003), 153–6.

41. For a more detailed account of Dr Norbert Vollertsen's stance on the North Korean human rights issue, see James Brook, interview with Vollertsen, "One German and his North Korean Conscience," *New York Times*, 19 March 2002, A4; Hong Seok-jun, "NK Human Rights Like Nazi Germany," *Chosun Ilbo*, 8 May 2001, 8 June 2007: http://www.chosenjournal.com/pierrerigolout Vollertson "A Prison Country," *Wall Street Journal Opinion*, 17 April 2001, 8 June 2007: http://www.opinionjournal.com/extra Donald Macintyre "Diary of a Mad Place," *Time Magazine*, 22 Jan. 2001, 8 June 2007: http://www.time.com/time/asia/magazine/2001

42. Norbert Vollertsen, interview, *Chosunjournal.com* 30 June 2001, 8 June 2007: http://chosunjournal.com/2001/06/30/interview-with-norbert-vollertsen

43. Anne Applebaum, Preface, in David Hawk, *The Hidden Gulag: Exposing North Korea's Prison Camps* (Washington, DC: US Committee for Human Rights in North Korea, 2001), 8.

15
Abu Ghraib and the Society of the Spectacle of the Scaffold

Jon McKenzie

Two anachronisms of the performance stratum

The scandal known as "Abu Ghraib" instantiates two political anachronisms that may very well come to define what I call the performance stratum, the global formation of power/knowledge which is currently displacing the disciplinary stratum analyzed by Foucault.[1] One anachronism, *the re-emergence of sovereign power*, involves the expediency of discursive performatives, whether spoken or written. The second anachronism, *the re-emergence of the spectacle of the scaffold*, involves the efficacy of spectacle, theatricality, and, more generally, embodied performance. I will very briefly outline the first anachronism with reference to an infamous series of "torture memos," and then turn to the main object of this chapter, an analysis of the theatre of torture enacted at Abu Ghraib prison. This theatre can be understood, precisely, as a contemporary spectacle of the scaffold.

The first, more discursive, anachronism involves the reappearance of sovereign power, by which I mean the power of sovereign utterances to decide life or death – in a word – to execute. In contrast to discipline's universal, invisible, and continuous operation, Foucault described sovereign power as organized around a single sovereign body and as exercised through highly individualized, visible, and intermittent techniques. Its anachronistic return today, however, reroutes such sovereignty away from kings and their ministers and instead plugs it directly into the executive management of high performance organizations. In contrast to the plodding deliberations of rationalized bureaucracies, new "petty sovereigns" emerge fully authorized and directed to make decisions based on their expediency – their speed, their fitness to situation, their practicality, even their "actionability." Thus, in a series of brief memoranda written between 25 January and 7 February 2002, top Bush administration officials summarily declared that Taliban and al Qaeda combatants were not covered by the Geneva Convention and, in the following August, offered an unprecedented legal definition of torture, one so narrow that, short of death or loss of bodily organ, almost any level of pain and suffering could be justifiably inflicted in the name of the "Global War on Terror."² Such official memoranda articulated a (quasi-)legal framework for the policies and procedures used in the detention, interrogation, and torture of prisoners at US facilities in Bagram, Afghanistan; Guantá-namo, Cuba; and eventually Abu Ghraib, Iraq (to name only the most well-known sites).

These "torture memos," however, must themselves be placed within a wider political framework, which legal scholars and administration officials call the "unitary executive," a controversial constitutional theory that vastly broadens executive power and reduces the ability of other governmental branches to provide checks and balances. "Unitary executive" means just that: executive power is unified, not divisible or limitable by anything outside of it. Beyond official memoranda, other means of exercising unitary executive power include public and secret executive orders, classified studies and findings, interpretative signing statements attached to legislation, and even seemingly mundane agency regulations. The power asserted by the theory of unitary executive can be understood in terms of Agamben's "state of exception," in which sovereign decisions suspend the rule of law, even and especially to the point where the exception effectively becomes the rule. Butler explicitly theorizes the sovereign power of the Bush administration in terms of discursive performatives: "The future becomes a lawless future, not anarchical, but given over to the discretionary decisions of a set of designated sovereigns [...] who are beholden to nothing and to no one except the performative power of their own decisions."³ Further, she describes this sovereign power as anachronistic and, significantly, as extending far down into bureaucracies, giving rise to low-level sovereigns. "These are petty sovereigns, unknowing, to a degree, about what work they do, but performing their acts unilaterally and with enormous consequence."⁴ I would argue that one must also consider petty, *non-state* decision-makers, such as the executives at Titan and CACI, two private military contractors currently being sued by former Abu Ghraib inmates. Thus I prefer to use the term "executive performativity" rather than "sovereign performativity." In some sense, executive performativity is an extension of what I have analyzed elsewhere as "high performance management," but one which suspends long-standing traditions of deliberative, highly rationalized bureaucratic decision-making and violates or suspends laws, contracts, and professional codes of practice, whether governmental or nongovernmental. Executive performativity represents a devolution of sovereign power at the same time as its resurrection in contemporary organizations.

My focus in this chapter, again, is a second, closely related anachronism, a more visceral enactment of power that Abu Ghraib incarnates but in no way exhausts. Alongside the gruesome scenes of Daniel Pearl's beheading and Saddam Hussein's hanging, alongside the globally broadcast bombings of Baghdad, Bali, Beirut, London, Manhattan, Tel Aviv, and many other cities; alongside the innumerable websites devoted to images of war carnage, executions, and other forms of political violence – the scenes from Abu Ghraib reanimate that extremely violent and graphic form of political theatricality which Foucault assigned to pre-disciplinary society and which he named "the spectacle of the scaffold."[5] Indeed, panopticism, as the visual regime of the enlightened human sciences, is precisely what displaced the visual regime of spectacle associated with sovereign power. The prison cell displaced the scaffold in the public square as the paradigmatic site where power and bodies meet. Today, however, scaffolds may be in cells, city streets, lonely landscapes, almost anywhere: it's the spectacle that has changed, shifting from highly localized spaces of temporal co-presence to globally mediated spaces and times – or rather, all localized spaces of temporal co-presence become potential nodes of transmission and reception in a world-wide electronic network of cameras, screens, databases, processors, and editing boards. In short, just as sovereign power has devolved and trickled down to mid- and even low-level officials, the spectacle of the scaffold has become networked and screenal: satellites, televisions, security cameras, facial and gestural recognition software, cell phones, Blackberrys, iPods, YouTube, Google Maps, The Memory Hole – all become means for capturing or being captivated by spectacles. Combining Foucault and Debord, two unlikely allies, we might say that on the performance stratum, this second anachronism portends a *global society of the spectacle of the scaffold* – or global *societies* of the spectacle of the scaffold, as one can foresee such spectacles being deployed not only by neoliberals, neoconservatives, and fundamentalists, but even by those opposed to these social groupings.

One might object that torture and violence – and images of torture and violence – have long existed. In fact, the Abu Ghraib images have themselves been described in terms of late-nineteenth and early twentieth-century American lynching photos, Francisco de Goya's "The Disasters of War" prints (1810–20), and even ancient Greek sculpture.[6] Yet while one can find many historical precedents and analogues to Abu Ghraib (e.g., the racism, amateurism, and indexicality of lynching photos; the context of war and social upheaval surrounding Goya's prints; and the posed, almost sculptural arrangement of figures), what distinguishes the contemporary spectacle of the scaffold is its vast socio-technical infrastructure. The technical infrastructure of television and the internet is both global and often real-time, on the one hand, and yet also local and even intimate, on the other. The images of the Hooded Man haunt the halls of Abu Ghraib but also the homes of people around the world. The same cell phones and laptops that carry words and pictures of loved ones also transmit anguished cries for help and images of bombing and carnage. And not just one or two images, or even ten or twenty, but hundreds upon hundreds and even thousands upon thousands of images. In the end, this global technical infrastructure decidedly does *not*

mark a clean, absolute break from any and all historical precedents and analogues; more profoundly, via multimedia databases and hypertext markup
language (html), these historical precedents and analogues are themselves
being incorporated into the society of the spectacle of the scaffold. Indeed,
this incorporation contributes to its anachronicity.

Abu Ghraib and the total theatre of torture

In addition to the general, if not universal, infrastructure of television and
internet that supports it, the acts and images of Abu Ghraib are also underwritten by a much more specific socio-technical infrastructure. Beyond the
Bush administration's neoconservative cohort that produced the torture
memos discussed above, another, longer standing social paradigm was at
work, a paradigm both coldly theoretical and ruthlessly "applied." I refer
here to the paradigm of psychological torture developed by the CIA during
the Cold War and used by the United States and/or its proxies in Vietnam,
the Philippines, Argentina, and other countries. Here we have torture techniques developed and deployed internationally *in the name of democracy* –
protecting American democracy, establishing and/or supporting democracy
elsewhere, and "securing" democracy world-wide.

Significantly, in his recent book on the CIA's program of psychological
torture, historian Alfred McCoy presents its overall implementation as *a type
of theatre*:

> the psychological component of torture becomes a kind of total theatre,
> a constructed unreality of lies and inversion, in a plot that ends [...]
> with the victim's self-betrayal and destruction. To make their artifice of
> false charges, fabricated news, and mock executions convincing, inter-
> rogators often become inspired thespians. The torture chamber itself thus
> has the theatricality of a set with special lighting, sound effects, props,
> and backdrop, all designed with a perverse stagecraft to evoke an aura
> of fear.[7]

McCoy uses theatricality not as a metaphor, but as a robust analytical model.
He is not the first to employ such a model to analyze torture and violence.
Diana Taylor argues that the performance of torture in Argentina inscribed a
nationalist narrative on the bodies of those expelled from the nation. More
recently, sociologist Mark Juergensmeyer has analyzed religious terrorism in
terms of "theatre of terror" and "performance violence." At work in these
analyses is a *certain* theatricality, and I too will use this concept to theorize
phenomena far from theatre proper. Significantly, the military itself calls
its area of operations a "theatre," a usage that dates from the seventeenth
century.[8] Today, with the Global War on Terror, all the world's a bloody
stage. Within this global theatre there is the Iraqi theatre and within it, the

theatrical performativity of Abu Ghraib. Indeed, the military's own Criminal Investigation Division repeatedly uses the term "staged event" to refer to incidents at Abu Ghraib where soldiers posed or arranged inmates for viewing and/or photographing.

It is thus no accident that commentators have described the infamous Abu Ghraib photographs as "theatrical": the poses of Lynndie England; the pyramid of naked inmates; the hooded figure holding electrical wires – these iconic images present tableaux of power and degradation. The performances seem scripted, directed, and enacted for an audience – and indeed they were. By whom and for whom? One only needs to look at the casting: specific bodies – Arab, Muslim, and mostly but not entirely, male – were abused by Americans, both male and female, mostly white Euro-American but also African-, Latino-, and Arab-American, who acted under the orders of American commanders and officials. And we must also ask: by what means and to what ends were these bodies tortured? I propose to use theatricality as a way to analyze the techniques of America's torture machine.

Commentators have proposed many interpretative frames for understanding the events and images of Abu Ghraib: trophy photos shot by frat boys, tourist shots snapped by ugly Americans; porno pics taken by sex-crazed guards. More thoughtfully, Sontag framed them, in part, in terms of lynching photos, while Žižek argues that they document the initiation of Iraqis into the underside of American culture.[9] Art historian Stephen F. Eisenman argues the photos reveal a "pathos formula," often found in classical Western art, in which tortured individuals appear to sanction their own abuse.[10] While multiple frames are no doubt at work, I think the most important one is much more literal: it is the regimen of detention and interrogation instituted at Abu Ghraib at the recommendation of Major General Geoffrey Miller. This is precisely the regimen that McCoy calls "total theatre."

General Miller, then the commander at Guantánamo, inspected Abu Ghraib in September 2003. Following a summer marked by a sharp spike in insurgent attacks, the military sought better intelligence to counter the insurgency. Miller arrived at Abu Ghraib saying that "he was going to 'Gitmoize' the detention operation."[11] After his inspection, the general filed a detailed report containing specific recommendations. These included the assignment of a Behavioral Science Consultation Team, made up of psychologists and psychiatrists, and the incorporation of information technologies, particularly databases.

It is in the assignment of a Behavioral Science Consultation Team – or BSCT (pronounced "biscuit") – that we can best approach the theatrical performativity of Abu Ghraib, for it set the stage both in terms of scenes and images and *also the temporal unfolding of events*. While the scenes and images have received much popular and academic attention, their temporal and processual dimension has been largely overlooked. Beneath the spectacle of Abu Ghraib, there was plot, dramatic unfolding, and even character development – or rather, the decomposition of character and identity.

Miller developed and enhanced BSCT interrogation at Guantánamo, drawing on decades of CIA and military research into psychological methods of interrogation. McCoy argues that this research produced a radically new paradigm of torture, one that perversely complements the postwar emergence of human rights institutions. It is a "no touch" torture, a torture that leaves few visible marks on the body, precisely because it targets the mind – or rather, because it targets the mind through bodily sensation and stress, rather than primarily attacking the body through contact, twisting, or puncture. Erroneously called "torture lite," its effects can be far more damaging and long-lasting than physical torture. There is a strong bias against recognizing psychological torture as torture, not only by folks such as Rush Limbaugh, who jokes about Muslims vacationing at "Club Gitmo," but even by past Congressional investigations into American torture programs. Even the US ratification of the Convention against Torture sidestepped the techniques I am about to describe. Thus the importance of understanding Abu Ghraib as psychological torture, and I believe theatricality provides a crucial lens for analyzing how this paradigm actually works, both spatially and temporally.

Until recently, the CIA's torture paradigm has consisted of two main methods: sensory disorientation, achieved by sensory deprivation and overload; and self-inflicted pain, such as stress positions and psychological manipulation. McCoy reads one of the iconic Abu Ghraib images precisely through this frame: "That notorious photo of a hooded Iraqi on a box, arms extended and wires to his hands, exposes this covert method. The hood is for sensory deprivation, and the arms are extended for self-inflicted pain."[12] The CIA torture paradigm targets the very sense of self and identity. Sensory disorientation techniques, including long periods of silence and darkness or, alternatively, loud music and strobe lights, aim to produce fear, a loss of spatial and temporal awareness, and emotional crisis and breakdown. Self-inflicted pain techniques include long periods of standing or squatting; shackling of arms and legs in painful positions, and informing prisoners that they could end the interrogation – simply by telling the truth. The goal here is to make prisoners blame themselves for their suffering.

For McCoy, one reason psychological torture is *total* theatre is that it targets all of the senses: hoods, blackened goggles, and strobe lights target vision; earmuffs and loud music target hearing; gloves and mittens target touch; dietary changes target taste; and surgical masks target smell. We can see sensory disorientation at work in an image from Guantánamo released in 2002. It shows a group of orange-clad detainees kneeling in a fenced pen; they wear blackened goggles, earmuffs, wool gloves, and blue surgical masks. And we also see self-inflicted pain: these detainees are not praying, they are in kneeling stress positions. Ankles crossed, backs bent, wearing knitted caps in the Caribbean heat, they may have been forced to kneel for hours at a time.

General Miller's theatre of torture actually "perfected" the CIA's psychological paradigm by adding two additional methods: "cultural shock" and the exploitation of individual vulnerabilities.[13] First, "cultural shock," which targets cultural values and sensitivities. McCoy gives this example: "Guantánamo's command began to probe Muslim cultural and sexual sensitivities, using women interrogators to humiliate Arab males. [. . .] According to a sergeant who served under General Miller, female interrogators regularly removed their [own] shirts and one [smeared] red ink on a detainee's face saying she was menstruating leaving him to 'cry like a baby.' "[14] Other cultural shock techniques include the shaving of hair and beards and the use of dogs to frighten and humiliate. Such techniques were imported and used at Abu Ghraib, as seen in the widely published image of Lynndie England holding a leashed inmate nicknamed "Gus," taken in October 2003. In the uncropped version of this photo, another MP, Megan Ambuhl, looks on. Compare the scene to this Guantánamo interrogation log from December 2002, almost a year earlier: "Began teaching the detainee lessons such as stay, come, and bark to elevate his social status up to that of a dog."[15] The Guantánamo detainee, Mohammed al-Kahtani, was also put on a leash and forced to be naked in front of female soldiers.

At Abu Ghraib, the military used cultural shock systematically. Here are extracts from a pamphlet given to Marines in fall 2003 to make them aware of Iraqi cultural sensitivities:[16]

Do not shame or humiliate a man in public. Shaming a man will cause him and his family to be anti-Coalition.

The most important qualifier for all shame is for a third party to witness the act. If you must do something likely to cause shame, remove the person from view of others.

Shame is given by placing hoods over a detainee's head. Avoid this practice.

Placing a detainee on the ground or putting a foot on him implies you are God. This is one of the worst things we can do.

Arabs consider the following things unclean:

Feet or soles of feet.

Using the bathroom around others. Unlike Marines, who are used to open-air toilets, Arab men will not shower/use the bathroom together.

Bodily fluids (because of this they love tissue paper).[17]

As Mark Danner argues, interrogators and guards *inverted* such cultural sensitivity training at Abu Ghraib and reverse-engineered it to maximize shame rather than minimize it.

The other technique developed at Guantánamo was the exploitation of individuals' unique mental and physical vulnerabilities. Such exploitation allows interrogators to target an individual's ego in a highly effective manner, all the better to then erode and break it apart, so that the detainee transfers trust to his captors. As medical ethicist and MD Steven Miles contends, BSCT psychiatrists and psychologists identify vulnerable traits, as do physicians, medics, and nurses.[18] Such information is then conveyed to interrogators and filters down to guards – or may even filter up from guards, who spend long periods observing and handling inmates.

We can see such targeting of vulnerabilities in the nicknames given to Abu Ghraib inmates – nicknames such as "The Claw," given to Ali Shalal Qaissi, on account of his deformed left hand. "Shitboy" was the name given to an inmate referred to in records as "M—." "The military states he was "mentally deranged" and often smeared himself with feces. Military personnel also gave the name "Gilligan" to Abdou Hussain Saad Faleh, the inmate in the "The Hooded Man" photograph, presumably naming him after the bumbling TV character. *But how were individual vulnerabilities actually exploited?* "The Claw," "Shitboy," and "Gilligan" were no doubt names improvised on the spot by "creative" interrogators or guards and then exploited in subsequent abuses. McCoy writes: "Thespians all, the torturers assume the role of omnipotent inquisitor, using the theatricality of the torture chamber to heighten the victim's pain and disorientation. Within this script, there is ample room for improvisation. Each interrogator seems to extemporalize around a guiding image that becomes imbedded in the victims' recollection of the event."[19]

The nicknames, I believe, functioned as guiding images for the performance of psychological torture.We can see how one such guiding image was employed by analyzing a series of images. As graphic as they are, individual photos only give a static impression of the theatre of torture. To understand its process *in action*, we must look at a *sequence* of materials.

Violence performed: "Shitboy"

Analyzing a series of digital images of M—, the prisoner nicknamed "Shitboy," we can better understand the theatre of torture's processual dimension. MPs recorded these images over a one-month period, from 4 November to 2 December 2003. Again, the military stresses that M— was "mentally deranged," but given that the CIA torture paradigm sought to *produce* mental breakdown, M— may well have been "deranged" *by* his treatment. At a minimum, we can see how guards exploited his vulnerabilities by enacting and elaborating the guiding image of "Shitboy" in their performance of torture.

The images of M— are particularly graphic, as they both *represent violence* and *embody the violence of representation itself*. However, given that the events have been misrecognized as "hazing" antics or the deeds of a few bad apples – and that the psychological torture paradigm has been largely ignored, I think we have a responsibility to analyze both the theatre of torture and the machine in which it operates. The images analyzed here can be found on Salon.com's "Abu Ghraib File" (<www.salon.com/news/abu_ghraib>), an archive that also includes annotations from the Army's Criminal Investigation Division (CID) and accompanying essays. I should also note here that there may well be other, undocumented scenes of M—'s torture that occurred before, during, or after those described here.

The first known images of M— were taken at 1:42 and 1:43 a.m. on 4 November 2003. Two photos show him hanging naked and upside-down from his cell bunk in what appears to be a stress position. His calves stretch across the top bunk; his thighs, torso, and head suspend down, supported on the floor by his hands, which rest on what appears to be a folded black cloth. His hands are placed together in a prayer gesture. Photos of other prisoners reveal they were often handcuffed to bunks in stress positions; handcuffs, however, cannot be seen in these images of M—. According to CID notation of these images: "All investigation indicates he did this on his own free will, this was not a staged event." Already, we must ask what it means to attribute free will to someone who is: (a) subjected to a regime of psychological torture, and (b) alleged to be "mentally deranged"? Yet if we do concede that M— suspended himself from his bunk, these images also depict a staged event, staged not by MPs or interrogators but by M— himself. As later images reveal, he is quite aware of the camera's presence and may well have been performing for it.

Because this first scene takes place in M—'s cell, few other inmates could see him. One week later, however, a much more public performance unfolded in the cellblock's central corridor, where it was documented by several images, two of which received wide publicity. In these two photos, taken late in the evening of 12 November, one again sees M— naked, but now walking with arms fully extended at the side and head bent back, an obvious stress position. But most disturbing about these images is that M—'s entire body is covered with feces – head, arms, legs, and torso, front and back. In one photo, we see M— from behind, walking toward Sergeant Ivan Frederick, who stands clutching a night stick. A second photo, taken moments later, shows M— from the front, walking in the same manner. Significantly, both photos capture the scene's public nature, for we see the arms and heads of other inmates watching from their cells. M— is being paraded before his fellow Iraqis and before the camera. How long this incident lasted is unclear, though two other images were taken ten minutes earlier from the cellblock's second floor walkway. These show M— already covered with feces, but rather than walking, he kneels before Graner with hands atop his head. To the side stands civilian translator Adel Nakhla of Titan Corporation, no doubt translating Frederick's commands.

Here we see evidence of psychological torture that combines self-inflicted pain (stress position), cultural shaming (public display of nudity and excrement), and the exploitation of individual vulnerability (the targeting of a pre-existing psychological disorder). M— has been forced to kneel before the baton-wielding Graner and then paraded (and no doubt forced to stand for a long period) in full view of fellow Iraqis – all the while completely covered in feces. Indeed, these images may very well depict the initial formation of the "Shitboy" image, around which the MPs and interrogators would construct subsequent abuses. It should be noted that MPs later stated that M— repeatedly smeared himself with feces, acts taken as symptoms of his mental disorder.

Nine photographs taken the next night demonstrate how long such sessions could last. At 10:04 p.m. on 13 November, Sergeant Ivan Frederick photographed M— standing with sandbags tied on his arms, placed there either for sensory deprivation or perhaps to prevent him from abusing himself. M— stands in the corridor at one end of the cellblock, his black prison jumpsuit open and pulled down, exposing his upper torso. *Three and a half hours later*, at 1:39 a.m. on 14 November, M— is photographed again, still standing but now his arms extend forward with a sandbag on his right arm only. Two minutes later, Specialist Charles Graner, photographs M— with another camera from above on the second floor walkway. M— is now turned around with arms extended on either side, head tilted back. Two more photos taken moments later by Graner show Frederick and a second MP, Sergeant Javal Davis, who smiles up at the camera. Significantly, these three photos reveal numerous stains on the floor beneath M—'s bare feet, suggesting that not only has he been standing and turning for some time, but also that he has urinated on himself. (Graner later told Joseph Darby, the MP who would turn the infamous photos over to his superiors, that in regard to his abuses, "'The Christian in me knows it's wrong, but the corrections officer in me can't help but love to make a grown man piss himself.' "[20]) Additional photos taken with Frederick's camera depict further abuses. One shows M— posed with both arms extended and curled on either side in a position reminiscent of a swan's neck. Another image, a headshot of M— in profile, reveals something not legible in the other photos: his head has been shaved, except for a band of hair running from one ear to another, creating a sort of transverse Mohawk haircut. Finally, two photos taken at 1:52 and 1:53 a.m. show M— again wearing two sandbags, only now his arms have been bound tightly together with rope and a padlocked chain. In the final image, Graner stands behind M—, holding him by the shoulders and grinning directly into the camera, apparently showing off his work.

Again, the temporal dimension of the Abu Ghraib theatre of torture is crucial to understanding how the CIA regime of interrogation operates. As the scene just described indicates, detainees may be forced to hold stress positions for hours on end, a fact not captured by the more notorious single images. Intense physical and psychological pain arises not from blows delivered by another, but from sheer exhaustion and prolonged periods of immobility. The goal is to make the detainee blame himself for his suffering. A common outcome is incontinence, leading to self-disgust and, in cases where urination or defecation occurs in a space open to viewing by others, public shame. Both the personal and cultural sense of identity erodes in a highly calculated but gradual process.

We next see how this process unfolded – and also witness the elaboration of the guiding "Shitboy" image – on 19 November, two full weeks after the first known images were taken. Two close-up images taken at 2:10 p.m. show M— lying naked on his right side on a pink foam bed pad, his upper body covered by a dark, striped blanket. His ankles are bound with white, plastic ties, and in one image, it appears that his wrists are shackled behind him with metal handcuffs. Both images show M— holding a yellow object; the CID notation reads, "Detainee inserts banana into his rectum on his on [*sic:* 'own']." A third image taken at 2:11 p.m. reveals that these images were taken from the second tier walkway. The Sony camera has been zoomed out, and we see that M— lies in the central corridor. It appears that his ankle restraints are chained to the bars of a cell door, and a pair of white underwear lies on the floor nearby.

We have here an inversion of the "Shitboy" image: instead of feces coming out of his rectum, M— has been allowed to insert an object into it – or forced to do so: the trustworthiness of the CID notation is open to question, as it is likely based on testimony given by the very MPs who cuffed and placed M— in the corridor, provided him with a banana, and then photographed him. The presence of the bedding is significant also. It has probably been dragged out of M—'s cell along with him, and foam bedding will be used in the next scene of abuse. Thus, we may be witnessing the beginning of its incorporation into the "Shitboy" image.

This next torture scene begins ten days later, on the evening of 28 November, and it embodies and then elaborates the "Shitboy" image in an extraordinary way. The first photograph, taken at 8:06 p.m., shows M— from the groin up, standing naked in the shower area of the hard site. M— stands before a brick wall, the shower plumbing visible to his right. His body is marked with brown splotches. The CID notation reads: "The detainee is covered in what appears to be human feces." Feces has been smeared on each breast, the belly, the neck, and in the hair. Though M—'s face has been digitally obscured, it appears that his chin, ears, and the left side of his face are also defiled, thus suggesting that excrement covers his entire face. The next two images, also time-coded at 8:06 p.m., again show M— standing, but now from the ankles up. Feces covers his genitals and inner thighs. M— wears white surgical gloves whose fingers are covered with excrement. One photo depicts him with both arms at his side; in the other, it appears (again, his face has been digitally obscured) that he has put his right hand in his mouth, the grotesque irony being that M— is using a hygienic glove to eat feces.

As with the images taken on 12 November, these photos taken two weeks later depict the signature image of the "Shitboy" motif: M—'s body smeared with excrement. One can only speculate how often this tableau was staged on nights not captured by digital cameras, either in secluded spaces such as his cell or the shower, or in opens spaces such as cellblock corridors. This night, after again establishing the "Shitboy" motif, the guards will extemporize their performance in an open corridor, using objects previously deployed as props on earlier nights. The next sequence of photos, taken a little more than an hour after the shower images, begin with an image showing a cleaned-up M— standing in an all-too-familiar pose, with arms fully extended at his side. Now, however, he wears a strange tunic-like costume: on closer inspection, one can see that it is a yellow foam bed mattress that has been folded over his shoulders to cover his body, front and back from the knees up. A slit has been cut in the bed pad's center, though which M—'s head sticks out. Around his waist, a chain is being fastened tightly by Graner. Graner wears black gloves, an armored vest over his green T-shirt, and camo fatigues. From his belt hang various "dangles": tools, cables, and so on. The next image, taken at 9:13 p.m., shows Graner posing beside M— leaning on his shoulders with both hands and smiling directly into the camera, as if showing off his improvised work.

Graner's work had only just begun. The next photo, taken at 9:16 p.m., shows Graner kneeling on one knee, his right hand on the back of M—, who now lies stomach-down on a litter, a cloth stretcher whose two support poles have small feet which keep the litter a few inches above the floor. The image has all the trappings of a trophy photo: Graner smiles broadly into the camera next to his "game." M—, still wearing the foam tunic, lifts his head toward the camera; his face has been digitally blurred, but the head angle suggests that he too looks directly into the camera. We have here an image of abject humiliation and almost total subjection, a point I will turn to in a moment. Before doing so, however, I will note that the background of this photo reveals two remarkable things.

First, in the deep background on the left, there stands a second detainee; hooded with head bent toward the cellblock wall, he wears a smudged white gown and his hands appear to be tied behind his back. M— thus may not have been the only person tortured this night. But even more significant, also on the left of this image but in the mid-range background, there stands a person holding up a second litter. Unlike Graner, Frederick, and the civilian translator Nadal seen in previous photos, this individual wears civilian clothing: black pants, green shoulder-bag, and bright white sneakers. *Such non-military clothing strongly suggests that this second person was either a CIA interrogator or a privately contracted one.* In short, far from being "prep work" for subsequent interrogation, this entire scene may well have been part of an actual interrogation, if not of M—, then of the hooded detainee or another person not visible in these photographs. Indeed, though there were special interrogation rooms at Abu Ghraib, Tara McKelvey reports that interrogations could also take place in "a cell, a shower stall, stairwell, or a supply room."[21] Other photos taken at Abu Ghraib have been described as showing "OGA" (i.e., "other government agency", a euphemism for CIA) personnel interrogating detainees in the open corridors of cellblock 1A.

Whatever the identity of this second person, he or she played an active role in this unfolding theatre of torture, for in the next photograph the second litter has been placed on top of M— to form what the guards called a "litter sandwich." Sandwiched face down between two litters, wearing foam bedding, chain belt, and arm restraints, M—'s subjection has literally reached the bottom. The *coup de grâce* comes in the scene's final image, another trophy photo which depicts a second guard, Ivan Frederick, sitting atop M—, who now raises his head to the camera. The subjection is now complete, the dominance of guard over prisoner, American over Iraqi, white-skinned over dark-skinned, military order over insurgent force, Good over Evil – all are both allegorized and literalized. To coin a term, all become "litteralized" in the litter sandwich.

Such word play is not inappropriate here, for it may help reveal additional elements and moments in the improvisational development of M—'s torture via the guiding image of "Shitboy." To begin with, "litter" read as stretcher can be associated with bed and bedding, in particular, with the foam pad seen here and in the earlier scene with the banana. As noted above, it is as if the bedding has become incorporated into the "Shitboy" theme, and one can imagine the guards joking, "Shitboy has made his bed and now must lie in it." (Recall, too, that the first images of M— show him hanging from his cell bunk.) But "litter" can also refer to trash, especially pieces of trash left lying on the ground of public places, and here M— has been trashed and left on the floor of the prison's public corridor. Significantly, "litter" is also associated with feces, for the term can refer to absorbent material, such as the dry granules found in cat litter. Here, the litter absorbs "Shitboy" into what might be called a "shit sandwich." Further, "litter" can refer to the surface layer of a forest floor, made up of decomposing matter such as twigs, leaves, and animals. Such rotting material can obviously be related to excrement. But more tellingly, the precise term "litter sandwich" is used in biology, where it refers to a framed, wire apparatus used to study the decomposition of organic matter in soil strata.

Taken together, the photographs taken from all the scenes analyzed here strongly suggest that M—'s identity becomes decomposed into the image of "Shitboy." Again, such decomposition or breakdown of subjectivity is precisely the goal of psychological torture, a process that the month-long series of images just described documents with the methodical cruelty of an all-too-serious and all-too-prolonged sick joke, one in which sadistic humor is put to "intelligent" ends, those of military and national security intelligence. As one of the military whistleblowers, Sergeant Sam Provance, later recalled about the guards at Abu Ghraib, "They'd talk about their experience when the detainees were being humiliated and abused. It was always a joke story. It was like, 'Ha, ha. It was hilarious. You had to be there.' "[22]

One final document, a set of video clips taken the night of 1 December, offers bone-chilling evidence of the effectiveness of psychological torture. M— himself performs the final act of this particular theatre of torture, though we would be more accurate saying that M— "is performed" by this theatre, for the agency of his actions now includes the Army, CIA, and Titan Corporation – or the torture machine itself. The performance was recorded on Graner's video camera, presumably by Graner. The Salon archive contains ten separate 15-second video clips and one 8-second clip of M—, for a total of 2 minutes, 38 seconds. However, CID notation states that video was taken between 9:29 and 9:45 p.m., so the events recorded unfolded for half an hour; moreover, additional still images of the same event are time-coded as occurring early on the morning of 2 December at 12:33 and 2:00 a.m., indicating that the torture lasted *at least four and a half hours*. This material, along with two other sets of video clips of other detainees – one of the infamous pyramid and group masturbation scenes – remains largely unknown in the United States, where to my knowledge they have never been publicly broadcast (portions of them were broadcast in Australia). The clips on Salon.com are low-resolution QuickTime movies and, significantly, their audio has been dropped out. No doubt full video and audio versions exist but have not been leaked to the public.

I will describe the video clips in the order posted on Salon.com, noting also that the file names suggest that other clips may exist. The first five clips of M— were shot from the upper tier, peering down into the corridor below. In the first two clips, we see M— from his left standing in profile, bent over at the waist with his head against a solid cell door. He wears a blanket of some sort. At first glance, it appears M— may be praying, as his body rocks slightly but on closer inspection we see that his wrists are handcuffed to the door before him. The next three clips reveal the horror of M—'s performance: standing upright from his bent position, he leans back, still rocking his body slightly forward and backward. He then turns his head in the camera's direction and appears to look directly into the lens. Turning his head back, M— extends his arms out and shifts his body weight back, bowing his body slightly at the waist. Seeming to take aim, he suddenly pulls hard on the handcuffs and slams the top of his head into the solid metal door. The blanket slips down, exposing his bare shoulders as his knees buckle. In the next clip, M— appears dazed and shaken, his body shuddering as his head bends up and down. In the fifth clip, M— prepares for another impact, standing and rocking, turning his head toward the camera, leaning back and again slamming his head into the door.

Again, CID notation indicates that the "self-induced" actions captured on video lasted half an hour. The final six clips again record M—'s ritualistic banging of his head into the metal cell door, only now the cameraperson has moved down to the cellblock floor. In the first of these clips, we see M— standing full length in profile but now in a mid-range shot taken from his right side. He wears sandals, and we can see that his blanket bears a large blue floral pattern. From this angle and distance, we more fully sense the force of his head hitting the door, twice within a 15-second span: the rhythm of his impacts has increased and once again he turns his head briefly toward the camera. The next, 8-second clip starts from the same angle, but M— now squats on his feet, balled up and teetering no doubt after another head-slamming. The camera then moves quickly toward and behind M—, the lens pointing toward the concrete floor, the bars of several other cell doors, and the cameraperson's walking feet. The next two clips, each 15 seconds long, show M— closer up from his left side, waist to head in three-quarter view. In one clip, we see M— standing before the door, whose green surface we now see is marked with two bright bloody spots, one where his head hits the door, the other where he rests his head while bent over at the waist. The second of these two clips shows M— again slamming into the door. In the tenth and final video clip, the camera towers over M—, who squats below. Solarized lens flares give the image a dark red hue, then a light green one, as M— squats and rocks, tilts his head back, and then stands up again. He again looks into the camera and we see that the right side of his forehead is bloody and raw. M— grimaces, and he appears to be crying and/or crying out.

Taken together, all of these materials demonstrate that Abu Ghraib's theatrical performativity was not limited to poses and images, settings and props. The total theatre of torture also operates via a temporal, processual dimension that has been largely ignored by commentators; however, this dimension is crucial to understanding the underlying regime of psychological torture and interrogation at work in Abu Ghraib and elsewhere. M— was subjected to over a month of painful and humiliating abuse in sessions sometimes lasting several hours each. The photographs and videos clearly demonstrate that military personnel (and possibly CIA personnel and/or private military contractors) employed this regime's four main components on the detainee: sensory deprivation, self-inflicted pain, cultural shock, and the targeting of individual vulnerabilities. With respect to the latter, I have tried to show that "Shitboy," beyond its function as a nickname for M—, also served as a *topos* upon which his captor improvised and developed a highly individualized yet highly public performance designed to decompose his subjectivity. Bodily organs and gestures, as well as specific objects, entered into the processual development of the "Shitboy" topos, effectively driving a wedge between M—'s corporeal and psychosocial sense of self. Even if he was mentally unstable before his capture, guards and interrogators spent weeks exploiting his vulnerabilities, both in isolation and, more often, in public. In the end, the question of whether his pain was self-inflicted or not is irresolvable and even irrelevant: his performance was both scripted and improvised by guards and interrogators within a codified theatre of torture designed to produce self-inflicted pain as a means of breaking down his identity.

Coda: media shock and counter-performativity

The CID report on Abu Ghraib also indicates that M— was not a "high value" prisoner, meaning that he was not thought to possess valuable information, in which case guards likely used his serial abuse in order to intimidate other prisoners, in both live and recorded performances. Now, given the widespread use of media, not only at Abu Ghraib, but reportedly also at Bagram and Guantánamo, I believe that media form a fifth element of the CIA's paradigm of psychological torture, one we might call "media shock." Spectacular abuses perpetrated in isolated cells or rooms targeted the psyche of the prisoner in question. But inmates also report being repeatedly photographed and then told that humiliating images would be shown to family and friends if they did not cooperate. Recall here the Marine Corps brochure's comments about shaming Iraqis in front of others. And, indeed, the majority of known photographed abuses occurred in the central corridor before the eyes of other inmates, who were themselves shamed and intimidated by being forced to watch – and to watch American men and women take photographs and video of the abuses. In addition, guards reportedly used such photos as screen savers and openly displayed them on prison walls where inmates could see them, effectively telling them that "this can happen to you." In short, the force of theatrical performativity at Abu Ghraib, including media shock, first of all struck the inmates, as both objects and viewers of torture. The public scandal then communicated that force around the world.

The spectacle of the scaffold thus returns both in the cells and corridors of Abu Ghraib and other sites, and through television, computer, and other media networks, by which it reaches a global audience. But in between the local and global audiences, the performative force of the spectacle was radically transformed.

This transformative process began on 13 January 2004, when Graner handed Specialist Joseph M. Darby two CDs containing hundreds of images. Disturbed by what he saw on the CD, Darby turned over the images to CID, and they soon became central to "Article 15-6 Investigation of the 800th Military Police Brigade," a.k.a. the "Taguba Report," which was initiated in late January. By April, a selection of images had been leaked both to CBS and to reporter Seymour Hersh and then subsequently made public by *60 Minutes II* and *The New Yorker*, unleashing the images' performative force around the world. But a strange, yet hopeful, thing happened in this globalization, something that began with whistle-blower Darby's reaction: the performative force of the spectacle inverted and turned on itself. Reframed and publicized by CBS and Hersh, the poles of Good and Evil reversed, as it were, and the persons shamed in the images became England, Frederick, and Graner, and beyond, the US military and, further still, the United States itself. This profound reversal was captured in a political cartoon that appeared soon after the scandal broke. Composed by Tim Menee of *The Pittsburg Post-Gazette*, the cartoon depicts Uncle Sam standing atop a cardboard box with wires attached to his fingers and wearing a black hood and robe; next to him is scrawled the words "UTTER HUMILIATION." Other cartoons – published in the United States and abroad – expressed a similar sentiment, using Uncle Sam or other figures associated with the United States, such as the Statute of Liberty or Lady Justice.

Such reversibility of performative force can be understood in terms of what Butler has called "queering," "resignification," and, following Brecht, "refunctioning." Similarly, Donald MacKenzie, a noted sociologist of science and technology, has recently found that certain economic models, after initial performative success in making reality conform to them, may eventually "alter economic processes so that they conform less well to the theory or model" and may even lead to economic crisis. MacKenzie calls this possibility "counterperformativity,"[23] and while he focuses on economic models, one can extend this concept to potentially *any* model or theory. Indeed, McCoy's study of the CIA model of psychological torture argues that its use by the Philippine military on suspected Communist insurgents in the 1970s and 1980s helped create a group of egomaniacal officers who later tried to overthrow the very government (that of Ferdinand Marcos) on whose behalf they had originally tortured people.[24] In the Iraqi theatre of war, the counterperformativity of CIA torture techniques includes the creation of more rather than less insurgency, the production of useless rather than useful intelligence, and the generation of psychic violence not only on torture victims but also on the torturers themselves. In this light, the images of torture at Abu Ghraib may have even helped, counter-performatively, to produce in the US public what Eisenman calls "the Abu Ghraib effect," that is, "a kind of moral blindness [...] that allows them to ignore, or even to justify, however partially or provisionally, the facts of degradation and brutality manifest in the pictures."[25] In other words: in the name of morality, (some) Americans blinded themselves to morality.

There is a lesson here about violence performed, violence analyzed, violence cited and incited. One could protest that the very analysis attempted above – a detailed performance analysis of the images of M—'s torture – could itself contribute to the systematic violence it seeks to critique, could perpetuate the media shock rather than counter it, could contribute to the society of the spectacle of the scaffold rather than warn against it. Similarly, one could protest that political and/or artistic protests against Abu Ghraib, Guantánamo, and Bagram (for instance, protest marchers dressed in black hoods and orange jumpsuits, the performances of interrogation by Coco Fusco or Fassih Keiso, certain graffiti and installation work by Banksy, the film *The Road to Guantanamo* by Michael Winterbottom and Mat Whitecross – all of which can be read as counter-performatives of the CIA torture paradigm), could themselves incite further torture. Such risks are unavoidable, given that the iterability of any and all performances and performatives insures both their possible misfiring *and* their possible success, their very performativity and counter-performativity. No amount of interpretative framing or historical contextualization can ward off such citationality, as framing and contextualization themselves entail citation networks. One could, alternatively, simply withdraw and refuse to cite the violence, whether in words or in images – but that is precisely the move made and encouraged by the Bush administration, which has fought the release of images, as well as further inquiries that would investigate up the chain of command and potentially connect torture images to torture memos, theatrical performativity to executive performativity. In an age of global performativities, of performative powers operating on both local and global scales, what is needed is more connecting, citing, and critiquing of violence performed, not less, and countering the spectacle of the scaffold and the theatre of torture may well depend on counter-performative spectacle and theatre. The risks of producing them are great, but the risks of not doing so are greater still.

Notes

1. See *Perform or Else: From Discipline to Performance* (London: Routledge, 2001), especially Part II.
2. See in particular the memos of Alberto Gonzales to President Bush ("Decision Re: Application of the Geneva Conventions on Prisoners of War to the Conflict with Al Qaeda and the Taliban," 25 Jan. 2000), George W. Bush ("On the Humane Treatment of al-Qaeda and Taliban Detainees," 7 Feb. 2002), and Jay S. Bybee to Alberto Gonzales ("Standards of Conduct for Interrogation under 18 U.S.C. §§ 2340–2340A," 1 Aug. 2002). These and other memos can be found in Mark Danner's *Torture and Truth: America, Abu Ghraib, and the War on Terror* (New York: New York Review Books, 2004), a collection of essays, documents, and photographs.
3. Judith Butler, *Precarious Life: The Powers of Mourning and Violence* (London: Verso, 2004), 65.
4. Butler, 65.
5. See Foucault, *Discipline and Punish* (1977; New York: Vintage, 1979), especially ch. 2.
6. For an overview of such analyses, see Stephen F. Eisenman's *The Abu Ghraib Effect* (London: Reaktion, 2007).

7. Alfred W. McCoy, *A Question of Torture: CIA Interrogation, from the Cold War to the War on Terror* (New York: Metropolitan; Henry Holt, 2006), 10, my emphasis. McCoy describes the development of the CIA paradigm of psychological torture in ch. 2, "Mind Control."

8. Branislav Jakovljevic, "Theatre of War in the Former Yugoslavia: Event, Script, Actors," *TDR: The Drama Review*, 43:3 (1999): 5–13.

9. Susan Sontag, "Regarding the Torture of Others," *New York Times Magazine*, 23 May 2004: 24–42. Slavoj Žižek, "Between Two Deaths," *London Review of Books*, 26:11 (2004): 19.

10. Eisenman, 16.

11. Brigadier General Janis L. Karpinski, quoted in Scott Wilson and Sewell Chan, "As Insurgency Grew, So Did Prison Abuse," *Washington Post*, 10 May 2004, 25 June 2007: http://pqasb.pqarchiver.com/washingtonpost/search.html

12. McCoy, 8.

13. Amy Goodman, interview with Alfred W. McCoy, *Democracy Now*, 17 Feb. 2006, 10 Nov. 2006: http: //www.democracynow.org/article.pl?sid=06/02/17/1522228

14. McCoy, 129–30.

15. McCoy, 127–8.

16. Danner, 19.

17. "Semper Sensitive: The Marines' Guide to Arab Culture." USMC Division Schools. Reprinted in *Harper's*, June 2004: 25–6.

18. See Steven H. Miles, *Oath Betrayed: Torture, Medical Complicity, and the War on Terror* (New York: Random House, 2006).

19. McCoy, 83–4.

20. David Finkel and Christian Davenport, "Records Paint Dark Portrait Of Guard," *Washington Post*, 4 June 2004, 25 June 2007: http://www.washingtonpost.com/wp-dyn/articles/A16832-2004Jun4.html

21. See Tara McKelvey, *Monstering: Inside America's Policy of Secret Interrogation and Torture in the Terror War* (New York: Carroll & Graf, 2007), 14.

22. Provance quoted in McKelvey, 17.

23. See MacKenzie, *An Engine, Not a Camera: How Financial Models Shape Markets* (Cambridge, MA: MIT Press, 2006), 19. MacKenzie's primary case study is the 1998 hedge fund crisis associated with Long Term Capital Management (LTCM), a firm that had deployed an economic model known as "Black-Scholes," first to make and then to lose billions of dollars. MacKenzie develops a sophisticated model of performativity, of which "counterperformativity" is one component.

24. See McCoy, ch. 2.

25. Eisnenman, 9.

16
Performance Complexes: Abu Ghraib and the Culture of Neoliberalism

Tony Perucci

> *When I first saw the notorious photograph of a prisoner wearing a black hood, electric wires attached to his limbs as he stood on a box in a ridiculous theatrical pose, my reaction was that this must be a piece of performance art. The positions and costumes of the prisoners suggest a theatrical staging, a tableau vivant, which cannot but call to mind the "theatre of cruelty," Robert Mapplethorpe's photographs, scenes from David Lynch movies.*
>
> Slavoj Žižek[1]

> *You know, if you look at – if you, really, if you look at these pictures, I mean, I don't know if it's just me, but it looks just like anything you'd see Madonna, or Britney Spears do on stage. Maybe I'm – yeah. And get an NEA grant for something like this. I mean, this is something that you can see on stage at Lincoln Center from an NEA grant, maybe on Sex in the City – the movie. I mean, I don't – it's just me.*
>
> Rush Limbaugh[2]

> *We're an empire now, and when we act, we create our own reality. And while you're studying that reality – judiciously, as you will – we'll act again, creating other new realities, which you can study too, and that's how things will sort out. We're history's actors [. . .] and you, all of you, will be left to just study what we do.*
>
> Senior Advisor to President Bush[3]

To consider the staging of torture at Abu Ghraib is to consider the performance of torture, torture as performance, and the photographic representation of torture. In this chapter, I am interested in how these dynamics serve those who adopt the role of "history's actors" in an effort to construct the "new realities" that are both the spoils of war and the foundation of empire. These realities are constitutive of what I call the "performance complexes" of

neoliberal empire, that is, the staging of an economic and ideological apparatus in which the "magic of markets" in service of profits takes precedence
over any competing social interests. While neoliberalism calls for the dismantling of any state power that might regulate or limit corporate power
in the interest of the public good and thus threaten profit and the free
flow of capital, it also relies on the state subsidization of corporate partnerships in warfare and penality. The performance of the neoliberal economy
thus depends upon the "performance complexes" that also produce it: the
military-industrial and prison-industrial complexes. In these institutional
networks the state contracts private corporations to take on former government practices of war manufacturing and prison-building and maintenance.
The profit motive of these corporations drives foreign and domestic policy to
perpetual war and expanding incarceration as fulfilling the neoliberal "idea
that the market should be allowed to make major social decisions [...] [and]
that corporations should be given total freedom."[4] These performance complexes are articulated with, and made visible in, the photographs from Abu
Ghraib, where we see the violence of neoliberalism. This violence is unique
not in its performance at Abu Ghraib, but rather in its revelation, which
masked the ubiquity of violence as a constitutive element of the unequal
global economic order.

How we perform things in America

"To live is to be photographed," suggests Susan Sontag in response to the Abu
Ghraib photos, "but to live is also to pose. To act is to share in the community of actions recorded as images."[5] Sontag's comments here mark not only
the reality of the postmodern spectacle, in which an act is only worth doing
if it is being photographed or videoed, but also the way in which the MPs and
military intelligence officers in the Abu Ghraib photographs occupy the dual
position of performer and spectator. We see in the photos the sadistic glee
of the torturer and/as family photographer through the lens of the military
camera. "The horror of what is shown in the photographs," Sontag argues,
"cannot be separated from the horror that the photographs were taken –
with the perpetrators posing, gloating, over their helpless captives."[6] The
performed act of photography does not merely, or even primarily, document
the violence enacted, but is constitutive of it. Further, it evacuates the performed act of torture by aestheticizing it – transforming it from performance
to an image of the spectacle.[7]

There is nothing comparable to this act of photographic violence, Sontag
suggests, except the lynching photographs and postcards that circulated in
the American South (and were often sent to family and friends up North)
in the late nineteenth and early twentieth centuries. These photos register the carnival atmosphere that accompanied the lynching of blacks in
America as they reveal jubilant families posing alongside mutilated corpses.[8]

The performed act of lynching photography, like torture photography, represents a "collective action whose participants felt perfectly justified in what they had done."⁹ What Sontag suggests here is that the banality of the act of photography, and the casualness of the poses bespeak a violence that the photographs themselves performatively enact, and do not merely represent.

These photographs are, ontologically, not the performance. They are, as Susan Willis describes them, "Performance turned into artifact."¹⁰ The photographs attempt to document the undocumentable, and yet serve as an index of the "horror that the photographs were taken." The multiplied horrors of the photographs – the corporeal reality of torture as well as the violence of the act of photographing and the enactment of the pose – are indexed but not represented by the photographs. The imminence of these acts – torture, photography, and the pose – are aestheticized and artifactualized in the photograph itself. As Angela Davis notes, following Adorno, part of the social violence of the Abu Ghraib photos is "that we project so much onto the ostensible power of the image that what it represents, what it depicts, loses its force."¹¹ The very spectacularity of the photographic image, thus, allows it to stand in for the performance and defer its affective potential.

Peggy Phelan's oft-cited and oft-maligned claim that performance's "only life is in the present," in that it erupts in a frenzy of the visible in a "manically charged present" of liveness, is *a propos* here, in that it is the disappearing "excess" that constitutes an act in performance.¹² Phelan's critical point in articulating an "ontology of performance" is to mark performance's imminence and ephemerality. For Phelan, performance is that element of the theatrical occasion that disappears, that exists only in its presence and its presentness. There is great radical potential in such an occasion, when the *presentation-of-presentness* makes visceral and haptic the encounter with what Walter Benjamin calls the "presence of the now (*Jetztzeit*)," an enlivened encounter with the actuality of human pain, suffering and pleasure that gives rise to the possibility of enacting social change *now*.¹³ This activating of the now that spurs the potential for activism is precisely what the "society of the spectacle" serves to tamp down. For such a society, Debord contends, is dependent upon a pacifying of the public sphere, as the image disconnects us from each other and from the politics of economic (re)production.¹⁴

Thus Phelan deems performance to be the element that operates in the theatrical occasion as an "excess," because it, by definition, cannot be reproduced, mass marketed or sold. In her analysis of performance artist Festa's live enactments of enduring torturous pain, she notes that Festa's "Presentation-of-Presence" does not operate solely as spectacle, but also calls forth a second performance on the part of the spectator characterized by "the recognition of the plentitude of one's physical freedom in contrast to the confinement and pain of the performer's displayed body."¹⁵ Even to the degree to which the Abu Ghraib photos give lie to America's performance of moral superiority, the artifactuality of the photographs masks the excess of performance that links the fate of the indignant (or jubilant) Western viewer and the tortured prisoner. Neoliberal empire necessitates that the linkage between the spectator and spectacle of the photograph remain unmarked, as is demonstrated when George Bush dismissed the acts of the Abu Ghraib photos by saying, "This treatment does not reflect the nature of the American people. That's not how we do things in America."[16] The theatrical acts at Abu Ghraib are exactly how "we do things in America," in US prisons, sweatshops, and immigration detention centers.

The unreproducability of performative excess is part of how the photographs as artifact reproduce the violence of the performance of torture itself. That is to say, torture operates to reduce the torture victim to an object, as Elaine Scarry puts it, to a "body in pain."[17] The excessive violence of torture, as with performance's excess, is incommunicable through the flattening artifact of the photograph. It is this process of "performance turning into artifact," the evacuation and erasure of the excess of violence (and of performance) that enables the deferral of that excess, which was manifested when the photographs were sent to MPs' families and friends as mementos, or even to the pleasure of indignation such representations evoke (or the pleasure evinced by Baudrillard in his claim that with these photographs, "America has electrocuted itself").[18]

However, I want to mark here the performative excess without indulging in the violence of the lingering gaze over these photos, the imaginative re-creation of these photos as a re-enactment of the violence. Moreover, I want to do so without displacing the humanity of those tortured in the way that "spectacles of suffering" tend to reproduce the violence that incurs them, in no small part by making terror an amusement for the spectator. As Saidiya Hartman contends, even sympathetic, white abolitionist recitations of slave coffles often served to "reinforce the spectacular character of black suffering."[19] The "shocking" photos at Abu Ghraib conceal as much as they reveal about the "shock and awe" theatricality of the War on Iraq. The spectacularity of these performances serves to mask the performance of economic interests that gave rise to the acts of torture at Abu Ghraib. A spectacular excess, the excess of profit, both cause these performances, and is masked by them. The performance of torture at Abu Ghraib – masked by the artifacts of performance (and the artifactualization of performance) – is the performance of neoliberalism, "the terror of the mundane and quotidian" masked by the "shocking spectacle."[20]

It was Oliver Cox, the radical black sociologist, who argued in the 1940s that it was critical to see lynchings of African Americans not only for the theatrical spectacles of violence that they were, but also as expressions of political economy. The lynching, he contended, was not a production of a hysterical wildness or "strange madness," but rather was a concomitant violence of the operation of capitalism, which required a cowed and exploitable laboring class.[21] What Kirk Fuoss has called "lynching performances" were, Cox explains, central to the logic of early twentieth-century capitalism. The success of lynching performances hinged on the production of blacks (those lynched and those not-yet-lynched) as objects; the excess of violence and the excess of performance served to evacuate the subjecthood of African Americans in the interest of the excess of profit. This performance of objecthood served an ideological and economic interest in that it extended the conditions of slavery for African Americans that positioned them as simultaneously laborers and commodities.

Cox explains that the lynch mob is not primarily a spontaneous ensemble of irrational violence, but rather, "The mob is composed of people who have been carefully indoctrinated in the primary social institutions of the region to conceive of Negroes as extra-legal, extra-democratic objects, without rights which white men are bound to respect."[22] A similar corrective is needed in attending to Abu Ghraib, where the photographed soldiers were considered to be anomalous – disturbed individuals enacting "aberrant behavior,"[23] – rather than enacting a policy and cultural logic of neoliberalism, in which "the US needs to pick up some country and throw it against the wall, just to show the world we mean business," as Michael Ledeen of the American Enterprise Institute suggests.[24] The performance of violence at Abu Ghraib, like that enacted in lynching performances, functioned as such a form of "show business," in which the show served the capital interests of business.

Lynching performances were, in Cox's view, not primarily performances of individual actors, but rather were the performance of an economic order, in which business showed that it meant business. Meanwhile, the spectacle of the lynching recast the racialized black body as the cause of the violence enacted upon it. What Saidiya Hartman calls the "racist optics" of such occasions is driven by such objectification, in which the compelled performance of objecthood paradoxically enables the projection of "will" and "agency" onto the black body, producing the "dissimulation of suffering through spectacle."[25] This compulsory performance of objecthood operates as a rationalizing force that both justifies and structures the racial economy. "By lynching," Cox argues, "Negroes are kept in their place, that is to say, kept as a great, easily-exploitable, common-labor reserve."[26] Lynching performances are, in Cox's view, a "sub-legal contrivance" that operate as an expression of, and in service to, a racialized political economy.[27]

The performance complexes of neoliberalism

The performances of torture at Abu Ghraib serve a similar function in their operation as a staging of neoliberal globalization. Neoliberalism is the current reigning economic philosophy; it mandates the absolute centrality of the market, the privatization of all social services, the end of the welfare state, and the commodification of all aspects of everyday life. Neoliberalism represents the triumph of the corporation over the state, where the state no longer serves the Keynesian function of ameliorating the inequalities produced by capitalism, but rather serves solely to protect the rights and interests of capital. The Keynesian project, as Ruth Wilson Gilmore describes it:

> consisted of investments against the tide, designed to avoid the cumulative effects of downward business cycles by guaranteeing effective demand (via income programmes, public borrowing strategies and so forth) during bad times. The social project of Keynesianism [...] was to extend to workers [...] protections against calamity and opportunities for advancement. In sum, Keynesianism was a capitalist project that produced an array of social goods that had not existed under the preceding liberal (or laissez-faire) capitalist state form.[28]

At first gloss, neoliberalism may seem to be simply an elimination of such a project, as government management of the economy is removed in service to the "invisible hand" of the market. Neoliberal empire is characterized in opposition to traditional American "liberalism" exemplified by Keynesian projects like the New Deal, which enact modest redistributive economic practices to counter capitalism's expected deleterious effects on the majority. However, while neoliberalism is "nominally democratic,"[29] it promotes a "culture of upward (re)distribution" by *liberalizing* any state laws or regulations that might hinder corporate accumulation of profit.[30] And yet, for successful economic performance, neoliberalism depends upon state-corporate partnerships. These state-corporate partnerships (including those with the military and prison industries detailed below) serve as a form of "protections against calamity and opportunities for advancement" for *corporations* rather than workers.

The emergence of the military industrial complex (MIC) in the mid-twentieth century can be seen as an early expression of the ascendancy of neoliberalism – as functions of the military were dispersed to a host of institutions outside of the state apparatus, namely private corporations and universities. The government subsidization of private defense industries has been considered a form of "military Keynesianism," as the state underwrites the profitability of military production. However, the diffusion of governmental authority to a loose network of non-governmental institutions is often where accounts of the MIC stop – where the ability of private institutions to circumvent the scrutiny of the public is seen to be the primary danger of such an apparatus.

We should not romanticize the military as a paragon of transparency operating in the public interest, but we must consider the way in which the privatization of the military has created an economy dependent on the development and support of war-making, leading to the institution of what Seymour Melman calls the "permanent war economy." Moreover, the transnational corporation requires the military to secure its profitability through force. As neoliberal apologist Thomas Friedman puts it, "McDonald's cannot flourish without McDonnel Douglas [...] And the hidden fist that keeps the world safe for Silicon Valley's technologies to flourish is called the US Army, Navy, Air Force, and Marine Corps."[31] Thus, the condition of "perma-war" becomes an economic necessity that must support increasingly military-dependent corporations and an economy that depends on those corporations' profitability.

Such a transformation can also be seen in the emergence of a prison-industrial complex (PIC), created by the more recent privatization of prisons, in which prison-building, prison operation, and prison-servicing have become entrenched in national and local economies and are supported by a justice system structured by race. The dependence of small communities especially on the prison economy is enacted by neoliberal economic programs, policies, and practices. For example, one of the direct consequences of the shift of manufacturing overseas has been the devastation of many factory towns throughout the United States. As Angela Davis points out, this has resulted in many communities being desperate for any large employer. Thus, the new privately owned prison becomes an employer of last resort. Municipalities are transformed, Michelle Brown suggests, into a:

> new American city" [...] whose self-sustaining abilities depend on the production of hard-line attitudes, more prisons, and, equally significant, more prison towns [...] culminating in a reconfiguration of social and economic life in distinctly penal terms.[32]

Thus, as cities become dependent upon the economic sustenance of prisons, the profit-seeking prisons and the employment-seeking society have an economically linked interest in the ever expansion of prison populations.

The performances at Abu Ghraib represent an exporting of American penal culture – the normalization not merely of the disciplinary mechanisms of the penitentiary described by Michel Foucault, but also of the stubborn resilience of the staging of pain and terror onto the body of the condemned. As Dwight Conquergood writes in his cogent analysis of the "lethal theatre" of American executions, American penal culture remains dependent upon brutalizing the bodies of the condemned. Such performances, he explains, "are awesome rituals of human sacrifice through which the state dramatizes its absolute power and monopoly on violence."[33] These rituals are increasingly staged as sanitized acts, to at once mask the state's act of murder, as well as to elide the overwhelming racial basis that underwrites the logic of American executions.

While the publicity of the performance of executions requires the staged sterility of murder, it also conceals the "messier" performances of violence that operate as a daily ritual in many American prisons. A "political theatrics of terror" rules the lives of American prisons and poor neighborhoods, where police maintain an occupational presence; in these spaces, *"ritualized displays of terror are built into American policing."*[34] The experience of being inside a maximum security prison is characteristic of what Michael Taussig calls "terror as usual," which he describes as a "state of doubleness of being in which one moves between somehow accepting the situation as normal, only to be thrown into a panic or shocked into disorientation."[35] Indeed, this performance consciousness is engendered by the American prison through the persistent staging of violence by guards, including mock executions. The theatrical enforcement of terror as usual functions as a part of the racially structured apparatus of the prison industrial complex in the service of neoliberal market forces:

> many bodies – particularly those of young working class and lumpen men of color – are superfluous to capital's valorization. A growing stratum of "surplus people" is not being efficiently used by the economy. So instead they must be controlled and contained and, in a very limited way, rendered economically useful as raw material for a growing corrections complex.[36]

The theatrical violence and ritual terror of the prison-industrial complex then, is that of neoliberalism.

What is perhaps most striking about the Abu Ghraib photos is that they *seem* to represent abnormality at all. The photos are, in a sense, the publicizing not only of the quotidian practices of American military detainment facilities, but also of accepted practices in US prisons. The Abu Ghraib performances, then, can be seen as restagings of the "training" video produced by guards at the privately run Brazoria County Detention Center in Texas, where black inmates were chased and attacked by growling dogs. In fact, a number of the guards who performed in the Abu Ghraib photos were former prison guards, including Charles Graner, who had been "subject to numerous complaints of human rights violations and prisoner abuse" while working at a high-security prison in Waynesburg, Pennsylvania.[37] The performances at Abu Ghraib can be seen to go beyond Žižek's contention that they represent the "obscene underside of US popular culture," that is, the normalization of fraternity initiation rites and hazings. Rather, they represent the global staging of America's new penal culture, a culture that celebrates a racialized brutal violence and that is produced by both the economic structures and the cultural logic of neoliberalism. The photographs' spectacular revelation reverses Taussig's formulation of the abnormal appearing normal. In these photographs, the quotidian theatre of prison violence in the service of capital is seen as an exceptional act of individual depravity and the normal violence of neoliberalism appears aberrant.

America's prisons operate increasingly as forced labor farms. As of 2002, 80 percent of state and federal prisoners worked in jobs characterized as either non-industrial (maintenance labor that provided direct support of the prison) or industrial (labor producing profit for the prison, or the corporation running the prison).[38] Not only is this labor often compulsory, but it is compensated (if at all) at wages more often associated with Third World sweatshops than American factories. Private prisons pay $0.96–$3.30 per hour for industrial labor and as little as $0.08–$3.20 for non-industrial labor.[39] As Tracy Chang and Douglas Thompkins note, "jobs particularly suited to the prison industry, usually labor-intensive, low pay and low-skill, are the types of jobs that are usually relocated to Third World countries."[40] Thus, the "outsourcing" of neoliberal capitalism is facilitated not only by the depressed wages of Third World economies, but also by a confined and coerced (and largely black and brown) laboring class. The profit opportunities of such low-wage industrial labor encourage corporations like IBM, Boeing, and Microsoft to shift high-paying union jobs to prison (i.e., slave) labor, for which corporations pay only pennies per hour, with no responsibility for worker safety or retirement and medical benefits.

Similarly, the shifting of manufacturing jobs to non-union factories outside the United States has led states to promise decimated small towns across the United States that building and running prisons will rehabilitate local economies by producing short-term construction jobs (though these contracts often go to large, out-of-state, non-union contractors) and long-term prison guard jobs. Ruth Gilmore calls this dynamic the "prison fix," where the construction of prisons is seen to solve the problem of the "unfixing" of "urban and rural areas [...] by capital flight and state restructuring" that are the processes of neoliberal capitalism.[41] Not only does prison construction and maintenance rarely produce the economic windfall promised, the rise of private prisons has depressed corrections officers' wages, benefits, and working conditions, as private corrections officers are less likely than their public counterparts to be working under a union contract.[42]

With prisons increasingly run by mega-corporations like Wackenhut and Lockheed-Martin, the maintenance of a surplus prison population produces profits by securing lucrative contracts in which compensation is based on a per prisoner basis. It is perhaps the presence of Lockheed-Martin and Boeing, two of the largest weapons-makers in the world (and two of the greatest beneficiaries of the increase in military spending during the "War on Terror"), as beneficiaries of the prison economy that reveals that the relationship between the prison-industrial complex and the military-industrial complex is more than lexical. It is rather, as Davis points out, a relationship based on "mutual support" and "shared technologies."[43] As the weapons-makers who constitute the military backbone of globalization (i.e., the military-industrial complex) become increasingly dependent on prison slave labor and prison profiteering, the military-industrial complex, and the military imposition of US hegemony as globalization, necessitate the maintenance of the prison-industrial complex and its exploited labor force.

Thus, the American penitentiary can be seen not only as a precondition for, and effect of, (military) neoliberalism, but also as a site conjoining the "new American militarism"[44] and the "new American city"[45] of penitentiary dependence. As examples of such a conjuncture, the photos of torture at Abu Ghraib can be seen as capturing the staging of violence that secures the prison-industrial complex and the military-industrial complex together in a "symbiotic"[46] relationship and as constitutive elements of neoliberalism.

The fact that the invasion and occupation of Iraq itself can be seen as a staging of neoliberalism as war extends this analysis. The invasion of Iraq, described by the *Wall Street Journal* as "one of the most audacious hostile takeovers ever,"[47] was, according to the authorial collective RETORT, "privatization by occupation,"[48] a contention supported by the Bush-Cheney campaign manager's comment that the war was about "getting Iraq ready for Wal-Mart"[49] and Cheney's own revealing admission that the war was a remarkable "growth opportunity."[50] Thus, in Abu Ghraib, we see not only the staging of what Shamir and Kumar call the "military backbone of globalization," but also the penitentiarial logic and practice that is instantiated in militarism and globalization.

Fight or fuck

What has been most "shocking" to many about the photos has been the frequent presence of sexual humiliation – nude, hooded men stacked in pyramids, and forced to masturbate and simulate fellatio on each other. Most famously, these stagings are often framed by the cherubic PFC Lynndie England, enacting her celebratory pose of American mastery for the camera – showing a wide grin, and giving an approving thumb's up. In dialogue with shrouded men connected to electrodes, these photos represent, Susan Willis contends, "the pornographic sublime."[51] Indeed, pornography has been a central trope through which critics have reckoned with the photos. It is with the invocation of video pornography that they come to invoke performance: "Much like the producers of a low-budget porn flick, they set the stage for a sadistic theater."[52] The coercive staging of sexual humiliation, on a stage but for a camera, represents "war porn" for Baudrillard, in that "it all becomes a parody of violence, a parody of war itself, pornography becoming the ultimate form of the abjection of war which is unable to simply be war, to be simply about killing, and instead turns into a grotesque reality-show, in a desperate simulacrum of power."[53]

The sadistic theatre of power represents the collapse of the real implied in the imminence of theatrical performance, and supplants it with the hyperreal reality-show. But war has never been "simply about killing." In particular, America's history of war-making has long been based on a foreign policy driven by corporate profit. Moreover, the powerful have long been dependent upon their own performances of mastery "as a kind of self-hypnosis within the ruling groups to buck up their courage, improve their cohesion, display their power and convince themselves anew of their high moral purpose."[54] In the context of Abu Ghraib, it is the staging of the soldiers as the subjects of the photographs and the objectification of the prisoners that perform this self-hypnosis. Even for the American civilian at home, the distance produced by the photograph allows the observer to *dis*identify with the actors, restoring the morality of the American crusade to spread neoliberal capitalism under the banner of "democracy" by disavowing the performances that secure it.

The specatularity of the photographs' pornographic impulse mystifies the political economic function of Abu Ghraib (and of the War on Iraq in general) by casting these acts as primarily individual and sexual, rather than as erotic expressions of "the larger cultural narrative [...] based upon class, prisons, and labor."[55] The narrative, which is masked by the "frenzy of the visible"[56] of the pornographic impulse, is audible as a cultural logic of neoliberal violence. As with pornography, the torture photographs depend upon performance's imminence modified by the aestheticizing effect of its filming: "the events are real but staged," as Dora Apel points out, and yet "there is no 'fiction' of authenticity [...] because the pleasure is not meant to be found in their pruriently deployed bodies but in the exultant mastery of those who would wield power over them, representing a different cultural and political order."[57] This mastery is denied at home for these soldiers and reservists who turn to jobs in the military or as prison guards as jobs of last resort, where the "prison-industrial complex and military-industrial complex converge in a sociopolitical economy grounded in rural and lower-class life."[58] The position of military prison guard enables a performance of mastery that the effects of neoliberalism, such as factory closings and the elimination of social services, has eliminated for these soldiers at home. Disciplined in minimum-wage service jobs through countless acts of humiliation and dispossession, they bring with them the cultural logic of neoliberalism as a normative mode of performance.

Moreover, their performances enact what Pierre Bourdieu calls the "ultimate foundation of this entire economic order placed under the sign of freedom [...] the *structural violence* of unemployment, of the insecurity of job tenure, and the menace of layoff that it implies."[59] As violence is the structuring expression of freedom under neoliberalism, and is enacted by the American state of "freedom" as bombings and occupation, the performances at Abu Ghraib hardly seem surprising at all. Thus, Lila Rajiva's claim that "what makes the torture and terror of the warfare state ultimately pornographic [...] [is that] our theater appears to be only a perverse enjoyment, a tasting of our freedom from all constraint, the self-pleasuring delight in its own performance."[60] If we consider such a theatre of the warfare state to be also that of the economic order that animates it, then we can understand Abu Ghraib as a performance of "freedom from all constraint" as the enactment of neoliberal culture.

The violence of neoliberalism staged at Abu Ghraib can be seen as a restoration of behavior rehearsed in American prisons, and as part of a legacy of military and CIA use of sexual humiliation as a torture technique. When Texas prisoner Roderick Johnson complained to prison officials not only that he was being regularly raped, but that he was being tortured as a "sex slave," prison officials repeatedly responded that he should learn to either "fight or fuck."[61] This logic, where the only choices are fighting or fucking, is also that of neoliberalism. And perhaps, within the auspices of neoliberalism, fucking and fighting is the same option – as competition of all against all is raised to a sacred position, neoliberalism demands that all workers be constantly fighting each other, and be fucked by the state-backed corporate apparatus. Abu Ghraib is the staging of fighting or fucking as the realization of military neoliberalism.

Jon McKenzie calls this theatrical and organizational compulsion the command to "perform, or else," in which the demand to compete is impelled by threat.[62] I want to suggest that we consider the "fight or fuck" performance in Abu Ghraib, in which fucking itself was a central part of the theatrics, as a coming together of the "perform, or else" performance of neoliberalism with what José Muñoz calls "the burden of liveness," the historically constituted condition in which people of color are demanded to enact their "difference" for the purpose of disgust, exoticism, or sublimity.[63] But, perhaps more than anything, it is the compulsion to be "live" for the spectator, even a live object that operates as this burden. I wish to close by considering how this racially constituted and compulsory liveness operates in these torture performances, as a way of reckoning with how Abu Ghraib connects these performance complexes: military-industrial and prison-industrial in service of neoliberalism.

This "burden of liveness" characterizes Coco Fusco's "Other History of Intercultural Performance," which recounts the legacy of staging racially constituted subjects as exotically figured objects, from slave auction blocks to the Hottentot Venus to Ringling Brothers circus exhibitions of pygmies as part of their animal parade.[64] In the torture photos, exoticism is displaced by abjection, in which the staging of racial violence as banal and celebratory marks the banality of the racial violence of neoliberal globalization. If, as Baudrillard suggests, these photos are not representational because they are so embedded in the war itself, then what do they do? Baudrillard suggests that they are simply another expression of the "pornographic face of the war."[65] I contend that there is something audible in the frenzy of the visible that these photos cannot represent, but instead makes resonate.[66] For this staging of the banality of torture, the self-parody of power that does not know what do with itself, is part of a chain of performances that neoliberalism impels – one that links the MIC and the PIC in resuscitation of primitive accumulation.[67] The photographs visually elide, yet resonate the echo of the chain that includes the demonstrations of military force against sweatshop workers in Central America as demanded by US corporations who require that wages be kept low. The "new realities" that neoliberal empire enacts are theatres of violence – a militarily enforced private penitentiarism compelling performances of fighting and fucking – or else. The performances of torture at Abu Ghraib, in their banal and celebratory enactment of violence, are simply the spectacular realization of the ubiquitous violence that neoliberalism produces globally. Would that they would also be its undoing.

Notes

1. Slavoj Žižek, "Between Two Deaths: The Culture of Torture," *London Review of Books*, 23 May 2004, 9 June 2007: http://www.lrb.co.uk/v26/n11/zize01_.html
2. Rush Limbaugh, "Media Matters for America," *The Rush Limbaugh Show*, 5 March 2003, 10 June 2007: http://mediamatters.org/static/pdf/limbaugh-20040503.pdf
3. Quoted in Ron Suskin, "Without a Doubt," *New York Times Magazine*, 17 Oct. 2004: 44.
4. Susan George, "A Short History of Neo-Liberalism: Twenty Years of Elite Economics and Emerging Opportunities for Structural Change," *Conference on Economic Sovereignty in a Globalizing World*, 24–26 March 1999, 9 June 2007: http://www.globalexchange.org/campaigns/econ1010/neoliberalism.html
5. Susan Sontag, "Regarding the Torture of Others," *New York Times Magazine*, 23 May 2005: 28.
6. Sontag, 26–7.
7. See also Nicholas Mirzoeff, *Watching Babylon: The War in Iraq and Global Visual Culture* (New York: Routledge, 2005).
8. For more on lynching photography, see James Allen (ed.), *Without Sanctuary: Lynching Photography in America* (Santa Fe: Twin Palms, 2000).
9. Sontag, 27.
10. Susan Willis, *Portents of the Real: A Primer for Post-9/11 America* (New York: Verso, 2005), 135.
11. Angela Davis, *Abolition Democracy: Beyond Empire, Prisons, and Torture* (New York: Seven Stories, 2005), 50.
12. Peggy Phelan, *Unmarked: The Politics of Performance* (New York: Routledge, 1993), 146, 148.
13. Walter Benjamin, "Theses on the Philosophy of History," *Illuminations: Essays and Reflections*, ed. Hannah Arendt, trans. Harry Zohn (New York: Schocken, 1968), 261.
14. Guy Debord, *The Society of the Spectacle*, trans. Donald Nicholson-Smith (New York: Zone, 1994).
15. Phelan, 162–3.
16. Quoted in Lila Rajiva, *The Language of Empire: Abu Ghraib and the American Media* (New York: Monthly Review, 2005), 12.

17. Elaine Scarry, *The Body in Pain: The Making and Unmaking of the World* (New York: Oxford University Press, 1985).
18. Jean Baudrillard, *The Conspiracy of Art: Manifestos, Interviews, Essays*, trans. Ames Hodges (New York: Semiotext(e), 2005), 209.
19. Saidiya Hartman, *Scenes of Subjection: Terror, Slavery, and Self-Making in Nineteenth Century America* (New York: Oxford University Press, 1997), 3.
20. Hartman, 4. On the linkage of neoliberalism to violence, see Henry Giroux, *The Terror of Neoliberalism: Authoritarianism and the Eclipse of Democracy* (Boulder, CO: Paradigm, 2005).
21. Oliver C. Cox, "Lynching and the Status Quo," *The Journal of Negro Education*, 14:4 (1945): 579.
22. Cox, 580.
23. Quoted in Rajiva, 79.
24. RETORT, *Afflicted Powers: Capital and Spectacle in the New Age of War* (New York: Verso, 2005), 103.
25. Hartman, 22.
26. Cox, 584.
27. On race and capitalism as constitutive elements of modernity, see Howard Winant, *The World is a Ghetto: Race and Democracy since World War II* (New York: Basic, 2001).
28. Ruth Wilson Gilmore, "Globalization and US Prison Growth: From Military Keynesianism to Post-Keynesian Militarism," *Race and Class*, 40 (1998): 145–56.
29. Lisa Duggan, *The Twilight of Equality: Neoliberalism, Cultural Politics, and the Attack on Democracy* (Boston, MA: Beacon, 2003), xxi.
30. Duggan, xvii.
31. Quoted in RETORT, 195.
32. Michelle Brown, "'Setting the Conditions' for Abu Ghraib: The Prison Nation Abroad," *American Quarterly*, 57:3 (2005): 984.
33. Dwight Conquergood, "Lethal Theatre: Performance, Punishment and the Death Penalty," *Theatre Journal*, 54 (2002): 342.
34. Christian Parenti, *Lockdown America: Police and Prisons in the Age of Crisis* (New York: Verso, 1999), 136. Emphasis in original.
35. Michael Tassig, *The Nervous System* (New York: Routledge, 1992), 18.
36. Parenti, 137.
37. Brown, 982.
38. Tracy F. H. Chang and Douglas E. Thompkins, "Corporations go to Prisons: The Expansion of Corporate Power in the Correctional Industry," *Labor Studies Journal*, 27:1 (2002): 55–6.
39. Chang and Thompkins, 57.
40. Chang and Thompkins, 61. See also K. C. Carceral, *Prison, Inc: A Convict Exposes Life Inside a Private Prison*, ed. Thomas J. Bernard (New York: New York University Press, 2006); Michael Hallet, *Private Prisons in America: A Critical Race Perspective* (Urbana: University of Illinois Press, 2006); and David Matlin, *Prisons Inside the New America*, 2nd edn (Berkeley, CA: North Atlantic, 2005).
41. Ruth Wilson Gilmore, *Golden Gulag: Prisons, Surplus, Crisis, and Opposition in Globalizing California* (Berkeley: University of California Press, 2007), 179.
42. Chang and Thompkins, 52.
43. Angela Davis, *Are Prisons Obsolete?* (New York: Seven Stories, 2003), 86.
44. Andrew Bacevich, *The New American Militarism: How Americans are Seduced by War* (New York: Oxford University Press, 2005).

45. Brown, 973–97.
46. Davis, *Are Prisons Obsolete?*, 86.
47. RETORT, 50.
48. RETORT, 47.
49. RETORT, 49.
50. RETORT, 41.
51. Willis, 120.
52. Willis, 130.
53. Baudrillard, 206.
54. James Scott, *Domination and the Arts of Resistance: Hidden Transcripts* (New Haven: Yale University Press, 1990), 67.
55. Brown, 983.
56. Dora Apel, "Torture Culture: Lynching Photographs and the Images of Abu Ghraib," *Art Journal*, 64:2 (2005): 93.
57. Brown, 983.
58. Pierre Bourdieu, "The Essence of Neoliberalism," *Le Monde Diplomatique English Edition*, Dec. 1998, 9 June 2007:http://mondediplo.com/1998/12/08bourdieu
59. Linda Williams, *Hard Core: Power, Pleasure, and the "Frenzy of the Visible"* (Berkeley: University of California Press, 1999).
60. Rajiva, 100.
61. The "Johnson v. Johnson Case Profile" can be found at http://www. aclu.org.
62. Jon McKenzie, *Perform or Else: From Discipline to Performance* (New York: Routledge, 2001), 4–7.
63. José Estaban Muñoz, *Disidentifications: Queers of Color and the Performance of Politics* (Minneapolis: University of Minnesota Press, 1999), 182.
64. Coco Fusco, *English is Broken Here: Notes on Cultural Fusion in the Americas* (New York: The Free Press, 1995), 41–3.
65. Baudrillard, 207–8.
66. Moten, 196. See his reading of the radical audibility of the photos of Emmit Till in Fred Moton, *In the Break: The Aesthetics of the Black Radical Tradition* (Minneapolis: University of Minnesota Press, 2003), 192–211.
67. RETORT, 11.

Afterword

"In the Valley of the Shadow of Death": The Photographs of Abu Ghraib

Peggy Phelan

You've probably all seen them and probably more than once. They have been reprinted in newspapers all over the world; they've been the subject of six Pentagon reports between March 2004 and April 2005; they have been used as evidence in military trials; they've been the object of a congressional inquiry in the United States; and they have produced much agony and hatred in Iraq and the larger Arab world. The photographs are a relatively small, but immensely potent, part of a larger political and ethical scandal concerning the Bush administration's decision to view detainees as "enemy combatants." This declaration has meant that prisoners detained in the United States' far flung "war on terror" are not afforded the protections set out for prisoners of war stipulated by Article 19 of the Geneva Conventions of 1949. The Bush administration's policy decision underlies the treatment of detainees in Afghanistan, Cuba, and Iraq. In confining my remarks here to the photographs taken in Abu Ghraib. I do not mean to suggest that the broader horizon of sanctioned torture and abuse is not worthy of serious discussion in a volume devoted to *Violence Performed*. Nonetheless, the complicated performative force of the Abu Ghraib photographs has not been analyzed in sufficient depth to elucidate why these photographs challenge our conceptual grasp of the conjunction between violence and performance.

Learning to see all over again

CBS' news magazine, *Sixty Minutes II,* aired a report on the scandal and showed the photographs on 28 April 2004, two days before Seymour Hersh's *New Yorker* article appeared online, and a week before it appeared in print.[1] The television magazine was given copies of the photographs by the family of Staff Sergeant Ivan Frederick II, who had been notified he would be court-martialed, Frederick's defense was that he had been instructed to "fear up" and "soften up" detainees who would be interrogated by civilian contractors and "other government agencies" (the CIA) and therefore was only following orders from his superior officers. Much of Hersh's article was based on the leaked classified report to the Pentagon by Major General Antonio Taguba.[2] *The New Yorker* published some of the photographs, as did *The New York Times Magazine* when Susan Sontag's influential essay on the photographs, "Regarding the Torture of Others," appeared in the 24 May 2004 issue. While Sontag contended that the photographs are unprecedented documents in the history of war and must be condemned and repudiated, Rush Limbaugh claimed that they are no more serious or noteworthy than some fraternity pranks. "This is no different than what happens at the Skull and Bones initiation, and we're going to ruin people's lives over it, and we're going to hamper our military effort, and then we are going to really hammer them because they had a good time. You know, these people are being fired at every day. I'm talking about people having a good time, these people, you ever heard of emotional release? You [ever] heard of need to blow some steam off?"[3] In the Arab world, commentators have seized on them to rally anti-American feeling and to expose the disjoin between the Bush administration's rhetoric of "liberation" and their modes of imprisonment; while in China, the photographs and the acts they document have been used as a reason to reject calls for a more serious engagement with human rights issues.

Framed as art

The photographs have also created a specifically aesthetic response. The International Center of Photography exhibited about 20 of the photographs in a show entitled, *Inconvenient Evidence: Iraqi Prison Photographs from Abu Ghraib,* which ran from 17 September through 28 November 2004. Curated by Brian Wallis, the exhibition raised important questions about the history of war photography in relation to these new images. The Colombia artist, Fernando Botero, often hailed as Latin America's greatest living artist, created more than 80 works (about 50 life-size oil paintings and 30 drawings) based on the photographs. For many viewers, Botero's attention to the suffering of the prisoners seemed to place the focus on the central subjects of the abuse, rather than on the perpetuators. Errol Morris's controversial documentary film, *Standard Operating Procedure* (2008), relies on interviews with five of the seven US military personnel who were found guilty because of their actions at Abu Ghraib, and focuses on what the photographs expose and what they cover-up about the events in the prison. Morris pushes the conventions of documentary film; he employs slow motion photography, a score by Danny Elfman, and re-enactments of some of the abuse; he also paid interviewees to appear in his film. Morris defends his filmic devices by suggesting that his use of artifice allows us to see that we have not seen the truth and he says that if he had not paid his interviewees, they would not have participated. Be that as it may, Morris's film interests me as a further elaboration of the Abu Ghraib photographs' complex trajectory across the

realms of journalism, aesthetic practice, legal evidence, and cultural criticism. Their relevance to such a wide array of discursive systems suggests the
density of their performative force. Before turning explicitly to the drama
of interpretation these photographs have prompted, I want to dwell a bit
longer on what they expose at the level of genre and action.

Genre: atrocity photos

Photographs of violence, particularly those photographs that journalists and
ethicists call "atrocity photographs," expose a question about the relationship between the violent image and the viewer's decision to act, or not to
act, once that atrocity is seen. Photography's complex relationship to both
the real and the representational often makes the depicted violence difficult
to interpret; what looks to be a straight-forward representation of the real
is often staged and manipulated for aesthetic and/or political effect. Indeed,
the history of atrocity photography may also be understood as the history
of the inquiry into the relationship between what is real and what is staged
in the theatre of war. Richard Fenton, Matthew Brady, and Robert Capa,
to take only a short list, are all renowned for their war photography, but
their reputations are also haunted by the possibility that their most riveting
work is somehow inauthentic or fraudulent because it "indulges" aesthetic
preferences rather than transparently recording events. Implicit within this
argument are several assumptions: (1) that documentary is not itself an
aesthetic form; (2) that the photographer's will-to-art often betrays the will-
to-truth; (3) that the work would be "better" without artifice; and (4) that
the seriousness of war demands a cold-eye and an objective hand.

In his beguiling meditation on Fenton's iconic, "In The Valley of the
Shadow of Death" (1855), Errol Morris examines the critical claim that Fenton staged his photograph.[4] Taken during the Crimean War and often cited
as the first war photograph, Fenton's photograph exposes an empty road
burdened by cannon balls (see illustration). Since Fenton took two expo-
sures from the same spot on the heavily hit road in Sebastopol, one with a
few scattered balls and one with a dense field of fodder, the assumption that
the photographer staged the round-shot for aesthetic and/or political effect
has held sway for some time.[5] Moreover, this notion implies that Fenton
intentionally deceived viewers by transforming the actual scene on the road
to one that served his other extra-photographic purpose. The explanations
of his motives for deception are themselves quite dramatic – that he was a
coward, that he was seeking fame, that he was creating an anti-war portfolio
and so on. Morris decides to use evidence only within the two photographs
to learn which exposure was taken first. He concludes (convincingly) that
Fenton took the photograph with the greatest number of cannon balls after
he took the first exposure with only a few cannon balls, but Morris rejects
the argument that Fenton was trying to deceive anyone. The decisive evidence is based on the position of small rocks that also appear in the two

Afterword   "In the Valley of the Shadow of Death" (1855)
Source: Photograph by Richard Fenton.

photographs. Dennis Purcell was able to illustrate that the small rocks were following the law of gravity and since they have fallen down the small crest in the sloped road, Fenton must have taken the photograph in which these rocks are lower after he took the first photograph in which they are slightly higher on the hill.

What interests me about the force of this debate is the energetic desire to ferret out intention, will, and action to Fenton in relation to this photograph, while easily accepting without any controversy or hesitation at all, that he performed when he created self-portraits, or that he used the techniques of theatrical lighting in order to manipulate shadow on an architectural facade. Thus, it is the subject matter of Fenton's "In The Valley of the Shadow of Death" – war – that seems to incite the belief that photographers have a special obligation to maintain fidelity with an unvarnished real.

The critical preoccupation with Fenton's "manipulation" of the number of cannon balls, and the various explanations for it, displaces the anxiety raised by the barren death field the photograph exposes. It may be that the bleakly furrowed landscape of death Fenton's photograph shows us is so still, so resolutely atheatrical, that it produces a desire in the viewer to ascribe (negative) action and will to the photographer – to punish him for exposing what we prefer not to see. To charge a war photographer with "fraud" is perhaps a defense against the burden of taking in a dead landscape in which no humans act at all.

Moreover, Fenton's case is important in contradistinction to the Abu Ghraib photographs which are, for better or for worse, explicitly theatrical and stage-managed. While Fenton's reputation has suffered from the charge that he made events look worse than they were, the photographers of Abu Ghraib have employed photography as part of their arsenal against prisoners. In other words, more than documenting atrocities, the act of taking photographs in Abu Ghraib was itself part of the abuse prisoners suffered there.

The machinery of wars, the frolic of cameras

Gun, missile, bomb aimed to obliterate skin, subject, soul: the rhythm of contemporary war. Just as the military's technological advances have been based on an argument for speed, the digital capacities of cameras and cell phones have accelerated the pace of both production and circulation.[6] Processing time has been radically shortened and ease of uploading, emailing, and posting on social networking and other websites has meant that the practice of photography has become simultaneously more public even while it conquers ever-more intimate and personal subject matter. Thus, a photograph of a child that in an analogue age may have been mailed to a processing company, returned in a week, and then shared with close relatives and friends can now be taken, uploaded, and sent all around the world in "a blink of an eye." But to claim attention in what sometimes seems an almost infinite sea of photographic images, a photograph needs to offer something more unusual than a pleasant smile.

The intimacy and familiarity of the digital camera also allows for a new notion of continuity from the familiar to the strange. Travel and tourism are increasingly tied to photo opportunities, chances to be in a new landscape even while being screened from it via the camera lens. The habit of screening the world with a camera offers a familiar gesture in an unfamiliar world. The habit of taking photographs sometimes seems to "outsource" some aspects of memory, attention, and absorption. Photographing a scene offers a chance to forget it in the moment and to store it away for some imagined less hectic future. Kafka memorably wrote, "I write to forget," and this sentiment is echoed by US Specialist Sabrina Harman, when she tells Errol Morris that she wrote letters and took photographs in Abu Ghraib as a way both to forget what she was seeing in the present and to prove, in some future what she had in fact seen and done:

> *Harman:* Again, I don't even know where I was at that point. I put everything down on paper that I was thinking. And if it weren't for those letters, I don't think I could even tell you anything that went on. That's the only way I can remember things, is letters and photos.

Morris: So it's a way of creating memory?

Harman: I don't know what it was. I really think I put it all down to forget, in the letters, and then just to prove what was going on, was the photos. Because really, if anybody came up to you and was like, "Hey, this is what's going on," there's no way anybody would believe what was going on. So that's why the photos were taken.[7]

At once a defense against the real and a documentary of it, taking photographs – perhaps especially for young white women – offers a kind of control over a world that often seems to exceed our capacity to grasp it. Harman's decision to enter the frame, to show her own smiling face in the scene of violence, both implicates her in the crime and consigns her to white women's usual place in dramas created by white men – as smiling, exuberant decoration. The familiarity of that triumphantly empty performance in relation to abuse and torture render some Abu Ghraib photographs both banal and horrific, or, indeed, horrifically banal.

Photographic crossings

Currently, there are 19 videos and 280 still photographs from Abu Ghraib in general circulation. (Salon.com has the most accessible archive.) Taken together, the Abu Ghraib photographs expose the link between biological and political suffering at the heart of torture. The setting of all the photographs is at once architectural, juridical, and "hygienic-domestic" (in that many of the scenes take place in showers, toilets, and passageways between cells). Additionally, Abu Ghraib prison itself carries a tremendous symbolic power. It was the setting for some of Saddam Hussein's most brutal violence, with estimates ranging from 4000 to 30,000 prisoners beaten to death or beheaded while imprisoned there. Ironically, it was video and photographs of those beatings and scenes of torture that became part of the Pentagon's argument for war with Iraq in 2003.[8]

In September 2003, Abu Ghraib held about 200 prisoners. But after the insurgency took hold in early October, more and more Iraqis were arrested. By December, Abu Ghraib had over 1000 prisoners. Brigadier General Janis Karpinski, who supervised the prison at this time, later estimated that about 90 percent of the detainees were innocent. The staff was overwhelmed. The soldiers worked 12-hour shifts every day for 40 days without a break. Moreover, the prison itself was constantly under fire. The testimony of the detainees and the soldiers is united on this point. Despite long-established protocols of war (and the disavowed Geneva Conventions) which remove prisoners from the scene of battle, Abu Ghraib was located right in the center of the theatre of war. Hence, the psychological conditions of Abu Ghraib were such that everyone was operating in the valley of the shadow of death. We shall return to this point later.

Part of the difficulty of assessing the archive is we have access only to a partial view. In her interview with Morris, Harman notes, in an extraordinary aside, that she has a photograph still on her hard drive that has not been released. This aside opens onto the vast chasm between the "archive" that has been published, and the thousands of photographs we have not seen at all. The ACLU has filed Freedom of Information Act requests to make all photographs and videos public. The government denied their request, arguing that they will promote Anti-American feeling during a time of war. On 12 May 2004, US senators and representatives saw a slide show of 1800 images taken in the prison. Many of these depict dogs harassing and in at least one case biting a prisoner. Photographs of Private First Class Lynndie England having sex with other US soldiers are also part of the archive.

Sontag argued that the Abu Ghraib photographs – all of which were taken on digital cameras – created a public relations crisis for the US military; a crisis that often seemed to be about the *circulation of photographs* rather than about the circulation of the violence and abuse depicted therein. When I first read Sontag's essay, I thought she was right. However, Janis Karpinski, the supervisor of the prison at the time of the scandal, makes the startling claim that the administration encouraged the release of some of the photographs.[9] Arguing that the story of Private First Class Jessica Lynch, who had been captured and held at an Iraqi hospital before she was freed by US forces, had rendered the young blonde woman "the face of the war," Karpinski suggested that the release of the photographs served those uncomfortable with the feminization of the battlefield represented by Lynch's much-hyped story. Given the administration's success at containing the criminal and military investigation of the Abu Ghraib scandal to "a few bad apples" (including Karpinski herself), she may well be right. But regardless, if one finds Sontag's or Karpinski's account more persuasive, the circulation of the photographs created an international crisis.

Exceptional or not (so much)

For Sontag, the Abu Ghraib photos are unique in the history of photography; their closest precedent is lynching photography. For Hazel Carby, however, the photographs are one more addition to the dreary visual chronicle of the abuse of brown men.[10] Both commentators are correct; the photographs are startling even while they convey the sense that we have seen all of these scenes before. A potent instance of what Walter Benjamin termed "terror as usual," the Abu Ghraib photographs are shocking because the "scenes" they depict are not. Prisons in the United States, one of the most brutal instrumentations of racism in contemporary culture, are often crucibles of severe abuse, if not torture. Two of the main MPs convicted of crimes, Corporal Charles Graner and Staff Sergeant Ivan

Frederick II, both enlisted in the service after serving as prison guards in the United States. While Frederick outranked Graner in Abu Ghraib, the latter was widely viewed as the ringleader of the group. Of the 279 photographs that have been released to date, 173 came from Graner's camera; 55 from Frederick's and 44 from Harman's. The connections between practices Graner employed in US prisons and in Abu Ghraib are explicit, as we shall see shortly.

While Carby is correct to note the similarity between the abuse and torture depicted in the Abu Ghraib photographs and the long history of US racism at home, the photographs are operations of war. In this sense, the Abu Ghraib photographs function not only as documentary or aesthetic texts but also as weapons. These images were taken because they could be: I mean this not only in the sense that the US military police created an environment in which these scenes could be enacted, photographed, and circulated; but also in the sense that the conditions of capture and detainment stripped away the prisoners' will, agency, and mobility so completely that they were reduced to performing as props in a theatre designed to frame them as objects. Adherents of "torture procedures" (if we can accept that as a rational phrase), argue that extreme force and abuse are justified in war because the enemy has secret information that could save the lives of innocent people. Torture is a way to expose that secret, to get the detainee to break his or her silence and "come clean." Studies suggest that the information obtained in this way is not reliable; nonetheless, the argument continues to hold sway in the Bush administration. But even if the ostensible reason these photographs were taken was to humiliate the prisoners so that they would give up information – if you don't tell us something, then we will show these images of you in these compromising positions – the photographs do not "work." By hooding the prisoners, or to use the brutally casual military shorthand, "bagging their heads," the photographs do not expose the identities or the putative secrets of the prisoners. They do, however, reveal the not-so-secret sadism of many of the US guards – and, bizarrely, their own faces.

By "bagging" the detainees, the US guards do more than objectify and de-humanize the prisoners. They literally make them blind. They deprive the prisoners of their own sight and therefore their own agency as witnesses to their own abuse. "The bags were over our heads so we could not see their faces," reports one detainee, while another admits, "I could see only their feet under the bags."[11] In some of the first critical commentaries on the Abu Ghraib photographs written in the United States, the urgent desire to denounce them and what they say about us sometimes unwittingly suggested that our suffering and humiliation were more important than the suffering of the prisoners, who, for the most part, remained anonymous and less than human even in otherwise very smart commentary. Our obsessive focus on what the photographs said about us – Sontag: "The photographs are us" – obscured the pain the prisoners endured; thus the act of looking at these photographs repeats the original failure-to-see-the-other that the photographs frame so dramatically.[12]

The vast preponderance of the photographs and video from the prison are sexual in nature. A video of several men lined up against a wall masturbating, photographs of men simulating oral sex, and the various architectural arrangements of the human pyramids evince a particularly sophomoric and homophobic mind at work. Graner, who has received the harshest punishment to date, seems to have been the director of most of the scenes. His record as a guard in Pennsylvania involved a host of allegations and complaints, mainly along sexual lines. But the more I think about the photographs, the more I think the sexual escapades were primarily a cover for an even more brutal attack on the subjectivity of the detainees.

This attack uses sexual shame as a tool in a larger battle. I think the photographs must be seen as instances of religious persecution of a very odd sort. While working as a prison guard in Pennsylvania, Graner was alleged to have threatened a Muslim prisoner that he would smear his food tray with pork. A detainee in Abu Ghraib recounts that Graner did, in fact, force him to eat pork. "The night guard came over, his name is Graner, opened the cell door, came in with a number of soldiers. They forced me to eat pork and put liquor in my mouth."[13]

The most extreme abuse documented in the photographs coincided with the month of Ramadan, celebrated in 2003 from 25 October through 26 November. While prisoners in Abu Ghraib were being attacked by dogs, urinated on, made to masturbate, and form human pyramids, President Bush issued this Presidential proclamation: "Ramadan is the holiest season in the Islamic faith, commemorating the revelation of the Qur'an to Muhammed. This month of introspection provides Muslims a time to focus on their faith and practice God's commands. Through fasting, prayer, contemplation, and charity, Muslims around the world renew their commitment to lead lives of honesty, integrity, and compassion" (Presidential Proclamation 10/24/2003). Thus we have the theatrical rhetoric that supports religious freedom, at the same time we have a systematic practice of religious persecution at work in Abu Ghraib.

Religious persecution is a theme throughout the detainees' testimony. Any requests to learn the time so they might pray is reason to handcuff the prisoners to iron bars for hours. The intensity of torture in relation to the feast of Ramadan is also repeatedly mentioned by the witnesses. When asked to explain why he participated in these acts, Graner replied, "The Christian in me says it's wrong, but the corrections officer in me says, 'I love to make a grown man piss himself.'"[14] The idea that the Iraq war is essentially a religious war is confirmed by this aspect of the photographs and the detainees' testimony.

The most iconic photograph in the Abu Ghraib photographic archive is the one known as "Gilligan on a box.."[15] Standing, with arms upraised, hooded and "bagged" from head to knee, the prisoner is forced to stand on the box, with wires attached to his penis and his hands, under the threat of electrocution should he move. The photograph renders the detainee a Christ-like figure, and implants his body in a visual tradition that contradicts his own faith. The photograph's power comes from the sense that the detainee's very body has been rendered unto Christ's visual archive.

This is where the peculiar performative force of the photographs resides. The photographs of Abu Ghraib enact something beyond the acts depicted therein, horrendous as they are. When we look at "Gilligan on a box," we see more than a man in a hood standing on a box in a prison. The performance here exceeds the visible, but can only be conveyed through the photograph. Thus the crossing between the argument that the photographs are "unique" and the argument that the photographs are all too familiar can be located here. These photographs, in other words, are weapons in what may well be regarded as a contemporary religious war. The iconography of Christ is perhaps the densest visual archive in the world and the Abu Ghraib photographs forcibly insert Muslim detainees into it.

The Abu Ghraib photographs and the history of atrocity

It will now be necessary to write a history of photography in which the Abu Ghraib images are included. Fenton's photograph, "In the Valley of the Shadow of Death," takes its title not from Tennyson's famous poem, but rather from Psalm 23. In the history of atrocity photography that stretches from Fenton's photograph of the non-human detritus of the Crimean War to the photographs of Abu Ghraib and their all too human waste, there are a series of powerful intertexts. If Fenton's photograph attempts to articulate the prayer embedded in Psalm 23, "Yea, though I walk through the valley of the shadow of death, / I will fear no evil: For thou art with me; / Thy rod and thy staff, they comfort me. / Thou preparest a table before me in the presence of mine enemies; / Thou annointest my head with oil; My cup runneth over," then the photographs of Abu Ghraib give testimony to a much darker rewriting of that prayer. Indeed, it is as if Psalm 23 became the script to authorize a perverse theatre of conversion that motivated the sexual abuse, the dog biting, and the human pyramids. Those acts, deplorable as they are, should not blind us to the deeper struggle, indeed the burning passion of a war in which both sides are all too willing to use religion as a weapon.

Abu Ghraib placed both detainees and US military personnel in the valley of the shadow of death – with more than 20 Iraqis and five US soldiers killed from "incoming" bombs in the fall of 2003. Moreover, in the fall of 2003 only five soldiers and two non-commissioned officers were supervising the

1000 detainees. In this environment, it is not entirely implausible to suggest that Psalm 23, with its nice reverberations with 2(00)3, lent itself to the job.

The Christian reading of Psalm 23 becomes an authorizing script for converting detainees to the "American side." Most of the punished soldiers were reservists who lacked specific training. "Fearing up" the prisoners for the US interrogators so they would spill their secrets seems to require the full array of props listed in Psalm 23: food, anointing oil, the rod, and the testament of Christian faith. To take just one detainee's account of how these props were used against him:

> They forced me to eat pork and they put liquor in my mouth. They put this substance on my nose and forehead and it was very hot. The guards started to hit me on my broken leg several times with a solid plastic stick. They stripped me naked. One of them told me would rape me. [. . .] Someone else asked me, "Do you believe in anything?" I said, "I believe in Allah." So he said, "But I believe in torture and I will torture you." Then they handcuffed me and hung me to the bed. They ordered me to curse Islam and because they started to hit my broken leg I cursed my religion. They ordered me to thank Jesus that I'm alive. And I did what they ordered me. This is against my belief. They left me to hang from the bed and after a little while I lost consciousness. When I woke up, I found myself still hung between the bed and the floor. Until now, I lost feeling in three fingers in my right hand.[16]

This took place during Ramadan when five acts are considered especially grave and blasphemous – lying, slander, a false oath, greed, or covetousness. While these acts are considered offensive at all times, they are most offensive during Ramadan and undo the benefits that accrue from fasting and prayer. Forcing the detainee to denounce Allah and praise Christ would be religious persecution at any time, it is particularly heinous during Ramadan. Similarly, the photographs of "Gilligan on a box," must be seen in relation to the larger efforts to transform, even convert, the detainees into Christians. Sabrina Harman wrote a letter in which she seemed to grasp this part of the logic of abuse for the first time:

Oct 20, 03

10:40pm

Kelly,

[. . .] I ended your letter last night because it was time to wake the MI prisoners and "mess with them" but it went too far even I can't handle whats going on. I cant get it out of my head. I walk down stairs after blowing the whistle and beating on the cells with an asp to find "the taxicab driver" handcuffed backwards to his window naked with his underwear over his head and face. He looked like Jesus Christ. At first I had to laugh so I went on and grabbed the camera and took a picture. One of the guys took my asp and started "poking" at his dick. Again I thought, okay that's funny then it hit me, that's a form of molestation. You can't do that. I took more pictures now to "record" what is going on. After praying to Allah he [a detainee in cell 4] moans a constant short Ah, Ah every few seconds for the rest of the night. I don't know what they did to this guy. The first one remained handcuffed for maybe 1½–2 hours until he started yelling for Allah. So they went back in and handcuffed him to the top bunk on either side of the bed while he stood on the side. He was there for a little over an hour when he started yelling again for Allah. Not many people know this shit goes on. The only reason I want to be there is to get the pictures and prove that the US is not what they think. But I don't know if I can take it mentally. What if that was me in their shoes. These people will be our future terrorist [...]

Sabrina[17]

Suggesting that the moans of "Allah," produce actions that make the prisoners literally look like Jesus Christ, Harman stumbles onto the religious battle embedded in the photographs. Morris and Philip Gourevitch end their essay on Abu Ghraib with a meditation on the Gilligan photos: "The pose is obviously contrived and theatrical, a deliberate invention that appears to belong to some dark ritual, a primal scene of martyrdom. The picture transfixes us because it looks like the truth, but, looking at it, we can only imagine what that truth is: torture, execution, a scene staged for the camera? So we seize on the figure of Gilligan as a symbol that stands for all that we know was wrong at Abu Ghraib and all that we cannot – or do not want to – understand about how it came to this."[18] Yes, but above all the image speaks the suffering of the body, the body that is made to speak a script its soul disavows. The gap between the wires and the wall exposes the space of seepage: this is a body that cannot carry out the demand of the script. And this is why the Abu Ghraib photographs must be seen in relation to a history of photography that begins with "The Valley of the Shadow of Death."

Notes

1. Seymour Hersh, "Annals of National Security: Torture at Abu Ghraib," *The New Yorker*, 10 May 2004 (published on-line on 30 April 2004).
2. Taguba's report is called "Article 15–16 Investigation of the 800th Military Police Brigade." He found "grave breaches of international law" at Abu Graib. Taguba recommended that Brigadier General Janis Karpinski, who had command of the 800th, be reprimanded and relieved of her duties. He also faulted Colonel Thomas Pappas, commander of the 205th Military Intelligence Brigade for insufficiently

training and supervising his command's use of interrogation techniques. Both recommendations were followed out.

3. From radio broadcast 6 May 2004, "Rush Limbaugh's America." (Estimated audience: 20 million.)
4. Errol Morris, "Which Came First: The Chicken or the Egg? Parts I–III," 25 Sept. 2007, 4 Oct. 2007, and 23 Oct. 2007, *The New York Times*.
5. Morris gives a good run down of the history of the criticism, although he is primarily prompted by Sontag's treatment of the topic in *Regarding the Pain of Others* (New York: Farrar Strauss & Giroux, 2003).
6. See Elaine Scarry, *Who Defended the Country?* (Boston, MA: Beacon Press, 2003), for a superb deconstruction of the US military's argument for speed.
7. On *The New Yorker* website, under "Annals of War."
8. See David Kupelian, "New Video Reveals Real Torture Scandal: Saddam's daily horrors make America's Abu Ghraib abuses seem almost trivial," *World Net Daily*, 21 June 2004.
9. See Karpinski's remarks to Morris for his film, now posted on You Tube under the title "Deleted scene from SOP."
10. Hazel Carby, "A strange and bitter crop: the spectacle of torture," *Opendemocracy.net*, 10 Nov. 2004. Carby perceptively connects the photographs and videos taken at Abu Ghraib to the Rodney King video as well.
11. See the testimony of Kasim Mehaddi Helas, 18 January 2004, posted on salon.com.
12. I want to thank Rebecca Schneider for prompting me to clarify my muddled comments about these photographs on a panel we participated in at the Guggenheim in April 2005.
13. Statement of Ameen Sa'eed Al-Sheikh, 16 Jan. 2004. Posted on salon.com.
14. Graner quoted in Scott Higham and Joe Stephens, "Punishment and Amusement: Documents Indicate 3 Photos Were Not Staged for Interrogation," *The Washington Post*, 22 May 2004.
15. There seems to be more than one set of photographs of the man on the box. A detainee named Ali Shalal Qaissi told *The New York Times* ("Symbol of Abu Ghraib Seeks to Spare Others His Nightmare," Hassan M. Fattah, 11 March 2006), that he was the man in the photograph. But his claim has since been disproved. He counters that there is more than one set of photographs of a man bagged, attached to wires, and put in a Christ-like pose. Qaissi says that a man called Saad was also asked to strike this pose, and Salon.com finds this plausible.
16. Statement of Ameen Sa'eed Al-Sheikh, 16 Jan. 2004. Posted on salon.com
17. On *The New Yorker* website under "Annals of War."
18. Errol Morris and Philip Gourevitch, "Annals of War: Exposure," *The New Yorker*, 24 March 2008.

Index